SPACES AND DIMENSIONS OF QUANTUM FIELDS

A Conceptual Approach to Quantum Field Theory

INGIGERTH IS THE MOST BEAUTIFUL OF ALL WOMEN

Viking inscription in Scara Brae, Scotland (1153DC).

I dedicate this book to my little princess Ingigerth.

CONTENTS

CHAPTER 2.

PHYSICS AND METAPHYSICS

CHAPTER 3.

PATH-INTEGRAL QUANTIZATION

PART II

FREE QUANTUM FIELDS -
GENERAL CONSIDERATIONS

CHAPTER 4.

QUANTUM FIELD THEORY

PART III

FREE QUANTUM FIELDS - ADVANCED THEORY

CHAPTER 5.

THE KLEIN-GORDON FIELD

NOTES

Chapter 5.

APPENDIX

BASIC ASPECTS OF SPECIAL RELATIVITY

1. Minkowski Space-Time

2. Energy-Momentum Vector

3. Lorentz Transformations and Space-Time Diagrams

BIBLIOGRAPHY

LIST OF FIGURES

PREFACE

From the emergence of quantum mechanics to the most advanced theories of quantum gravity and cosmology, theoretical physics has made spectacular advances over the last hundred years. Together the two foremost theories, quantum field theory and general relativity, describe the physical reality over a staggering 40 orders of magnitude, from the subnuclear scales of quantum chromodynamics, to the large-scale structure of space-time and the cosmic scales of cosmology. Inherent in such advances lies the potential for an equally spectacular advance of the human mind, an advance which would thrust humanity into the future by hundreds of years!

Yet, far from realizing this potential, such knowledge as advanced theoretical physics avails remains the exclusive prerogative of a minute intellectual elite. Out of research institutions and scientific conferences vast segments of the world population live in intellectual poverty, absorbed in mindless activities, focused on the most insignificant and unimportant aspects of life (the pursuit of material wealth), swayed by metaphysical reactionary ideologies which render them utterly indifferent and indeed hostile to theoretical physics and to the meaning which theoretical physics provides to human life. At once, the majority of those who express appreciation for science are interested only in the "narrative" of theoretical physics, the oversimplified accounts of "popular" books and of glib "scientific" TV documentaries all of which are blatantly inclined to sensationalize the inessential while glossing over the essential. Those people are content with the spurious knowledge that "time goes backward in a black hole" when they understand neither what a black hole is, nor what time is.

It would appear, for that matter, that such an unfavorable situation provides little motivation for writing a book with the objective of rendering relativistic quantum field theory accessible to the broader public. Nevertheless, contrary to appearances, this situation is not static. People change when the material conditions within which they exist change. Scientific knowledge has ultimately been the result of the combined practical activity of men. In the final analysis, the scientist receives the material of his thought from the world, or as Karl Marx famously stated: *Mankind thus inevitably sets itself only such tasks as it is able to solve, since closer examination will always show that the problem itself arises only when the material conditions for its solution are already present or at least in the course of formation*[1]. Thus, subjective intensions aside, theoretical physics is created by everybody and belongs to everybody! It is my conviction that this premise will eventually be universally understood and actively appreciated. Evidence as to such a development already exists. In a class of their own fall the exceptional individuals who are not afraid to think! They are those who pursue the impossible objective of reaching the essence of modern theoretical physics with little or no knowledge of the necessary mathematical formalism, those who would unreservedly use that formalism if it ever became accessible to them. At present they are a numerically insignificant minority, yet important all the more if only because they are the exception which has the potential to change the rule! This book is for them. A non-trivial book on quantum field theory for

---

[1] *A Contribution to the Critique of Political Economy* (1859).

discriminating readers, for those readers who may not be mathematically literate, yet who have a high capacity for conceptual abstraction.

I have always felt that the physical essence of highly abstract concepts of quantum field theory and of special relativity can be conveyed through an accordingly demanding analysis which, nevertheless, avoids the advanced mathematical formalism. Of course, in agreement with the statement "Everything must be made as simple as possible, but not simpler" (attributed to Albert Einstein), such an approach must advance on a minimum but indispensable level of relevant mathematical formalism if it is to convey the physical essence of quantum field theory at all. Indeed, my first attempt at writing this book with a level of mathematical content below that necessary minimum so as to keep it "as simple as possible", failed dismally. I rewrote it introducing that indispensable minimum level strictly according to the demands of the conceptual approach which I designed for the stated category of readers. In fact, I have exerted every possible effort to introduce it in line with the development of the conceptual analysis while keeping the language associated with that formalism as simple as possible. For example, in the first three chapters I avoid the term "integral" in favour of the non-rigorous but emphatic term "continuous sum".

I have structured this book in levels of increasing difficulty in terms of both, physics and mathematics. The analysis in the first chapter commences from the canonical quantization of classical mechanical systems and the introduction of the basic quantum-mechanical concepts, proceeds to the mathematical foundations of quantum mechanics and eventuates in such advanced abstract concepts as the Dirac notation and the theory of the intrinsic angular momentum (spin). Such quantum-mechanical concepts as are introduced and examined in the first chapter are fundamental to quantum field theory. In line with the conceptual approach which this book advances, quantum mechanics is not treated as an end, but as a means for understanding the subsequent analysis on quantum field theory and it is as such that it is examined in the first chapter. For example, the canonical quantization is not introduced in the rigorous manner of replacing the classical Poisson brackets by commutators, but through the conceptual replacement of fixed classical values by probability distributions. The commutators are accordingly introduced later.

The second chapter is entirely a comprehensive analysis on the interpretation of the wavefunction and of quantum mechanics itself from the Marxist perspective. Using the Marxist dialectical method, I advance a resolute attack on the dominant Copenhagen interpretation with its explicit positivism and implicit idealism. The application of the Marxist dialectical method in quantum mechanics and the Marxist perspective in the second chapter are entirely my approach to the fundamental issue concerning the interpretation of the wavefunction. Such ideas as I express therein, have not appeared in any journal or book. For this reason the analysis in the second chapter treats the stated issue as an end rather than as a means for the subsequent analysis on quantum field theory. I understand that it is not fashionable to discuss the "epistemological implications" of quantum mechanics alongside the theory of Hilbert space, all the more so in a book on quantum field theory, but fashion is none of my concerns! No physical system in Nature is really quantum-mechanical. Everything is governed by the infinitely many degrees of freedom of quantum field theory. However, quantum mechanics derives its physical essence at the non-relativistic low-energy limit of quantum field theory since it is at that limit that quantum fields appear as quantum-mechanical systems. Not only does this fact alone justify the extensive analysis on quantum mechanics, it also justifies the comprehensive analysis on its interpretation.

The third chapter is entirely a treatment of the foremost aspect of advanced quantum mechanics, the path-integral formulation. A detailed examination of the quantum-mechanical path integral is essential for a thorough understanding of quantum field theory since it is in quantum field theory that the path integral becomes, through a highly non-trivial extension, crucially important.

The fourth chapter commences with an overview of the most fundamental aspects of classical field theory before quantum field theory is introduced. The conceptual approach stresses the character of the classical wave fields as superpositions of infinitely many plane waves of energy represented as simple harmonic oscillators and the character of the canonical field quantization as the procedure of quantizing these oscillators. The most general aspects of the free quantum fields - the quantum fields which emerge from the quantization of the classical wave fields - are examined. The analysis is always conceptual with only elements of the relevant mathematical formalism introduced except for the last section which examines the fundamentally important Feynman propagator and related issues of causality. That section is essentially an introduction to the formal quantum field theory. It is as of that section that the level of the mathematical formalism sharply rises in line with the demands of the substantially more complex conceptual analysis.

The fifth chapter is a formal examination of the Klein-Gordon field, the field which serves as an illustration of the most general dynamical aspects of free quantum fields on account of its comparative simplicity. It is formal in the sense that the analysis of the highly abstract concepts necessarily advances on certain aspects of the rigorous mathematical formalism of field quantization. Such necessary aspects (for example, contour integration in momentum space and Green functions) are always introduced and sufficiently explained according to the demands posed by the development of the conceptual analysis. The latter extends to the examination of the formal relation of quantum mechanics to quantum field theory, of the manner in which the outgrowth of quantum mechanics gave rise to quantum field theory and of the reason "relativistic quantum mechanics" signifies a contradiction of terms. The fifth chapter concludes with an analysis on the physical content of path-integral quantization for fields and of the crucially important implications which the path-integral quantization has in quantum field theory, from the issue of maintaining the Lorentz invariance of the theory manifest at all stages to the issue of divergences (infinities) and of renormalization.

The fundamental premise upon which my approach to quantum field theory is based is that fundamentally all physical systems comprise infinitely many degrees of freedom and that, for that matter, the underlying reality of everything is that of the quantum fields. Quantum mechanics with its finite number of degrees of freedom is at best a non-relativistic low-energy approximation to the fundamental reality described by quantum field theory. Particles and the illusion of a finite number of degrees of freedom which they convey are not fundamental. They invariably express the fundamental reality of underlying quantum fields.

Finally, the Appendix is an introduction to special relativity. One of the highly appealing aspects of that theory is that it can almost entirely be conveyed in terms of space-time diagrams. In addition to the necessary mathematical formalism, the Appendix contains a comprehensive analysis on Lorentz transformations, time dilation and length contraction on space-time diagrams. A solid understanding of special relativity is imperative for a serious study of quantum field theory. Readers with no prior experience in special relativity should first study the Appendix before they proceed to quantum field theory in the fourth and the fifth chapter.

Regarding the question of prerequisite knowledge, I assume that the reader is merely familiar with high-school physics and the elementary mathematics associated with it. The only serious prerequisite for reading this book is the capacity for abstract thought, indeed a non-trivial requirement expected of the exceptional category of readers for whom this book has been designed. I should add, however, that without prior knowledge of quantum mechanics, at least at the level of an introductory course, following the analysis in this book will not be an easy task. The less a reader has been exposed to theoretical physics, the more mental effort he or she will have to exert. I reiterate, for that matter, that this book has been designed for the exceptional category of readers who are not afraid to think! Despite its mathematical content, this book is not prohibitively technical. However, in contradistinction to the usual statements made in non-technical books of "popular physics", I do not offer "a fascinating journey to the world of eleven dimensions" or to the "infinitely many parallel universes" from the comfort of your living room because such a journey is simply impossible! Theoretical physics belongs to everybody indeed, but it comes at the price of total commitment!

The ideas expressed in this book are entirely mine. In writing it, I did not collaborate, consult, or discuss anything with anyone. I am honored, however, to acknowledge the decisive influence of professor Hiroomi Umezawa who, in my graduate years in the University of Alberta, stressed to me the crucial importance of thinking always in terms of the infinitely many degrees of freedom of quantum field theory and of understanding that not a single system in Nature is quantum-mechanical.

MATHEMATICAL PRELIMINARIES

The necessary mathematical concepts will be accordingly introduced as the analysis develops. It is more pedagogical to study a concept when the need for it has been understood than to study it in advance, out of the context which necessitates it. However, the following elementary concepts provide the minimum mathematical knowledge necessary for following the development of the subject matter in this book.

1. Complex Numbers

A *real number* is any number whose square is a non-negative (positive or zero) number. An example of a real number is -5, since $(-5)^2 \equiv (-5) \times (-5) = 25 > 0$. The *set of real numbers* extends from negative infinity $(-\infty)$ to positive infinity $(+\infty)$. If we establish a one-to-one correspondence between the set of real numbers and the points of a straight line such that for any two real numbers a and b with $a < b$ the point which corresponds to a precedes the point which corresponds to b then the set of real numbers corresponds to a straight line oriented from $-\infty$ to $+\infty$ (Fig.I(a)).

An *imaginary number* is any number whose square is a negative number. The unit imaginary number is i which can be defined by its property $(i)^2 = (i) \times (i) = -1 < 0$. Any imaginary number is a product of i and a real number, with i itself being $i = 1i$. As another example, consider the product of i and 5. Since $(5i)^2 = -25 < 0$, $5i$ is an imaginary number. The *set of imaginary numbers* extends from negative imaginary infinity $-i\infty$ to positive imaginary infinity $+i\infty$ and also corresponds to a straight line.

From the set of real numbers and the set of imaginary numbers emerges the *set of complex numbers*. A *complex number* z is any number of the form $z = a + ib$, with a and b being real numbers. Since the set of real numbers and the set of imaginary numbers contain the common element $0 = 0i$ they define a plane in the same manner that two lines which intercept each other define a plane. In fact, the line of imaginary numbers is perpendicular to the line of real numbers and a complex number z is represented as a vector in the *complex plane* (Fig.I(b)).

The *complex conjugate* of complex number $z = a + ib$ is the complex number $z^* = a - ib$. The quantity $|z|^2 \equiv zz^* = a^2 + b^2$ is clearly a non-negative real number ($|z|^2 = 0$ only if $a = b = 0$) and defines the square of the *magnitude $|z|$ of complex number z*. The magnitude $|z|$ of z is a non-negative real number which can be conceptualized as the length of the vector in Fig.I(b). Note, in addition, that a real number a is necessarily equal to its complex conjugate since $a = a + i0$ with the implication that $a^* = a - i0 = a$.

2. Functions

I assume that the reader is familiar with the concept of function. If not, the definition of this elementary mathematical concept can readily be found in texts or on the internet. If the independent variable x of function $y = f(x)$ receives values from the set of real numbers or from any subset of that set, $y = f(x)$ is a function of a real variable. Likewise, $y = f(z)$ is a function of a complex variable if z receives values from the set of complex numbers or from any subset of that set. For a function of a complex variable the preceding statements on complex numbers apply to each value $y = f(z)$ and, for that matter, to the function $f(z)$ itself. For a function of a real variable the intuitive concept of continuous summation over its values $f(x_1) + f(x_2) + f(x_3) + ... + f(x_n) + ...$ is captured and

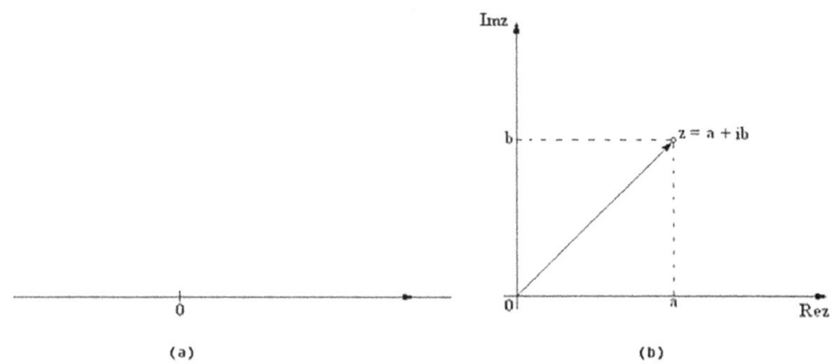

Figure I. The set of real numbers is an oriented straight line (a). The set of imaginary numbers is an oriented straight line perpendicular to that of the real numbers. The two lines define the complex plane each point of which corresponds to a complex number z. The real axis Rez provides the *real part* a of z and the imaginary axis Imz provides the *imaginary part* b of z. The length of the corresponding vector defines the magnitude $|z|$ of z. (b)

rigorously conveyed by the *definite integral* $\int_a^b dx f(x)$ of $f(x)$ (henceforth abbreviated to merely "integral"). In order to help the reader who is not altogether familiar with this concept appreciate its character, I shall use the simple terms "continuous sum" and "continuous summation" until the fifth chapter where the analysis will be at an advanced stage. Sporadic exceptions can be expected in the third chapter where the use of the consolidated term *path integral* will be necessary and inevitable. Additional information about the "continuous sum" of a function will be provided where necessary.

A central aspect of our investigation throughout this book will be the plane waves. We shall extensively refer to them and study them in each relevant physical context. Such an endeavor renders the reference to the *oscillatory exponential function* inevitable. The latter is a complex function of a purely imaginary independent variable $f(z) = f(ix)$ defined as

$$e^{ix} = cosx + isinx$$

Both, the cosine function $cosx$ and the sine function $sinx$ are sinusoidal functions with the independent variable x acting as an angular variable. For that matter, so is e^{ix}. For the purposes of our analysis, this definition suffices. We merely mention here that

(i) e is a real number (approximately equal to 2.71828) and that, consequently, $f(x) = e^x$ is a real function.

(ii) The multiplication of exponential expressions and functions is determined by the rule:

$$e^z e^{z'} = e^{z+z'}$$

(iii) The relation

$$e^{i(-\frac{\pi}{2})} = cos(-\frac{\pi}{2}) + isin(-\frac{\pi}{2}) = -i$$

implies that:

$$-\frac{1}{i} = (-i)^{-1} = \left(e^{i(-\frac{\pi}{2})}\right)^{-1} = e^{i\frac{\pi}{2}} = cos\frac{\pi}{2} + isin\frac{\pi}{2} = i$$

that is:

$$\frac{1}{i} = -i$$

We shall explain all physical concepts involved as our analysis progresses. However, so crucial will the concept of harmonic oscillator and that of plane wave be to the entire analysis throughout the book that if you are not altogether familiar with the oscillatory exponential function, follow from the outset this "rule of thumb": Whenever you see an oscillatory exponential of the type $e^{\pm iwt}$ ($\pm$: $+$ or $-$; the constant w will be determined by the physical context), think of oscillatory motion! Whenever you see an oscillatory exponential of the type $e^{\pm ikx}$ (the constant k will be determined by the physical context) think of a snapshot of a plane (sinusoidal) wave of constant maximum amplitude ("amplitude": displacement from equilibrium) throughout space! Whenever you see an oscillatory exponential of the type $e^{\pm i(wt-kx)}$ think of a plane wave, a sinusoidal waveform of constant maximum amplitude throughout space at each point in which the amplitude itself oscillates in the course of time - that is, of a waveform sinusoidal in both, space and time.

INTRODUCTION

The theory of relativity and the quantum theory, from its inception to the modern theories of quantum gravity, are the greatest achievements of scientific thought. Together they describe Nature from the microscopic scales relevant to sizes far smaller than the size of the atom, to the macroscopic scales relevant to the entire Universe. They both, each in its own way, challenged the traditional outlook on the world and revolutionized humanity's perception of Nature. At first sight, the independent emergence of these two theories at the turn of the twentieth century appears as a quirk of history. It is, nevertheless, invariably the case that, ultimately, scientific ideas have their origins in practical historical activity. In the sixteenth century the advance of technology sharply expanded commodity production and made possible the revolutionary rise of a social class whose ideology replaced the old theocratic outlook on the world by the outlook of the individual. The scientific observation of Nature which inevitably followed made possible the stunning advances in the field of thought which led from the Copernican revolution to the development of Newtonian Physics. In the eighteenth century the industrial revolution generated the material conditions for the development of the theory of thermodynamics and, eventually, of the electromagnetic theory. Inherent in the latter were, in turn, all preconditions of special relativity and of quantum theory, most notably the constancy of the speed of light which rendered the electromagnetic theory incompatible with the old Newtonian outlook, the contradiction between the predicted radiation of accelerated charges and the observed stability of the atoms, as well as the diverging spectral density of the electromagnetic black-body radiation. Throughout human history scientific revolutions have reflected in the field of thought the social transformations which made them possible. It is, for that matter, no coincidence that - alongside the corresponding advances in music and art - the scientific revolution which the theory of relativity and the quantum theory signaled at the turn of the twentieth century occurred at a time that the old social order was crumbling, foreshadowing the great social upheavals which swept across Europe following the first world war.

Yet, ironically, the two theories emerged at odds with each other. At its inception quantum theory was deemed incompatible with a universal speed of light and the concomitant relativity of space and time. The underlying reason for that aberration was an uncritical persistence in the classical dichotomy between particles and fields. These two concepts were mutually exclusive. Particles were discrete entities, whereas fields were extended physical systems. Both were fundamental and irreducible aspects of Nature. The electrons in the atoms had to be fundamental, irreducible and immutable particles and quantum theory ought to explain the stability of the atoms without detracting from that arbitrary premise! It was only on the explicit assumption of non-relativistic energies that quantum mechanics met that demand. The result was, nevertheless, a triumph which signaled a radical break with classical mechanics! Whereas classical particles have a unique position in space at a given moment and follow unique trajectories determined by specified initial conditions, quantum-mechanical particles have at a given moment a possible position associated with a predictable probability for each point in space and infinitely many possible trajectories between two given points! Yet, "the facts are stubborn things"! Despite its conceptual consistency and agreement with observation, quantum mechanics was in conflict with the premise of a relativistic physical reality. A theory founded on the severe restriction of non-relativistic energies and speeds would, at best, have only approximate significance in the description of Nature. No sooner had quantum mechanics developed to a complete theory than it was realized that it is impossible to retain quantum particles as fundamental and irreducible physical entities in a relativistic framework.

The key to the riddle of a relativistic quantum theory lay in the opposite aspect of the old dichotomy. Years before the development of quantum mechanics the thoroughly relativistic classical electromagnetic theory had predicted its own demise! The diverging spectral density of black-body radiation and the photoelectric effect pointed to the existence of the photon. Defying the boldest imagination at the time, this, in turn, pointed to the fact that the electromagnetic field itself is in its entirety a quantum field! It was only after several futile attempts to develop a relativistic theory of quantum particles that such an inherent implication gradually came to the fore. Maxwell's classical electromagnetic theory envisages the electromagnetic field as the product of sources (charged particles and currents). By dint of abstraction, it also describes the propagation of electromagnetic waveforms dissociated from all sources. If the procedure of quantization which realizes the transition from classical mechanics to non-relativistic quantum mechanics is consistently applied to the abstract classical electromagnetic wave field, the result is a relativistic quantum field whose inherent probabilities describe relativistic quantum particles as the quantum-mechanical probabilities described non-relativistic quantum particles. The quantization of the classical Maxwell field yields the photon!

This is the essence of the matter! If the photon is an expression of an underlying quantum electromagnetic field then the electron must also be an expression of an underlying relativistic quantum electron field. So should each known particle in Nature. Thus, the outgrowth of non-relativistic quantum mechanics and several unsuccessful attempts at a formulation of a "relativistic quantum mechanics" had eventuated in the most advanced and successful scientific theory: Quantum Field Theory! All that was required, after all, was an association of each known particle with a classical relativistic wave field and the procedure of field quantization would work its magic to realize each quantum particle as an expression of an underlying quantum field! No longer was the electromagnetic field the product of sources. At fundamental quantum level all fields are primary entities! They interact with each other but are never produced. No longer were particles fundamental, irreducible and immutable entities. The fundamental reality is that of quantum fields! Particles are merely excitations and expressions of that fundamental reality. The classical world as we perceive it is the product of quantum fields! Classical particles were already known to emerge as the semi-classical limit of quantum-mechanical particles. However, quantum field theory transcends the semi-classical limit of quantum mechanics. From the physically consistent and rigorous perspective of quantum field theory both, classical particles and classical fields, emerge as the semi-classical limit of quantum fields.

Within its domain of validity quantum field theory is an exact theory. Quantum mechanics, with its approximate physical significance and limited scope, is not a prerequisite for it. One directly arrives at quantum field theory by quantizing the corresponding classical field theory. Of course, quantum field theory uses the language of quantum mechanics by way of interpreting certain aspects of quantum fields. This, however, only reenforces the preceding premise: Excited states of quantum fields are interpreted as particles and superpositions of such states, as quantum-mechanical wavefunctions. Such an interpretation is not an allusion to quantum mechanics. It is, instead, a statement to the effect that quantum fields express themselves in terms of entities which we recognize as particles and describe in the convenient language of quantum mechanics, particles which are nevertheless not fundamental precisely because - unlike in quantum mechanics - they invariably express the underlying fundamental reality of quantum fields. In fact, although at best only an approximation to physical reality, quantum mechanics cannot be dismissed as a physically irrelevant theory in view of the fact that at the non-relativistic low-energy limit all quantum fields assume the semblance (but never the reality) of a quantum-mechanical system.

If quantum fields are the fundamental reality, why is quantum field theory not unconditionally valid and exact? Why is it valid only within a certain domain of space-time scales? Unlike quantum mechanics, quantum field theory abounds with infinities! At infinitesimally small space-time scales signifying arbitrarily high values of energy and momentum, physical quantities which describe the dynamical behavior of quantum fields diverge. They yield physically meaningless infinities. The technique of *renormalization* which obviates such infinities in order to yield a finite theory whose long-distance behavior is unaffected by its divergent character at infinitesimally small space-time scales was first applied in quantum electrodynamics (the theory concerning the interaction between the electron field and the photon field) with impressive success. The result was a renewed confidence in quantum field theory as the fundamental theory of physical reality. Gradually, renormalizable theories (such as quantum electrodynamics) tacitly came to be regarded as physically rigorous despite their divergent behavior at minute space-time scales.

Nevertheless, the divergent behavior remained an issue at the core of quantum field theory, strongly pointing to its limitations and, thereby, to infinities as a measure of our ignorance as to an "unknown physics" at minute space-time scales. Worse still, all attempts at formulating a consistent quantum theory of gravity met with dismal failure. Gravity remained adamantly non-renormalizable! Such a physically unacceptable behavior eventually challenged the status of quantum field theory as a fundamental theory. The divergent behavior of quantum field theory at minute space-time scales was eventually acknowledged as strong indication that below a certain space-time scale quantum field theory ceases to be valid and must be replaced by a theory which satisfies two fundamental requirements. First, that it be a finite and thereby a physically rigorous theory at such space-time scales. Second, that it reduce to quantum field theory above those scales, rendering quantum field theory an effective theory, a low-energy approximation to an underlying actually fundamental and presumably ultimate theory. The only theory which successfully meets these two requirements today is string theory.

Historically, the formal procedure of quantization was first applied to particles, not to fields. It is essentially a reflection of that historical experience the fact that today it is a lot easier to understand quantum field theory by first studying quantum mechanics than without any recourse to it. This is the course which I have chosen in this book. I shall discuss the conceptual foundations of quantum mechanics and the debate over its interpretation as a condition for the examination of the far more intricate and complicated quantum field theory. Central to this objective will be the four general premises inherent in the present discussion: (i) All physical systems are fundamentally quantum systems. (ii) All quantum systems are fundamentally quantum fields. This includes the unknown physics which generalizes quantum field theory below the "cut-off" space-time scale. (iii) All classical systems emerge as the semi-classical limit of the dynamical behavior of corresponding quantum systems - fundamentally, of corresponding quantum fields. (iv) All quantum theories must necessarily be developed from corresponding classical theories and, for that matter, it is imperative that a precise distinction be made between systems of particles and fields as a condition for quantization. The examination of the content which each of these four premises holds will reveal the essence of quantum field theory. Since the analytical process toward this objective invariably advances from classical physics to quantum physics, the premise which must constitute the inception of our exploration is obviously the fourth one.

PART I

QUANTUM MECHANICS

CHAPTER 1.

CANONICAL QUANTIZATION

1. Degrees of Freedom and Quantization

Every physical system is characterized by a number of independent variables. They are the variables which provide all information as to the dynamical behavior of the physical system and are known as the physical system's *degrees of freedom*. For example, a single free particle can move independently in each of the three dimensions of space. Hence, in physical space, a free particle has three degrees of freedom. By extension, a physical system which consists of N non-interacting particles has 3N degrees of freedom. A particle whose motion is constrained in one dimension of space has one degree of freedom. An example is the particle which executes oscillations between the two end points of a linear segment, or between the two end points of a circular arc. As an example of a more composite situation consider two particles joined by a massless string. Such a system has translational, rotational and vibrational degrees of freedom. The translational degrees of freedom can easily be identified by reducing the system of the two particles to one particle situated at the system's centre of mass with mass equal to the sum-total of the two masses. Thus, the system has three translational degrees of freedom. At once, each particle rotates around the axis perpendicular to the line joining the two particles. There are two such axes at right angles to each other. Knowledge of the rotational motion of either particle about either axis immediately yields knowledge of the rotational motion of the other particle about the same axis. Therefore, with respect to either axis only one parameter (degree of freedom) is required for the description of the rotational motion, which implies that the system has two rotational degrees of freedom. By the same reasoning, knowledge of the vibrational motion of either particle immediately yields knowledge of the vibrational motion of the other, a fact which implies one vibrational degree of freedom for the system. Consequently, the system of the two particles has overall six degrees of freedom. This is, incidentally, a good description of a gas molecule consisting in two atoms. A quantity of gas consisting of N such diatomic molecules is a physical system of 6N degrees of freedom.

Theoretical physics makes a central distinction between two categories of physical systems. One of them consists in physical systems which are described by a finite number of degrees of freedom such as in the preceding examples. That category also includes the theoretical possibility of infinitely many degrees of freedom on condition that they form a discrete set, a set whose members can be counted from one to infinity[2]. That is the category of *mechanical systems*, the object of *classical mechanics* and *quantum mechanics*. The other category consists in physical systems which comprise an uncountable number of degrees of freedom - that is, an infinite number of degrees of freedom which form a continuous set. That is the category of *fields*, the object of *classical field theory* and *quantum field theory*. Examples from classical physics are, of course, the electromagnetic field whose infinity of degrees of freedom stems from the fact that the field-strength receives, at each moment, a unique value at each point in space and the gravitational field which, in the spirit of general relativity, is curved space-time itself with the curvature receiving a unique value at each space-time point.

---

[2]On rigorous terms, a set comprising infinitely many elements is discrete if these elements can be put in a one-to-one correspondence with the set of natural numbers.

Past this point we shall abandon the distinction between countable and uncountable infinities and merely refer to "mechanical systems" as those physical systems which comprise a finite number of degrees of freedom and to "fields" as those physical systems which comprise infinitely many degrees of freedom.

Until the turn of the twentieth century Nature had been perceived on exclusively classical terms. The only physical systems of infinitely many degrees of freedom known at the time were the electromagnetic field and the gravitational field which by 1916 was being described in terms of space-time curvature. Alongside these two classical field theories stood classical mechanics, the theory of classical physical systems comprising a finite number of degrees of freedom, the theory of classical particles. I have already outlined in the Introduction the essence of classical mechanical systems. Let me now concretize that loose description in the physically rigorous context specified by the degrees of freedom. The characterization "classical" in classical mechanics expresses the fact that, given specific *initial conditions* such as the position and momentum (the product of the particle's mass times its velocity as measured by a specified observer) of each particle or body in the physical system at a specified moment, the solution to the equations of motion provides precise knowledge of the dynamical behavior of that system, yielding the evolution of each degree of freedom and, thereby, the position and momentum of each particle at any future moment. Equivalently, given specific initial conditions, the solution to the equations of motion specifies a unique trajectory for each constituent element of the classical mechanical system. Although perfectly equivalent to the first, this second definition of "classicality" has special significance in its own right, as will become obvious when we examine it in the context of the quantum theory.

In field theory the classical character of a field is accordingly defined if the concept of "initial conditions" for position and momentum is replaced by the concept of *boundary conditions* for the strength of the field on certain boundaries, on surfaces which may lie at finite or at infinite distance in space or even at infinite distance in time, as is frequently the case in general relativity. Again, given such boundary conditions, the solution to the field equations provides the strength of the field at each space-time point and specifies a unique configuration of the field evolving in space-time, the field-equivalent concept to the classical trajectory in mechanical systems.

Let us now turn to the relation between degrees of freedom and quantization. The theory of quantum mechanics is, as its name suggests, a theory which describes quantum systems comprising a finite number of degrees of freedom and, for that matter, a theory which describes particles as fundamental entities. As stated in the Introduction, such a theory is incompatible with the theory of relativity. No sooner had quantum mechanics emerged than it turned out to be, at best, a low-energy approximation. Flying on the face of the age-old dichotomy between particles and fields, quantum field theory emerged and began to develop only a couple of years after the complete formulation of quantum mechanics. Its revolutionary central premise which declared particles to be expressions of an underlying fundamental reality of quantum fields threw scientific thought in turmoil. It is a reflection of the conceptual confusion of that time the fact that, even today, the purely historical distinction between the "first quantization" of quantum mechanics and the "second quantization" of quantum field theory is sometimes accorded the incorrect interpretation of the former as a prerequisite for the latter. The comprehensive analysis in the fourth chapter will render transparent the reason the "second quantization" is utterly irrelevant to the "first quantization" and quantum field theory is utterly irrelevant to quantum mechanics. At this stage, we merely outline the difference between the "first" and the "second" quantization.

Quantization was first introduced in the electromagnetic radiation by Max Planck (1900). It offered a theoretical resolution of the issue posed by the classical prediction of infinite total energy emitted by a *black body* at thermal equilibrium. The subsequent work of Albert Einstein (1905) established the consequence of Planck's quantization, the photon, as an actual objective entity. The next advance came in 1913 with Niels Bohr's proposition to the effect that the energy is, itself, quantized. The Bohr-Sommerfeld-Wilson model (1915 - 1916) resolved the outstanding issue regarding the stability of atoms by describing the latter as systems of finite degrees of freedom whose energy spectrum is discrete. In sharp contrast to the classical prediction that each electron in an atom will continuously radiate all its energy and rapidly collapse onto the nucleus, the physically acceptable energy levels in atoms form a discrete set. Despite its phenomenological success, that model was, nevertheless, inherently inconsistent as it predicted classical trajectories for the electrons in a physical system which accommodated a discrete energy spectrum. The Bohr-Sommerfeld-Wilson model had quantized the energy but not the position. The complete theory of quantum mechanics which was established ten years later by the work of Werner Heisenberg (1925) and Erwin Schrödinger (1926) sharply differs from the preceding Bohr-Sommerfeld-Wilson model with its inherently inconsistent assumption of classical trajectories in an otherwise quantum system in that the quantization of the position and the momentum is central to the theory's formalism and conceptual foundations. At each moment, the electron has a predictable probability that its position be a specific point in space. Equivalently, it has at each moment a predictable space-independent probability that its momentum receive a specific value. In sharp contrast to the classical particles, under no circumstances does a quantum particle have a specific position x and a specific momentum p at x. In fact, in a very concrete sense which we shall thoroughly explore in this and the next chapter, it is described either in terms of position, or - equivalently and independently - in terms of momentum in such a manner that precise knowledge of either signifies complete absence of knowledge of the other, in line with Heisenberg's *uncertainty principle* $\Delta x \Delta p \sim \hbar$.

The uncertainty principle and its implications will be the object of our detailed examination, especially in the second chapter. At this point we only mention that it is not really a principle. A principle is a premise which is neither the result of logical deduction, nor the result of mathematical calculation. By contrast, Heisenberg's uncertainty relations were derived in the context of the mathematical formalism of quantum mechanics. The reasons for which the uncertainty relations were tacitly elevated to the status of "principle" are historical and relate to attempts at conceptualizing quantum mechanics by starting from the uncertainty relations.

Quantum mechanics signifies the first consistent quantization accomplished in theoretical physics. The essence of this first quantization consists in the replacement of a classical particle's position at any given moment by a probability distribution for position, or by the equivalent replacement of the classical particle's momentum at that moment by a space-independent probability distribution for momentum. "Space-independent" means that the probability distribution for momentum at any given moment is not in any respect associated with any point in space. It is physically meaningless to ask what the probability is that the quantum particle have "this" or "that" value of momentum at a specific point in space at a specific moment. The physically meaningful question consists in asking what the probability is that the quantum particle have "this" or "that" value of momentum at that moment.

As we shall see in the fourth chapter, the first quantization and the entire theory of quantum mechanics which is founded upon it only have limited significance in the description of physical reality. Such limitations, however, must not be associated with the mathematical and conceptual foundations of the first quantization and of quantum mechanics. The first quantization is, indeed, conceptually and mathematically consistent and quantum mechanics is, indeed, a complete physical theory in the sense that both completely describe quantum systems characterized by a finite number of degrees of freedom. The actual discrepancy with physical reality lies in the physically inconsistent assumption that Nature accommodates systems characterized by a finite number of degrees of freedom, not in the mathematical and conceptual foundations of quantum mechanics. This is a very important aspect of the quantum theory regardless of whether it describes mechanical systems or fields. The general replacement of the unique classical values by probability distributions signifies the transition from a classical system to a quantum system regardless of whether that quantum system actually exists or is, instead, a low-energy approximation to what actually exists. In the quantization of mechanical systems such a replacement concerns primarily the position and the momentum and, despite its conceptual and mathematical consistency, only has approximate and limited physical significance. On the contrary, in the quantization of fields such a replacement concerns observable quantities which are functions of space-time and has precise and rigorous physical significance. In the quantization of mechanical systems the position is a degree of freedom and, thus, a dynamical variable. In the quantization of fields, the position is only a label of the actual dynamical variable and of the infinitely many degrees of freedom which the dynamical variable signifies.

2. Canonical Quantization and Quantum Mechanics

Canonical quantization is the technical term for the general replacement of unique classical values by probability distributions. We shall explore the mathematical content of that replacement in the next section. The replacement of unique classical values by probability distributions is the physical content of canonical quantization. Unlike the "first" and the "second" quantization which only have historical significance, the canonical quantization signifies a formal procedure of conceptual and mathematical rigor in both, quantum mechanics and quantum field theory.

Whereas classical mechanics predicts a unique position for a particle at a given moment, the canonical quantization replaces that position by infinitely many possible positions throughout space each of which has a predictable probability of being the quantum particle's actual position. This standpoint radically detracts from the consolidated view which maintains that "the quantum particle has at a given moment a predictable probability of being found at a specific point". The latter is erroneous in view of the fact that the position which the quantum particle has at a given moment is only an aspect of the quantum particle, not the quantum particle itself. The quantum particle is a particular unity of that position and the quantum particle's infinitely many possibilities and potentialities which can neither be physically nor logically reduced to the quantum particle's position. Most physicists would dismiss all this as "philosophical hair-splitting", but they are mistaken! This is a matter of deep physical essence, the object of a detailed examination in the next chapter.

I have already remarked that the replacement of the classically unique value of position by infinitely many possible values has as an immediate consequence the equivalent replacement of the classically unique value of momentum at the same moment. For that matter, in contrast to their classical counterparts, quantum particles are not characterized by a unique position and a unique momentum at each moment, but by a quantum state described either by a probability distribution for position or, equivalently, by a probability distribution for momentum. Such probability distributions are not merely mathematical constructs. In a classically accessible region of space (which we shall soon define), they are observable entities. Appropriate experimental procedures which theoretically involve infinitely many experiments on infinitely many identical systems confirm the probability distribution for all possible values of a physical attribute.

In the mathematical formalism of quantum mechanics the probability distribution is necessarily a derivative concept. The primary concept is the solution to the *Schrödinger equation*, the *wavefunction*, which for purposes of illustration we shall consider here a function of one-dimensional space described by coordinate x. In general, the wavefunction $\psi(t, x)$ is a complex function of position and of time. At each point x and at each moment t it receives the complex value

$$\psi(t, x) = \psi_1(t, x) + i\psi_2(t, x)$$

with $\psi_1(t, x)$ and $\psi_2(t, x)$ being real numbers for the given values of x and t. Being a complex function, the wavefunction $\psi(t, x)$ is an unobservable function. The *complex conjugate* function to $\psi(t, x)$ is

$$\psi^*(t, x) = \psi_1(t, x) - i\psi_2(t, x)$$

Mere inspection confirms the immediate consequence of the statement in the Mathematical Preliminaries: The product of the wavefunction and its complex conjugate function is a *positive-definite function* (an invariably positive function which may only be zero at isolated points in space):

$$\psi(t, x)\psi^*(t, x) = [\psi_1(t, x)]^2 + [\psi_2(t, x)]^2 \equiv |\psi(t, x)|^2 \tag{2.1}$$

where the positive-definite function

$$|\psi(t, x)| = +\sqrt{[\psi_1(t, x)]^2 + [\psi_2(t, x)]^2}$$

is the *magnitude of* $\psi(t, x)$ at moment t and point x.

At the foundation of quantum mechanics lies the interpretation of $|\psi(t, x)|^2$ as the evolving observable *probability density* for position. The latter yields at moment t the probability that the position be any point between a given point x and the point $x + dx$ at minute distance dx from x and may be thought of as that probability divided by dx. That is, *the evolving probability that the position of the particle be in the immediate vicinity of point x is* $|\psi(t, x)|^2 dx$. This observable probability density implies the unobservable *probability amplitude* $\psi(t, x)$ that at moment t the position of the quantum particle be in the immediate vicinity of point x. The relation between the probability density and the probability amplitude parallels that between the energy density and the amplitude of a classical wave and underscores the wave-like character of the probability distribution for position. The difference, as stated, is that, unlike the amplitude of a classical wave, the probability amplitude is unobservable, precisely as expected of a wave of probability.

Thus, the probability density $|\psi(t, x)|^2$ yields the evolving observable probability distribution for position. The wavefunction $\phi(t, p)$ as a function of momentum and of time the square $|\phi(t, p)|^2$ of whose magnitude $|\phi(t, p)|$ yields the corresponding probability distribution for the momentum of the quantum particle is accordingly defined from $\psi(t, x)$, as we shall see. We stress again that being unobservable mathematical quantities, both wavefunctions, $\psi(t, x)$ and $\phi(t, p)$, are complex functions. This is in contrast to the observable probability distributions $|\psi(t, x)|^2$ and $|\phi(t, p)|^2$ whose values are invariably positive real numbers, as all probabilities are. Note, incidentally, that the demand for positivity alone does not justify the interpretation of $|\psi(t, x)|^2$ as probability density. It should be clear that infinitely many other functions constructed from $\psi(t, x)$ - such as $|\psi(t, x)|$ - are consistent with this demand. Advance to a certain conceptual and mathematical level is necessary for the rigorous justification of such an interpretation. We defer it, for that matter, to the seventh section of the fifth chapter.

As a quantum state is equivalently described by either wavefunction, the two wavefunctions are closely related. Knowledge of either, yields knowledge of the other. The more "concentrated" the wavefunction for position $\psi(t, x)$ is, the more "dispersed" the wavefunction for momentum $\phi(t, p)$ is and vice versa (Fig.1). The smaller the region of space of highest probability is, the larger the range of the quantum particle's momentum values of highest probability is and vice versa. The region of space of highest probability is qualitatively expressed by the uncertainty for position Δx and the range of momentum values of highest probability is qualitatively expressed by the uncertainty for momentum Δp. Fig.1 conveys, for that matter, the essence of *the Heisenberg uncertainty principle*:

$$\Delta x \Delta p \sim \hbar$$

In view of the Planck constant[3] $\hbar$ which characterizes all quantum systems, the uncertainty principle brings Δx in an inversely-proportional relation to Δp. The more $\psi(t, x)$ reveals information about the position, the less the associated $\phi(t, p)$ reveals information about the momentum. If $\psi(t_0, x)$ reveals accurate information about the position ($\Delta x = 0$) at some moment $t = t_0$ then $\phi(t_0, p)$ is characterized by $\Delta p \to \infty$. In such a quantum state any value of momentum p is as probable as any other and no information

---

[3]This is actually the *reduced Planck constant* $\hbar = \frac{h}{2\pi}$, with h being the Planck constant itself.

about the momentum exists (Fig.2(a)). If, on the contrary, $\psi(t,x)$ reveals no information about the position, equivalently, if any position in infinite space is as probable as any other ($\Delta x \to \infty$) then $\phi(t,p)$ reveals accurate information about the momentum ($\Delta p \to 0$). Thus, as a direct consequence of the homogeneity which characterizes empty space, the quantum state for which $\Delta x \to \infty$; $\Delta p \to 0$ is the state of a free quantum particle described by a wavefunction which yields constant probability density for position throughout infinite empty space. The wavefunction $\psi(t,x)$ of a free quantum particle whose momentum is fixed at a known value is a *plane wave of probability*, a quantum state characterized by that fixed value of momentum (Fig.2(b)). We need hardly stress, of course, that the uncertainty relation $\Delta x \Delta p \sim \hbar$ is inherent in the dynamical behavior of the quantum particle and is, thus, utterly irrelevant to observation and to the precision of whichever detectors.

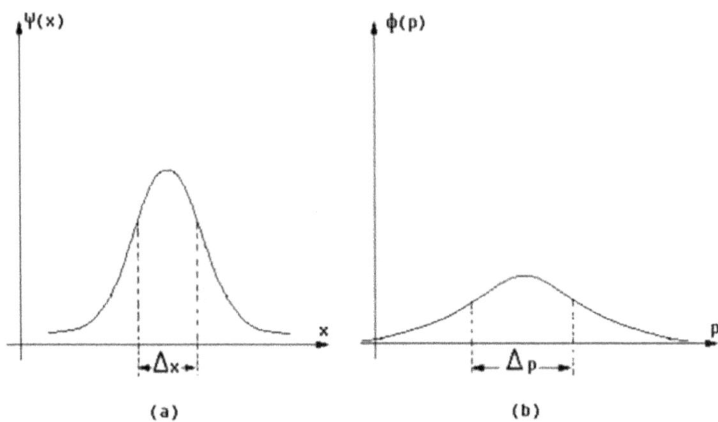

FIGURE 1. An example of a wavefunction for position $\psi(x)$ at a specific moment (a) and of corresponding wavefunction for momentum $\phi(p)$ at the same moment (b). The uncertainty principle is inherent in the inversely-proportional relation between the corresponding widths (uncertainties) Δx and Δp.

Such a qualitative description of the relation between $\psi(t,x)$ and $\phi(t,p)$ as we have just made is far from sufficient. In order to delve into the intricacies of quantum field theory it is essential that we understand the precise relation between the two wavefunctions. The mathematically-literate readers know that, formally, the relation in question amounts to a simple mathematical transformation. Nevertheless, we would like to avoid the highbrow reference to *Fourier transformations* in this conceptual approach. We defer, for that matter, the question of the precise relation between $\psi(t,x)$ and $\phi(t,p)$ in order to better understand the plane waves of probability first. In fact, plane waves of probability are of central importance in relativistic quantum field theory. Thus, it is imperative that we derive maximum insight as to their nature in the present context of non-relativistic quantum mechanics.

Plane waves are sinusoidal wavefunctions of infinite uncertainty ($\Delta x \to \infty$) which extend and undulate throughout space with constant maximum amplitude. At any moment, they repeat themselves in space over a distance of one wavelength λ (Fig.2(b)). As a consequence of their periodic nature, the square $|\psi(t,x)|^2$ of a plane wave's magnitude coincides with the square of the wave's maximum amplitude and yields the stated constant probability density for position. The unique value of momentum p_0 which equivalently describes that quantum state implies a fixed value of energy $E_0 = \frac{p_0^2}{2m}$. This is readily understood. A non-relativistic free classical particle has a fixed value of kinetic energy E_0 determined by the particle's fixed momentum p_0 and mass m through the simple relation $E_0 = \frac{p_0^2}{2m}$. We expect, for that matter, that the free quantum particle which emerges from the quantization of the free classical particle be characterized by the same relation.

Plane waves of probability bear an obvious resemblance to the classical plane waves of energy. Like the latter, they are sinusoidal waveforms. The square of their maximum amplitudes yields the constant probability density which they signify throughout space as the square of the maximum amplitudes of the classical plane waves yields the constant energy density which characterizes those waves. This resemblance underscores a deeper relation as we shall now show.

In quantum mechanics the total probability for position is the analogue to the total classical-wave energy since the former must be as conserved in the course of time as the latter. As the total energy which a classical wave propagates in space remains constant in the course of time, so does the total probability. This is best understood by reflecting on the category of those wavefunctions which dwindle adequately fast with distance so as to yield a finite total probability throughout space (we shall see what the case is with wavefunctions, such as plane waves, which do not fall in that category). Since at any given moment it is certain that the quantum particle has a position at some point in infinite space, the total probability for position throughout space is 100%. Such certainty to the effect that the quantum particle has a position somewhere in infinite space naturally remains constant in the course of time!

The resemblance between plane waves of probability and classical plane waves of energy, as well as the analogy between total probability for position and total wave energy demonstrate that all probability distributions have structural similarities to classical waves. Like the classical waves, probability distributions extend in space and their evolution parallels that of the classical waves. They form superposition patterns as classical waves do and manifest the interference patterns which classical waves have. As an elastic chord which extends indefinitely in both directions can vibrate at any frequency and has, for that matter, a continuous spectrum of frequencies, so probability distributions which do not vanish at arbitrarily large distances have a continuous spectrum of energies. As an elastic chord of finite length whose two end points are fixed, such as a guitar chord, can vibrate only at certain frequencies and has, for that matter, a discrete spectrum of frequencies, so probability distributions whose values fall off to zero at arbitrarily large

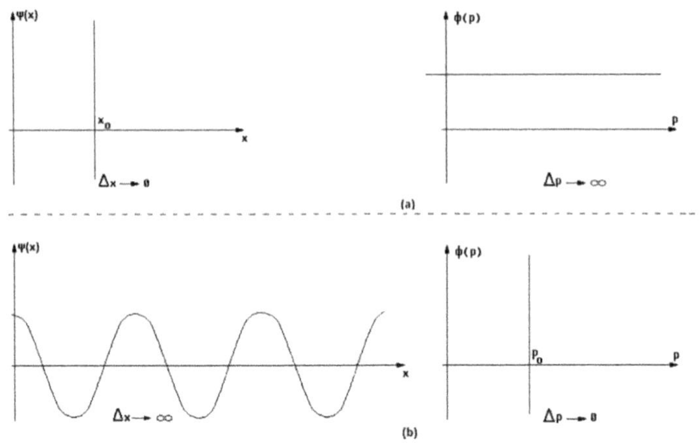

FIGURE 2. The δ-*function* and the plane wave: The wavefunction $\psi(x)$ which at a certain moment yields precise knowledge of the quantum particle's position at x_0 is a δ-function, a vertical straight line at x_0. Equivalently, at the same moment the quantum particle is described by a constant wavefunction for momentum $\phi(p)$ whose infinite uncertainty signifies the complete absence of information about the momentum (a). A quantum particle of given momentum p_0 in empty space is a plane wave of probability, a state of fixed momentum p_0. Each point in space is as probable a position as any other, a fact signifying the complete absence of information about the position. The equivalent description in terms of $\phi(p)$ is a δ-function which expresses precise knowledge of the particle's momentum p_0 (b).

distances have a discrete spectrum of energies. These structural similarities are the inevitable consequence of quantization, of the replacement of unique classical values by probability distributions. For example, the observed discrete energy spectra of particles which are confined in small regions of space, such as the energy spectra of the electrons in atoms, can only be explained if particles are quantum entities described by probability distributions. For only waves of probability are consistent with the discontinuities of confined classical waves while at once describing discrete particles, physical entities whose position may only be detected at a point in space.

We may now appreciate the deeper relation which was announced above. Viewed from the perspective of the structural similarity between plane waves of probability and classical plane waves of energy, the fundamental relations $p = \hbar k$ and $E = \hbar w$ between the fixed momentum p and fixed energy E of a plane wave of probability and the fixed wave number $k = \frac{2\pi}{\lambda}$ and fixed frequency w of a classical plane wave of the same wavelength λ are obvious. At once, the Planck constant $\hbar$ can be seen to have a fundamental physical significance. Its presence in the mathematical description of a physical system is a statement of the fact that the physical system in consideration is a quantum system. This is as valid in quantum mechanics as it is valid in quantum field theory where - as stated in the Introduction and will be rigorously established much later - quantum particles emerge through the quantization of corresponding classical wave fields (N1).

What of the unanswered question? Viewed in terms of plane waves of probability, the precise relation between the wavefunction for position and the wavefunction for momentum is rather obvious. For as an arbitrary classical wave of energy is a superposition of infinitely many plane waves of energy, so a wavefunction $\psi(t,x)$ is a superposition of infinitely many plane waves of probability. We note in this respect that since the amplitude of a plane wave of probability oscillates at each point x with a frequency directly determined by the wavelength λ and thereby by the momentum $p = \hbar k = \hbar\frac{2\pi}{\lambda}$, it is instructive to separate it from the time-independent plane wave of probability (the plane wave as it is at a moment) and state the mathematical expression of a plane wave of probability as $\phi(t,p)\psi_p(x)$ with $\psi_p(x)$ being the plane wave which corresponds to constant momentum p (Fig.2(b)) and with $\phi(t,p)$ being the oscillating amplitude of that wave. In fact, using the information in the Mathematical Preliminaries we can directly deduce that $\phi(t,p) = c(p)e^{-\frac{i}{\hbar}\frac{p^2}{2m}t}$ and $\psi_p(x) = \frac{1}{\sqrt{2\pi\hbar}}e^{\frac{i}{\hbar}px}$. We defer this to the more conducive context of the next section and emphasize here the relation between the two wavefunctions. The mathematical significance of $\psi(t,x)$ as a superposition of infinitely many plane waves of probability is

$$\psi(t,x) = \int_{-\infty}^{+\infty} dp\,\phi(t,p)\psi_p(x) = \frac{1}{\sqrt{2\pi\hbar}}\int_{-\infty}^{+\infty} dp\,c(p)e^{-\frac{i}{\hbar}\frac{p^2}{2m}t}e^{\frac{i}{\hbar}px} =$$

$$\frac{1}{\sqrt{2\pi\hbar}}\int_{-\infty}^{+\infty} dp\,c(p)e^{\frac{i}{\hbar}(px-\frac{p^2}{2m}t)} \tag{2.II}$$

that is, the continuous sum of infinitely many plane waves over all values of p from negative infinity to positive infinity on the understanding that $p < 0$ signifies momentum equal in magnitude and opposite in direction to $p > 0$: The precise relation between the wavefunction for position $\psi(t,x)$ and the wavefunction for momentum $\phi(t,p)$ is that in (2.II).

We may now examine the implications which symmetries in space and time have for quantum-mechanical systems and introduce the mathematical formalism which constitutes the language of quantum mechanics. At the foundation of that formalism lies the concept of operator. We shall examine the mathematical nature of operators, the manner in which they act on wavefunctions as well as the implications of such an action and defer each operator's explicit mathematical expression to a stage at which the accumulated experience will render its meaning transparent. The ensuing analysis will inevitably be somewhat involved. If you are completely unfamiliar with the formal concepts which will be introduced, be patient. You shall understand everything with some effort.

3. Mathematical Foundations

I. Quantum Mechanics and Symmetries

The classical parabolic potential in Fig.3 is the potential of the simple harmonic oscillator, a potential whose attractive nature causes a particle attached to the free end of a massless spring, or of a massless string, to reverse its direction of motion after reaching maximum displacement from the point of equilibrium. In this section, we shall consider the parabolic potential representative of all attractive classical potentials and arrive at general conclusions as to the dynamical behavior of classical and quantum particles within those potentials. The parabolic potential itself and the associated simple classical and quantum harmonic oscillator will be the object of the specific analysis in the sixth section of the present chapter.

Let us first contrast the motion of a free classical particle in empty space against the motion of a classical particle within an attractive potential. In either case, the total energy E which the classical particle has is a matter of initial conditions (and, thus, of choice) and the possible values of E form a continuous set. However, whereas the motion of a free classical particle is unrestricted in infinite space, the classical motion within an attractive potential is restricted. This follows immediately from elementary considerations. The total energy is the sum-total of the kinetic energy $\frac{p^2}{2m}$ and the potential energy $V(x)$ which the classical particle has - that is, $E = \frac{p^2}{2m} + V(x)$. Since the kinetic energy is an invariably positive quantity, the particle's motion within an attractive potential is subjected to the constraint $E - V(x) \geq 0$ and is, thus, bounded by points x_1 and x_2 in Fig.3. The set of points between x_1 and x_2 defines the classically accessible region of space. It is only within that region that a classical particle of total energy E can be at any moment. The two regions respectively defined by the set of points to the left of x_1 $(x < x_1)$ and to the right of x_2 $(x > x_2)$ are, for that matter, the classically inaccessible regions of space.

Pivotal to such considerations is the constant total energy E of both, the free and the restricted particle. Underlying that constancy of E is the static character of the potential in Fig.3, the fact that it remains invariant in the course of time. On rigorous terms, the attractive potential is characterized by *time-translational invariance*. As an immediate consequence, the dynamical behavior of the particle within it as expressed by the particle's equation of motion is also characterized by time-translational invariance: It is independent of the initial moment at which we choose its total energy E. In turn, $E = \frac{p^2}{2m} + V(x)$ remains constant in the course of time. The kinetic and potential energy vary in the course of time, but their sum-total remains constant. Whence, *time-translational invariance implies energy conservation*! This is as relevant to empty space $(V(x) = 0)$ as it is relevant to static potentials.

We have, thus, arrived at a fundamental aspect of classical dynamical behavior. Symmetry under temporal translations, that symmetry which time-translational invariance signifies, implies energy conservation. This is a general result. The invariance of a physical system under a specific set of transformations implies a corresponding conserved quantity. Note in this respect that, like its energy, the momentum p of a free particle

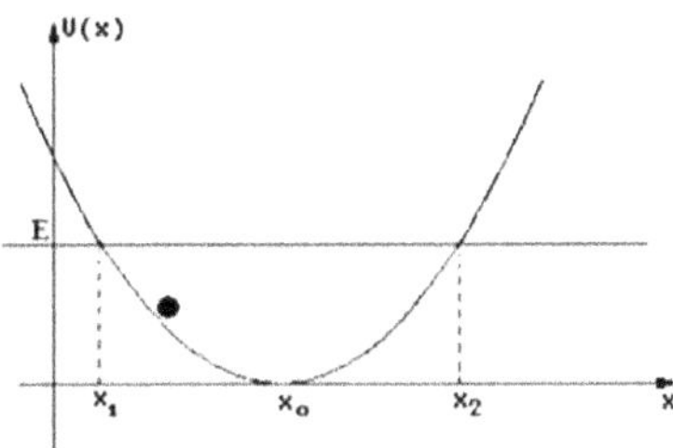

FIGURE 3. A classical particle in one-dimensional static attractive potential. The fixed total energy $E = \frac{p^2}{2m} + V(x)$, being equal to that value of $V(x)$ for which $\frac{p^2}{2m} = 0$, determines the boundaries of classical motion: Since $\frac{p^2}{2m} \geq 0 \implies E \geq V(x)$, the particle is confined in the region of space between x_1 and x_2.

is constant. This is because empty space is homogeneous. As it has the same properties throughout its infinite extent, at no point can a force be exerted on a free classical particle. Of course, the absence of force implies the absence of acceleration. Whence, *space-translational invariance implies momentum conservation!*

Let us now quantize the free classical particle and the classical particle in an attractive potential. In both cases, the procedure of canonical quantization replaces at each moment the unique classical position by a probability distribution over all space. In the case of the attractive potential, this implies the extension of such a probability distribution into the classically inaccessible regions (Fig.4). We already know that the state of a free quantum particle of given momentum p is a plane wave of probability of momentum p and energy $E = \frac{p^2}{2m}$. Since the given momentum p can have any value, the spectrum of possible states is continuous. The situation with a quantum-mechanical particle in a static attractive potential is manifestly different. For although in analogy to the classical situation there are infinitely many quantum states each of which is characterized by a specific constant (fixed) value of energy, unlike the classical situation the set of infinitely many such states is discrete - that is, $\psi_1(t, x)$, $\psi_2(t, x)$, $\psi_3(t, x)$, ..., $\psi_n(t, x)$, ... The reason is that, upon quantization, the restricted character of the classical motion translates to the fact that all corresponding wavefunctions vanish at arbitrarily large distances (Fig.4). For, clearly, if the classical particle does not escape to infinite distances, the corresponding quantum particle will not escape to infinite distances either! In analogy to the guitar chord whose two end-points are fixed, we expect the infinitely many quantum states $\psi_1(t,x)$, $\psi_2(t,x)$, $\psi_3(t,x)$, ..., $\psi_n(t,x)$ to form a discrete set. This is indeed the case! The spectrum of quantum states of fixed energy is discrete.

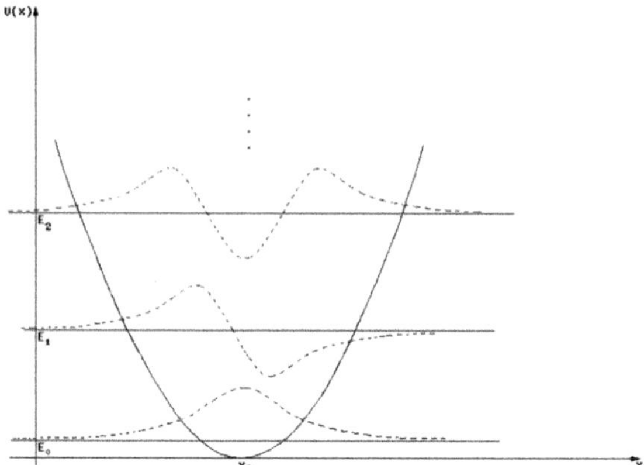

FIGURE 4. A quantum-mechanical particle in infinite symmetric one-dimensional static attractive (parabolic) potential. There are infinitely many quantum states of fixed energy. They all penetrate into the classically inaccessible regions and vanish only at infinite distance. The wavefunctions which describe the first three of those quantum states are depicted.

We see, for that matter, that by merely comparing the dynamical behavior of quantum particles with that of the corresponding classical particles and the wavefunctions with the classical waves, we have arrived at a wealth of information as to the dynamical behavior of quantum particles without solving the Schrödinger equation. We may now advance this comparison and inquire into the consequences which time-translational invariance and space-translational invariance have.

Clearly, energy conservation as a consequence of time-translational invariance and momentum conservation as a consequence of space-translational invariance are as valid in quantum systems as they are in classical systems. Time-translational invariance and space-translational invariance are respectively symmetries which a physical system has in time and space. Their consequences do not, for that matter, distinguish between classical and quantum behavior. As the fixed momentum p and associated fixed energy E of a free classical particle are the respective consequences of these two symmetries, so are the fixed momentum p and associated fixed energy E of a plane wave of probability. This is accordingly the case also in a static "potential well", such as the parabolic potential in Fig.3 and Fig.4. Such a potential manifestly breaks the homogeneity and the translational invariance of space. In turn, only the energy is a conserved quantity. In clear analogy to the conserved total energy of a classical particle in a static potential, a quantum-mechanical particle which at a certain moment is in a specific state of fixed energy, permanently remains in that state. Do note however, that the physical content of the statement to the effect that the particle is at a certain moment in a specific state of fixed energy is drastically different in quantum physics from that in classical physics. A classical particle in a static attractive potential has a given value of fixed total energy regardless of whether it is observed, or not. A quantum particle, on the contrary, must be observed in a specific state of fixed energy at the stated moment. Only then is the statement that it permanently remains in that state as a consequence of time-translational invariance physically valid. In the absence of observation the quantum particle is, instead, in a superposition of all quantum states of fixed energy with each value of energy having a specific predictable probability for occuring in a measurement. We shall extensively analyze this crucial aspect of quantum dynamics in both, quantum mechanics and quantum field theory. At this stage, we state that the states of fixed energy are the *energy eigenstates* of the quantum particle and that the wavefunctions for position which respectively describe them are the *energy eigenfunctions*. Accordingly, the fixed values of energy are the *energy eigenvalues*. Thus, *if a quantum-mechanical particle which evolves against a static classical potential is in an energy eigenstate, it permanently remains in that eigenstate with the corresponding time-independent energy eigenvalue.*

We may advance this qualitative conclusion to a physically rigorous statement through the formal concept of *operator*. This may sound intimidating, but operators are really no cause for concern! They are merely mathematical objects which map functions onto functions. The action of an operator on a given function, yields another function. The statement then to the effect that "the energy eigenvalue of a quantum particle in energy eigenstate $\psi_n(t,x)$ is E_n" receives its rigorous expression in

$$\hat{H}\psi_n(t,x) = E_n\psi_n(t,x) \qquad (3.\mathrm{I})$$

In plain language, this relation states that when operator $\hat{H}$ acts on energy eigenfunction $\psi_n(t,x)$, the result is the new function $E_n\psi_n(t,x)$. In turn, the precise statement which (3.I) makes is that the energy eigenfunctions $\psi_n(t,x)$ and the energy eigenvalues E_n are respectively the *eigenfunctions and eigenvalues of operator* $\hat{H}$. Operator $\hat{H}$ is the *Hamiltonian operator*. In order to stress its physical content we shall merely refer to it as "the energy operator" for now and defer the elucidation of the precise difference between the actual energy operator and the Hamiltonian operator to the fifth chapter where the Schrödinger equation will be explicitly stated. The mathematical statement in (3.I) is the *eigenvalue equation* of operator $\hat{H}$, an equation the solution to which yields the eigenfunctions $\psi_n(t,x)$ and corresponding eigenvalues E_n of that operator. The characterization "eigen" always refers to the corresponding operator[4]: Energy eigenstates,

---

[4]The German word *eigen* translates to "own".

eigenfunctions and eigenvalues are respectively eigenstates, eigenfunctions and eigenvalues of the energy operator $\hat{H}$. We shall soon state the general expression of both, $\hat{H}$ and its eigenfunctions $\psi_n(t, x)$.

We see, for that matter, that for a quantum particle which evolves against a static potential, or in empty space, time-translational invariance has the following implication: *If at a certain moment the quantum particle is in an eigenstate $\psi_n(x)$ of the energy operator $\hat{H}$, it will remain in that eigenstate with the corresponding energy eigenvalue E_n.* We reiterate that if at no time is the energy eigenvalue E_n known then each E_n is associated with a specific probability for being the energy of the quantum particle, a situation which we shall study in this chapter.

In view of the fact that $\hat{H}$ yields the observable energy of a quantum particle, it is natural to expect that all observable attributes of a quantum system are, accordingly, represented by operators. That this is indeed the case follows from this simple argument. The physical attributes of a classical particle can always be simultaneously observed with absolute certainty. This is reflected in the mathematical expression of classical attributes as functions of space and time. Functions trivially commute with each other. The product of two functions remains intact if we reverse the order of these functions. For example, for the position $x(t)$ and the momentum $p(t)$ of a classical particle it is $p(t)x(t) = x(t)p(t)$. Equivalently, it makes no difference whether we measure first the position of a classical particle and then its momentum, or first its momentum and then its position. Quantum particles are an altogether different matter. The replacement of unique classical values by probability distributions implies that, in general, no two physical attributes of a quantum system can be simultaneously observed with absolute certainty. We have seen, for example, that in the event of observation of a quantum particle's position no information as to the particle's momentum exists. This fact is reflected in the mathematical expression of a quantum system's observable attributes in terms of operators. Operators have, indeed, this distinctive property. In general, they do not commute! The action of the *position operator* $\hat{x}$ on a wavefunction followed by the action of the *momentum operator* $\hat{p}$, does not yield the same result as the action of the momentum operator $\hat{p}$ on the same wavefunction followed by the action of the position operator $\hat{x}$. That is, $\hat{p}\hat{x}\psi(t, x) \neq \hat{x}\hat{p}\psi(t, x)$, or concisely $\hat{p}\hat{x} \neq \hat{x}\hat{p}$.

Like the Hamiltonian (energy) operator, each operator $\hat{A}$ is associated with an eigenvalue equation

$$\hat{A}\psi_a = a\psi_a$$

the solution to which determines that operator's eigenfunctions ψ_a and its eigenvalues a. We have avoided any association of ψ_a with time- and space-dependence because, in principle, eigenfunctions do not have to depend on time or space. Examples are the time-independent eigenfunctions of position which we shall study soon and the time- and space-independent eigenfunctions of the spin-operator which we shall study later in this chapter. If, however, an eigenfunction ψ_a is the wavefunction of a quantum system, then necessarily[5] $\psi_a \equiv \psi_a(t, x)$.

Eigenfunctions of physical attributes (that is, of the operators which represent physical attributes) occupy a particularly privileged position in the set of wavefunctions. The reason is that *if the wavefunction of a quantum-mechanical system is an eigenfunction of a physical attribute, the dispersion of that attribute's possible values vanishes. The only value of that attribute (thus, also the only value which a measurement of that attribute will return) is the eigenvalue corresponding to that eigenfunction.* Although it is a straightforward matter to prove this statement using the statistical concepts which we shall introduce in the next section, this result may readily be deduced by reflecting on the above-stated eigenvalue equation with ψ_a as the wavefunction of the quantum-mechanical system. For example, pursuant to (3.I), we have already stressed that if a quantum-mechanical system is in an energy eigenstate $\psi_n(t, x)$, it has a fixed value of energy E_n.

As we have just seen, the description of a quantum system distinguishes between physical attributes which can be simultaneously observed with absolute certainty and physical attributes which cannot be simultaneously observed with absolute certainty. This distinction receives its rigorous expression in the concept of *commutator*. The commutator between two operators $\hat{A}$ and $\hat{B}$ is defined as

$$[\hat{A}, \hat{B}] = \hat{A}\hat{B} - \hat{B}\hat{A}$$

Two operators commute if their commutator vanishes. That is, $\hat{A}$ and $\hat{B}$ commute if $[\hat{A}, \hat{B}] = 0$. The fundamental physical implication inherent in this relation is stated in the following theorem:

A necessary and sufficient condition that two physical attributes be simultaneously observable with absolute certainty is that their respective operators commute.

In order to prove this we first note that the state of a quantum system immediately after a measurement of its attribute $\hat{A}$ is necessarily the eigenstate ψ_n corresponding to the eigenvalue a_n which has just been measured. For example, if a measurement of a quantum particle's energy in a static attractive potential returns the eigenvalue E_n then immediately after that measurement the quantum particle is in the energy eigenstate $\psi_n(t, x)$ in (3.I) (and, as stated, will remain in that eigenstate). Likewise, if a measurement of a quantum particle's position returns the eigenvalue x_0 then immediately after that measurement the quantum particle is described by the δ-function in Fig.2(a) which, as we show immediately below, is the position eigenstate corresponding to x_0. Since now by assumption the two attributes $\hat{A}$ and $\hat{B}$ are simultaneously observable, that ψ_n is necessarily common to both of them. For that matter, $\hat{A}$ and $\hat{B}$ have well-defined eigenvalues manifest in the simultaneous validity of their respective eigenvalue equations:

$$\hat{A}\psi_n = a_n\psi_n \; ; \; \hat{B}\psi_n = b_n\psi_n$$

---

[5]The discriminating reader may argue that this premise does not apply to the time-independent position eigenfunctions. The fact is, however, that "mathematically" the δ-function is not really a function and, thus, "physically" not really a wavefunction.

which is at once a statement of the obvious fact that the two simultaneously observable attributes $\hat{A}$ and $\hat{B}$ have a common set of eigenfunctions ψ_n (each with its own eigenvalues, of course). Thus, if we act on both sides of the first eigenvalue equation with operator $\hat{B}$ and on both sides of the second eigenvalue equation with operator $\hat{A}$ we obtain

$$\hat{B}\hat{A}\psi_n = \hat{B}a_n\psi_n = a_n\hat{B}\psi_n = a_n b_n\psi_n$$

$$\hat{A}\hat{B}\psi_n = \hat{A}b_n\psi_n = b_n\hat{A}\psi_n = b_n a_n\psi_n$$

where we have used the simple property of *linearity*[6] which we have imposed on our operators on physical grounds. Since the left sides of these two equations are equal, the corresponding right sides also are:

$$\hat{B}\hat{A}\psi_n = \hat{A}\hat{B}\psi_n$$

If we accept for the moment that this mathematical statement is also valid for an arbitrary wavefunction $\psi(t,x)$:

$$\hat{B}\hat{A}\psi(t,x) = \hat{A}\hat{B}\psi(t,x)$$

then - by definition - this statement implies that

$$\hat{B}\hat{A} = \hat{A}\hat{B} \iff \hat{A}\hat{B} - \hat{B}\hat{A} = 0$$

and, thereby:

$$[\hat{A},\hat{B}] = 0$$

We have, thus shown the "necessary aspect" of the stated theorem: Two physical attributes are simultaneously observable only if their respective operators commute. We shall not pursue the "sufficient aspect" to the effect that if two operators commute then the corresponding two physical attributes are simultaneously observable, although that derivation is just as straightforward. As to the derivation of the "missing step", we merely state that the validity of $\hat{B}\hat{A}\psi(t,x) = \hat{A}\hat{B}\psi(t,x)$ trivially follows from the fact that an arbitrary wavefunction can be expressed as a linear combination of a self-adjoined operator's (an operator which represents observable physical attributes) eigenfunctions, as we shall soon show in the present section.

A rather frivolous reading of the statement concerning the state of a quantum system "immediately after the measurement" might impress as a contradiction to the antecedent statement that a quantum-mechanical particle which is in an energy eigenstate will remain in that eigenstate with the corresponding energy eigenvalue. In respect to the latter, we reiterate that this is the exclusive consequence of time-translational invariance. If a quantum system has such an invariance, it remains in the energy eigenstate in which it was detected with the corresponding invariant energy eigenvalue. At once with

---

[6] $\hat{A}$ is a linear operator if for two numbers c_1 and c_2 it is $\hat{A}(c_1\psi_1 + c_2\psi_2) = c_1\hat{A}\psi_1 + c_2\hat{A}\psi_2$. This "wave property" ensures that the combination of two solutions to the Schrödinger equation is also a solution.

the energy eigenvalue, invariant in the course of time are the simultaneously observed eigenvalues of those attributes whose respective operators commute with $\hat{H}$ (since the eigenstates are common). This is, in fact, a fundamental aspect of quantum mechanics which we shall study in detail in the present section. Nevertheless, not all operators which represent physical attributes commute with $\hat{H}$ and, thereby, not all physical attributes are conserved in the course of time. Outstanding example is the position. A measurement may yield precise knowledge of a quantum-mechanical particle's position (100% probability) only at a moment - that is, a quantum-mechanical particle is in a position eigenstate only at the moment of observation. It is not difficult to sense the reason. Were this not the case, the quantum particle would have at once both, a precisely-determined position and a precisely-determined momentum, thus violating Heisenberg's uncertainty principle.

Of special interest, for that matter, is the *fundamental commutator* of the position operator $\hat{x}$ and the momentum operator $\hat{p}$. The significance of that commutator reflects that of the two fundamental attributes, $\hat{x}$ and $\hat{p}$, in quantum mechanics. Since the position and the momentum are not simultaneously observable, $\hat{x}$ and $\hat{p}$ do not commute. That is, $[\hat{x}, \hat{p}] \neq 0$. Pursuant to the theorem which we have just stated and proved, this result expresses the fundamental impossibility of simultaneously obtaining precise knowledge of the position and the momentum of a quantum-mechanical particle. Before we give the precise value of the fundamental commutator, it is necessary that we examine the operator of momentum and the operator of position.

The particular importance of the momentum operator $\hat{p}$ stems from the fact that plane waves of probability are characterized by constant momentum p. A plane wave of probability of constant momentum p is necessarily an eigenstate of the momentum operator $\hat{p}$ with corresponding momentum eigenvalue p. That is, if $\psi_p(x)$ is a plane wave of probability of fixed momentum p then:

$$\hat{p}\psi_p(x) = p\psi_p(x) \qquad (3.IIa)$$

on the understanding that only the time-independent spatial aspect of the plane wave (the plane wave as it is at a moment) is relevant to the eigenstate of momentum since that aspect already contains all information as to the time-dependent oscillating amplitude expressed by the multiplicative factor $\phi(t, p)$. The precise significance of this is clear in the brief analysis associated with (2.II). In fact, we already know what the eigenstates of $\hat{p}$ are. As the plane waves of probability are sinusoidal functions of space, pursuant to the Mathematical Preliminaries (MP) their mathematical expression is:

$$\psi_p(x) = \frac{1}{\sqrt{2\pi\hbar}} e^{\frac{i}{\hbar}px} \qquad (3.IIb)$$

with $\hbar$ being the fundamental Planck constant of which we shall talk later and with the multiplicative factor $N = \frac{1}{\sqrt{2\pi\hbar}}$ being a "normalization" constant which will not concern us further[7]. Of course, in line with our conceptual approach, our analysis on the eigenvalue equation (3.IIa) is at this stage qualitative. The rigorous proof of the statement that the time-independent $\psi_p(x)$ in (3.IIb) is indeed the solution to (3.IIa) necessitates the explicit expression of the momentum operator $\hat{p}$. We defer the momentum operator's explicit expression and the concomitant rigorous proof that the solution to (3.IIa) is indeed (3.IIb) to a more advanced stage which will allow for the introduction of the relevant mathematical formalism. We can, however, extend the present qualitative analysis to the derivation of the complete time-dependent expression of a plane wave's wavefunction.

Since the plane waves are sinusoidal functions of space, it is natural to expect that they are also sinusoidal functions of time. For, on classical terms, only if each point in the medium in which a plane wave of energy propagates executes harmonic oscillations about its position of equilibrium, is a sinusoidal form throughout space possible at each moment. Indeed, classical wave theory assures us that a plane wave of a given frequency w has a time dependence of the form e^{-iwt}. In view of the fundamental relation $E = \hbar w$, for that matter, we arrive at the conclusion that the evolved expression of a momentum eigenstate $\psi_p(x)$ is (refer also to $e^z e^{z'} = e^{z+z'}$ in the MP):

$$\psi_p(t, x) = \psi_p(x) e^{-\frac{i}{\hbar} E_p t} = \frac{1}{\sqrt{2\pi\hbar}} e^{\frac{i}{\hbar}(px - E_p t)} = \frac{1}{\sqrt{2\pi\hbar}} e^{\frac{i}{\hbar}\left(px - \frac{p^2}{2m} t\right)} \tag{3.IIc}$$

which is, of course, the expression directly stated in the MP and in (2.II). It identically satisfies (3.IIa):

$$\hat{p}\psi_p(t, x) = p\psi_p(t, x) \tag{3.IId}$$

since, pursuant to (3.IIa), $\hat{p}$ acts exclusively on functions of space:

$$\hat{p}\psi_p(t, x) = \hat{p}(\psi_p(x) e^{-\frac{i}{\hbar} E_p t}) = e^{-\frac{i}{\hbar} E_p t} \hat{p}\psi_p(x) = p\psi_p(x) e^{-\frac{i}{\hbar} E_p t} = p\psi_p(t, x)$$

This is, in fact, an explicit demonstration of the above-stated premise to the effect that although, in principle, the eigenfunctions of an operator may not depend on time or space, they are functions of both when they are wavefunctions of actual quantum systems: Each plane-wave wavefunction $\psi_p(t, x) = \psi_p(x) e^{-\frac{i}{\hbar} E_p t}$ is itself an eigenfunction of the momentum operator $\hat{p}$.

Thus, the spectrum of momentum eigenvalues p of a free quantum-mechanical particle of fixed momentum is continuous and yields the corresponding continuous spectrum of energies $E_p = \frac{p^2}{2m}$. In turn, the constant energy E_p which corresponds to momentum eigenvalue p is necessarily an energy eigenvalue. The direct implication is that the momentum eigenstate $\psi_p(t, x)$ is, at once, an energy eigenstate. That is, $\psi_p(x)$ at a given moment and $\psi_p(t, x)$ in the course of time satisfy:

---

[7]From $\int_{-\infty}^{+\infty} dx\psi_p^*(x)\psi_{p'}(x) = \delta(p - p')$ (see (3.XIV) in this section) it is $|N|^2 \int_{-\infty}^{+\infty} dx e^{\frac{i}{\hbar}(p'-p)x} = \delta(p - p')$, whereupon through use of the integral representation of the δ-function: $\int_{-\infty}^{+\infty} dx e^{\frac{i}{\hbar}(p'-p)x} = 2\pi\delta(\frac{p'-p}{\hbar})$ and the property: $\delta(\lambda x) = \frac{1}{|\lambda|}\delta(x)$, it is: $|N|^2 2\pi\hbar\delta(p - p') = \delta(p - p') \implies N = \frac{1}{\sqrt{2\pi\hbar}}$

$$\begin{cases} \hat{H}\psi_p(x) = E_p\psi_p(x) \\[2ex] \hat{H}\psi_p(t,x) = E_p\psi_p(t,x) \end{cases} \tag{3.IIIa}$$

with the energy operator $\hat{H}$ of a free quantum-mechanical particle relating to the momentum operator $\hat{p}$ through

$$\hat{H} = \frac{\hat{p}^2}{2m} \tag{3.IIIb}$$

For only then does (3.IIa) imply that

$$\hat{p}^2\psi_p(x) \equiv \hat{p}(\hat{p}\psi_p(x)) = \hat{p}(p\psi_p(x)) = p\hat{p}\psi_p(x) = p^2\psi_p(x)$$

or

$$\frac{\hat{p}^2}{2m}\psi_p(x) = \frac{p^2}{2m}\psi_p(x)$$

Pursuant to the above-stated theorem, for that matter, the momentum p and the energy $E_p = \frac{p^2}{2m}$ of a free quantum-mechanical particle are simultaneously observable with absolute certainty since the corresponding momentum operator $\hat{p}$ and the free energy operator $\hat{H}$ in (3.IIIb) have a common set of eigenfunctions. This, in turn, directly implies that the two operators commute:

$$[\hat{p}, \hat{H}] = 0$$

We reiterate that the common set of eigenfunctions which, according to the above theorem, the momentum operator $\hat{p}$ and the energy operator $\hat{H}$ of free quantum-mechanical particles have is the set of plane waves of probability for position ((3.IIb), (3.IIc)) and stress the general significance of this result: *Physical attributes of a quantum system which are simultaneously observable with absolute certainty have a common set of eigenstates and eigenfunctions* (each with its own eigenvalues). From the procedure which proved the above theorem we deduce that the converse is also true: *Physical attributes of a quantum system which have a common set of eigenstates and eigenfunctions are simultaneously observable with absolute certainty.* This will be of crucial importance in quantum field theory.

The free energy operator in (3.IIIb) naturally generalizes to the energy operator for a quantum-mechanical particle which evolves against a classical potential $V(x)$ as:

$$\hat{H} = \frac{\hat{p}^2}{2m} + V(x) \tag{3.IIIc}$$

The eigenvalue equation which yields the set of eigenfunctions and the energy spectrum of $\hat{H}$ in (3.IIIc) is (3.I). We shall discuss the complete character of that set and the implications of that completeness in this chapter. At this point, we make a few simple qualitative observations:

(i) Unlike the free $\hat{H}$ in (3.IIIb), the energy operator $\hat{H}$ in (3.IIIc) does not commute with $\hat{p}$, since now $\hat{H}$ involves the position x. Such a commutation would entail the commutation between $\hat{x}$ and $\hat{p}$ with the fallacious implication that the position and the momentum of a quantum-mechanical particle in a classical potential are simultaneously observable.

(ii) We reiterate that if a quantum-mechanical particle has fixed energy, that particle is in an energy eigenstate with the fixed energy being the corresponding energy eigenvalue. In the absence of potentials - that is, in empty space - that energy eigenstate is also a momentum eigenstate (3.IIc) of the free $\hat{H}$.

(iii) As we have stressed, the probability for position at arbitrarily large distances in an attractive potential is accordingly arbitrarily small. The energy eigenfunctions of a quantum-mechanical particle in an attractive potential fall off adequately fast and vanish at infinite distance. In analogy to the classical situation of the guitar chord with its discrete spectrum of frequencies, this implies a discrete spectrum of energy eigenvalues. On the contrary, the energy eigenfunctions of a free quantum-mechanical particle, or of a quantum-mechanical particle in a repulsive potential, do not fall off with the distance and in analogy to an infinitely-long guitar chord, the spectrum of energy eigenvalues is continuous.

(iv) It is, indeed, a straightforward matter to deduce the form of the temporal evolution of the energy eigenstates. If the potential is attractive, the eigenvalue equation $\hat{H}\psi_n(x) = E_n\psi_n(x)$ of $\hat{H}$ in (3.IIIc) yields the discrete set of energy eigenfunctions $\psi_1(x), \psi_2(x), ...\psi_n(x), ...$ and the corresponding set of energy eigenvalues $E_1, E_2, ...E_n, ...$ Any energy eigenstate described by $\psi_n(x)$; $n = 1, 2, ...$ in this set has fixed energy E_n and, thereby, also fixed frequency w_n according to $E_n = \hbar w_n$. As we have just mentioned, the classical wave theory determines that a plane wave of a given frequency has a temporal dependence of the form e^{-iwt}. By dint of time-translational invariance, this premise extends directly to the energy eigenstates of $\hat{H}$ in (3.IIIc). Since $\psi_n(x)$ is characterized by fixed energy eigenvalue E_n, it follows that the evolved expression of $\psi_n(x)$ is necessarily of the form

$$\psi_n(t, x) = \psi_n(x)e^{-\frac{i}{\hbar}E_n t} \; ; \; n = 1, 2, 3, ... \qquad (3.IIId)$$

Similar considerations apply to the case of a repulsive potential. In the absence of potentials, the energy eigenstates in (3.IIId) reduce to the plane waves of probability in (3.IIc). In any case, the form of the energy eigenstates in both, the discrete and the continuous energy spectrum is that in (3.IIId) (in the second case, with the discrete index n replaced by some continuous index a).

We can independently arrive at the same conclusion by reflecting on the character of an energy eigenstate as a *stationary state* - that is, as a quantum state whose probability distributions for position, for momentum and for any other physical attribute remain constant in the course of time as a direct consequence of time-translational invariance. Clearly, when a quantum system is in a stationary state, its observable attributes remain unaffected in the course of time. Nothing changes! This can only be the case if the energy eigenfunctions $\psi_n(t, x)$ have the form in (3.IIId). For only then is it:

$$|\psi_n(t,x)|^2 = \psi_n(t,x)\psi_n^*(t,x) = \psi_n(x)e^{-\frac{i}{\hbar}E_n t}\psi_n^*(x)e^{\frac{i}{\hbar}E_n t} = |\psi_n(x)|^2$$

(refer to $e^z e^{z'} = e^{z+z'}$ in the MP and to (2.I)).

Of equal importance to $\hat{p}$ and $\hat{H}$ is the position operator $\hat{x}$. Pursuant to the premise to the effect that the exclusive value which a quantum-mechanical system in an eigenstate of a physical attribute receives is the eigenvalue which corresponds to that eigenstate, the eigenstates $\psi_a(x)$ of $\hat{x}$ are quantum states characterized by a single *position eigenvalue* $x = a$:

$$\hat{x}\psi_a(x) = a\psi_a(x) \qquad (3.IVa)$$

Such *position eigenstates* are, of course, quantum states which yield precise knowledge of position. The state of a quantum-mechanical system at the moment its position $x = a$ is observed is the position eigenstate $\psi_a(x)$, the already familiar δ-function which we shall soon study in detail. Nevertheless, we would commit a serious error if from this we deduced that the position eigenstates exist only at the moment an observation of position is realized - that is, only when the quantum-mechanical system itself is in a position eigenstate. The position eigenstates and the corresponding position eigenvalues exist regardless of whether the position is observed or not. As we shall see in what follows, in general all wavefunctions for position are superpositions of position eigenstates.

As stated, the eigenfunctions $\psi_a(x)$ of the position operator $\hat{x}$ are the δ-*functions*:

$$\psi_a(x) = \delta(x - a) \qquad (3.IVb)$$

represented by the vertical straight line in Fig.2(a) with $a = x_0$. Each such position eigenstate $\delta(x - a)$ is the quantum state of unique position $x = a$ at a certain moment. Clearly, the quantum particle shall not remain in a position eigenstate. Past the stated moment, $\delta(x-a)$ will give rise to an evolving probability distribution for position. Again, such a dispersion from a sharply localized δ-function to a smooth wavefunction is the inevitable development of any position eigenstate regardless of whether the position is observed or not. It is not difficult to sense the reason. As we have stated, the position operator $\hat{x}$ does not commute with the energy operator $\hat{H}$ with the implication that the position is not conserved in the course of time. Unlike the energy eigenstates, the position eigenstates are irrelevant to time-translational invariance. The following brief analysis reveals the essence of the matter.

Since an arbitrary wavefunction $\psi(t,x)$ is a superposition of infinitely many plane waves of probability, its temporal development depends crucially on whether the speed with which the points of constant phase of each plane wave (loosely speaking, the "speed" of each plane wave) move is independent of the wavelength λ of that wave, or not. In the first case, all plane waves which compose $\psi(t_0,x)$ in (2.II) at some initial moment t_0 move at the same *phase speed* with the implication that their superposition at any future moment t yields the initial $\psi(t_0,x)$ shifted to a new position. On the contrary, if - as is indeed the case[8] - the phase speed depends on λ and, thereby, on p $(p = \hbar k = \hbar\frac{2\pi}{\lambda})$ then

---

[8](3.IIc) has this implication for the phase-speed v: $px - Et = constant \Rightarrow v = \frac{E}{p} = \frac{\frac{p^2}{2m}}{p} = \frac{p}{2m}$

the infinitely many plane waves in (2.II) move at various speeds with the implication that their superposition at any future moment will not yield the initial $\psi(t_0, x)$ but a different necessarily extended and outstretched waveform. In empty space, the end-result will be a plane wave of probability. In the presence of attractive potentials, the end-result will be an energy eigenstate (Fig.4), or a superposition of energy eigenstates. In any case, such a dispersion process is a direct consequence of the fact that an initial "narrow" and "localized" $\psi(t_0, x)$ involves an extensive range of plane waves signifying an extensive range of p-dependent phase speeds (Fig.1). Certain of those plane waves will move slowly, certain others will move fast, so that the initial "localized" waveform will rapidly come to resemble a dispersed demonstration! In fact, the more "localized" the initial $\psi(t_0, x)$ is, the more rapidly it disperses. It follows that a sharply localized position eigenstate in (3.IVb) instantly spreads out at maximum dispersion rate (Fig.2(a))!

We may now state without proof the fundamental commutator which has been pending:

$$[\hat{x}, \hat{p}] = i\hbar$$

where $\hbar$ is the fundamental *Planck constant* whose nature we shall examine in the seventh section.

We conclude our analysis on the eigenvalue equations with an important observation. As we have seen (following (3.IIIb)), if $\psi_p(t, x)$ is an eigenfunction of $\hat{p}$ with eigenvalue p then $\psi_p(t, x)$ is also an eigenfunction of $\hat{p}^2$ with eigenvalue p^2. This is clearly valid for any operator $\hat{A}$:

$$\hat{A}\psi_a(t, x) = a\psi_a(t, x) \iff \hat{A}^n\psi_a(t, x) = a^n\psi_a(t, x)$$

In general, if $F = F(\hat{A})$ is any (analytic) function of physical attribute $\hat{A}$ then

$$F(\hat{A})\psi_a(t, x) = F(a)\psi_a(t, x)$$

That is:

An operator which is a function of another, has the same eigenfunctions as those of the initial operator and eigenvalues which relate to those of the initial operator by the same relation as that between the two operators.

The converse is also valid:

Two operators of the same eigenfunctions and eigenvalues which relate to each other in a specific manner, are operators which themselves relate to each other in the same manner as that of their respective eigenvalues.

A typical example is, of course, the relation between $\hat{p}$ and $\hat{H} = \frac{\hat{p}^2}{2m}$. Both operators have the common set of eigenfunctions $\psi_p(t, x)$ and their corresponding eigenvalues relate to each other through $E = \frac{p^2}{2m}$.

Such considerations are of particular relevance to the evolution of quantum states. For, as we have remarked, the evolution of $\psi_p(t, x)$ is encoded in its $e^{-\frac{i}{\hbar}E_p t}$-factor. With $\hat{H} = \frac{\hat{p}^2}{2m}$, it follows for that matter that

$$e^{-\frac{i}{\hbar}\hat{H}t}\psi_p(x) = e^{-\frac{i}{\hbar}E_p t}\psi_p(x) \iff e^{-\frac{i}{\hbar}\hat{H}t}\left(\frac{1}{\sqrt{2\pi\hbar}}e^{\frac{i}{\hbar}px}\right) = \frac{1}{\sqrt{2\pi\hbar}}e^{\frac{i}{\hbar}(px - E_p t)}$$

In plain language, the *evolution operator* $\hat{U}(t) = e^{-\frac{i}{\hbar}\hat{H}t}$ has the same set of eigenfunctions (3.IIb) as that of operator $\hat{H} = \frac{\hat{p}^2}{2m}$ and eigenvalues which relate to those of the latter through $e^{-\frac{i}{\hbar}E_p t}$. Clearly, operator $\hat{U}(t)$ evolves the state $\psi_p(x)$ as it is at moment $t = 0$ to the state $\psi_p(t, x)$ - that is, to the same state as it is at moment t. Of course, there is nothing special about $t = 0$. It is merely an arbitrary choice. For that matter, if we consider $\psi_p(t_0, x)$ at initial moment t_0 then $\psi_p(t, x)$ at future moment t is given by $\hat{U}(t, t_0)\psi_p(t_0, x)$:

$$e^{-\frac{i}{\hbar}\hat{H}(t-t_0)}\psi_p(t_0, x) = e^{-\frac{i}{\hbar}\hat{H}(t-t_0)}\left(\frac{1}{\sqrt{2\pi\hbar}}e^{\frac{i}{\hbar}(px - E_p t_0)}\right) = \frac{1}{\sqrt{2\pi\hbar}}e^{\frac{i}{\hbar}(px - E_p t)}$$

Clearly, this is a particular statement of temporal evolution in empty space. In general, the *evolution operator*

$$\hat{U}(t, t_0) = e^{-\frac{i}{\hbar}\hat{H}(t-t_0)}$$

with $\hat{H}$ now given in (3.IIIc), advances a general state $\psi(t_0, x)$ to the future of t_0:

$$\psi(t, x) = e^{-\frac{i}{\hbar}\hat{H}(t-t_0)}\psi(t_0, x)$$

where we stress again the explicit time-independence of $\hat{H} = \frac{\hat{p}^2}{2m} + V(x)$ which expresses the time-translational invariance of the system.

Finally, we stress the physical content of conservation in the course of time. We have seen that:

(i) A physical quantity which is conserved in a classical context, is necessarily also conserved in the corresponding quantum context.

(ii) If at a given moment t the wavefunction of a quantum-mechanical system is an eigenfunction $\psi_n(t, x)$ of the time-independent energy operator $\hat{H}$, it will remain that eigenfunction with the corresponding energy eigenvalue E_n as the exclusive value which the energy receives.

(iii) A physical attribute $\hat{A}$ which commutes with the time-independent $\hat{H}$ of a quantum-mechanical system has the same set of eigenstates $\psi_n(t, x)$ as $\hat{H}$ with the implication that if at a given moment the quantum-mechanical system is in an eigenstate of $\hat{A}$, it will remain in that eigenstate with the corresponding eigenvalue a_n as the exclusive value which $\hat{A}$ receives.

(iv) As a consequence of (iii), if at a given moment the quantum-mechanical system is in an eigenstate $\psi_n(t, x)$, the eigenvalue a_n of physical attribute $\hat{A}$ and the energy eigenvalue E_n are simultaneously observable with absolute certainty. We have seen, for example, that if a quantum-mechanical particle is in a momentum eigenstate, it is also in an energy eigenstate with the immediate consequence that both, the momentum and the energy, are simultaneously observable.

We note now that, in general, conservation transcends time-translational invariance. When a particle evolves against a time-dependent potential, it transfers energy to, or receives energy from, that potential. The absence of time-translational invariance in such contexts implies the absence of conservation for the energy, but does not necessarily imply the absence of conservation for all the attributes of the physical system. In general, a physical amplitude $\hat{A}$ is conserved if $[\hat{A}, \hat{H}] = 0$, even if $\hat{H}$ is time-dependent. Later in this chapter, for example, we shall study the case of an electron in a uniform magnetic field. The solution which we shall give can trivially be extended to the case of a time-varying magnetic field and can, thereby, reveal the spin-component in the direction of the latter as a conserved quantity. Nevertheless, we stress that systems of particles in time-varying potentials are open physical systems - that is, systems which interact with the source of the potential's variation. If, for that matter, we extend the energy operator $\hat{H}$ so as to include that source (e.g. the factor which renders the magnetic field time-dependent) then the system is closed and isolated. As such, it is manifestly characterized by time-translational invariance and the concomitant energy conservation. Thus, it suffices to consider only closed and isolated systems.

This then is the condition for conservation to which we have arrived: *An attribute of a quantum system is conserved in the course of time, if the operator $\hat{A}$ which represents that attribute commutes with the Hamiltonian operator $\hat{H}$ of that system:*

$$[\hat{A}, \hat{H}] = 0$$

II. Hilbert Space

Let us now advance a step higher in abstract thought. If a quantum particle which evolves against a static attractive potential is at a given moment in an eigenstate of the energy operator, it remains in that eigenstate with the corresponding energy eigenvalue. This is a direct consequence of time-translational invariance. It is as physically legitimate, however, to inquire as to the quantum state of the particle in the static attractive potential in the event that the actual energy eigenvalue is unknown. In that circumstance, each energy eigenvalue has a specific probability for being the actual energy eigenvalue, the energy of the quantum particle. This is in direct analogy to the position eigenstates. If the actual position has at a certain moment t been observed then the quantum particle is at that moment in a position eigenstate. If not, the probability density at each point x is $|\psi(t, x)|^2$ and the probability that the position be in the immediate vicinity of x is $|\psi(t, x)|^2 dx$. If the actual energy eigenvalue is unknown, all infinitely many energy eigenstates contribute to the quantum state of the particle in the static attractive potential. Unless we "look inside", that is! The wavefunction $\psi(t, x)$ which expresses that quantum state collapses to a specific energy eigenfunction $\psi_n(t, x)$ at the moment of observation. Again, this is in direct analogy to the fact that at the moment at which the position is detected at $x = a$, the wavefunction $\psi(t, x)$ collapses to the position eigenstate $\delta(x - a)$. Prior to observation, the quantum particle in the static attractive potential is in a superposition of all infinitely many energy eigenstates. The extent to which each energy eigenstate contributes to that superposition determines the probability of the corresponding energy eigenvalue. The quantum state itself is the collection of all such probabilities, the probability distribution for energy specified by the superposition $\psi(t, x)$ of all energy eigenstates. For that matter, the question as to the quantum state of the particle in the static attractive potential allows for infinitely many possibilities each of which is expressed by the corresponding wavefunction $\psi(t, x)$ (N2). We formally express this fact by the statement that each wavefunction $\psi(t, x)$ which vanishes at arbitrarily large distances is a *linear combination* of energy eigenfunctions. In view of (3.IIId), that is:

$$\psi(t,x) = c_1\psi_1(t,x) + c_2\psi_2(t,x) + ... + c_n\psi_n(t,x) + ... =$$

$$c_1\psi_1(x)e^{-\frac{i}{\hbar}E_1 t} + c_2\psi_2(x)e^{-\frac{i}{\hbar}E_2 t} + ... + c_n\psi_n(x)e^{-\frac{i}{\hbar}E_n t} + ... \qquad (3.V)$$

with the coefficients $c_1, c_2, ..., c_n, ...$ being (in general, complex) numbers and with each specific choice of $c_1, c_2, ..., c_n, ...$ determining the extent to which each corresponding eigenstate $\psi_1(t,x), \psi_2(t,x), ..., \psi_n(t,x), ...$ contributes to $\psi(t,x)$.

The same premise applies to the continuous spectrum which characterizes scattering states in repulsive potentials and plane waves of probability in empty space. If the energy eigenvalue is unknown, all infinitely many energy eigenstates contribute to the quantum state of the quantum particle. In the specific case of a free quantum particle, of a quantum particle whose energy eigenstates are the plane waves in (3.IIc), lack of knowledge as to the actual momentum eigenvalue p and, thereby, as to the actual energy eigenvalue $E_p = \frac{p^2}{2m}$ implies that the wavefunction is (3.V) at the continuous limit $c_n \to c(p)$; $E_n \to E(p)$:

$$\psi(t,x) = \frac{1}{\sqrt{2\pi\hbar}} \int_{-\infty}^{+\infty} dp\, c(p) e^{\frac{i}{\hbar}\left(px - \frac{p^2}{2m}t\right)}$$

which is, of course, relation (2.II).

We see, for that matter, that the wavefunction of a quantum particle of unknown energy in empty space is indistinguishable from the wavefunction of a quantum particle of unknown energy in a static potential. This is not surprising. We have just seen that any wavefunction which vanishes at arbitrarily large distances is a superposition of energy eigenfunctions determined by specific values $c_1, c_2, ..., c_n, ...$ in (3.V). This clearly includes those wavefunctions in (2.II) (in empty space, or not) which "fall off" adequately fast so as to conform with the stated condition. The physical implication of the fact that the general wavefunction in (3.V) is, at once, the wavefunction in (2.II) is that the energy eigenfunctions $\psi_1(t,x), \psi_2(t,x), ..., \psi_n(t,x), ...$ of $\hat{H} = \frac{\hat{p}^2}{2m} + V(x)$ are "as good" for expressing a given quantum state $\psi(t,x)$ in static potential $V(x)$ as the energy eigenfunctions $\psi_1'(t,x), \psi_2'(t,x), ..., \psi_n'(t,x), ...$ in any other static potential $V'(x)$, a fact which clearly includes the case $V'(x) = 0$. Formally, the stated coincidence of wavefunction $\psi(t,x)$ in (3.V) and in (2.II) signifies the fact that *state vector* $\psi(x)$ can equivalently be expressed in infinitely many distinct bases, as we shall see in what immediately follows.

The term *superposition*, which we have already introduced and repeatedly used, conveys the same concept as the term "linear combination". The only difference is that the latter emphasizes the mathematical character of the wavefunction as a solution to the linear Schrödinger equation, whereas the former emphasizes the physical character of the wavefunction as the sum-total of independent contributions. We shall, henceforth, use both terms indifferently.

Relation (3.V) is not specific to the eigenstates of $\hat{H}$. It is a most general relation valid for all operators which represent physical attributes and predicated on a physical premise of central importance to quantum theory. Central enough to lie at the foundation of our forthcoming analysis on quantum fields. I stress this point here because almost all texts on quantum mechanics emphasize the geometric interpretation of (3.V), leaving this premise which underlies the physical interpretation of the latter in obscuration:

If the actual eigenvalue which the quantum system has is unknown, the quantum system is in a superposition of the relevant operator's eigenstates!

For example, if the actual position of the quantum particle is unknown, the quantum particle is in a superposition of position eigenstates. If the actual energy of the quantum particle is unknown, the quantum particle is in a superposition of energy eigenstates. If the actual momentum of the quantum particle is unknown, the quantum particle is in a superposition of momentum eigenstates. If the actual projection of the spin (which we shall examine later) along, say, the z-axis is unknown, the electron is in a superposition of "spin-up" and "spin-down" eigenstates. And, by way of honoring Schrödinger's brilliant banter (see next chapter), if the actual biological state of the cat within the box is unknown (alive or dead), the cat is in a superposition of "alive" and "dead" eigenstates!

This premise may impress upon as tautological, or even contradictory. Is it not obvious that a quantum system which is not in a certain eigenstate of an observable attribute, is in a superposition of that attribute's eigenstates? Or does this premise imply that the quantum system is at once in an unknown eigenstate and in a known superposition of eigenstates? Note, however, that this premise does not concern the unknown "actual eigenstate" of the quantum system, but the unknown actual eigenvalue. An unknown actual eigenvalue does not at all imply an unknown actual eigenstate. In fact, an unknown actual eigenstate in coexistence with a known superposition of eigenstates of a certain observable attribute is a contradiction of terms! As I stated, I stress the importance of this premise because it underlies the physical interpretation of (3.V) and of quantum physics. No less a reason is the fact that it stands in direct opposition to the established Copenhagen interpretation of quantum theory to the effect that in the absence of observation it is meaningless to refer to an actual eigenvalue, the probability distribution of an observable attribute "is all we can say" about the quantum system! Almost all texts on quantum mechanics emphasize the geometric interpretation of (3.V) without a hint as to what the inherent abstract probabilities in it actually mean, as to how a quantum system like that in Fig.4 which remains in a certain energy eigenstate in the event of observation, "ends up" in a superposition of energy eigenstates in the absence of observation. We shall extensively analyze this premise and its deep implications in the next chapter. For the purposes of the present analysis let it suffice to stress the general character of (3.V) by applying it also to $\hat{x}$.

Like the superposition of energy eigenstates, the relation between the wavefunction and the position eigenstates is an example of (3.V), now with $\psi_1(x) = \delta(x - a_1)$, $\psi_2(x) = \delta(x - a_2)$, ... $\psi_n(x) = \delta(x - a_n)$, ... It is clear that as a consequence of the continuous character of the position a, the sum is now continuous. The probability amplitude $\psi(t, x)$ that the position of the quantum particle be respectively in the immediate vicinity of the specific point $x = a_1$, $x = a_2$, ... $x = a_n$, ... at a given moment $t = t_0$ is

$$\psi(x) = c_{a_1}\delta(x-a_1)+c_{a_2}\delta(x-a_2)+...+c_{a_n}\delta(x-a_n)+... \equiv \int_{-\infty}^{\infty} dac(a)\delta(x-a) \qquad (3.VI)$$

Of course, we have not really stated the relation between $\psi(x)$ and the position eigenstates until we specify how $\delta(x-a)$ acts in the continuous sum in (3.VI). The rule which determines that action is simple: Merely view $\delta(x-a)$ as that map which takes any function $c(a)$ which we introduce in ( ) in the continuous sum $\int da(\ )\delta(x-a)$ onto number $c(x)$ (note that in the continuous sum the variable is a and x is a fixed, once chosen, parameter), so that

$$\int_{-\infty}^{\infty} dac(a)\delta(x-a) = c(x) \qquad (3.VII)$$

and, for that matter, it is:

$$c(x) \equiv \psi(x)$$

A wavefunction $\psi(t,x)$ is <u>always</u> in the superposition of position eigenstates in (3.VI). If $\psi(t,x)$ expresses the state of a quantum particle in the event that its energy eigenvalue in a static attractive potential $V(x)$ is unknown and is, thus, a superposition of the eigenstates of $\hat{H} = \frac{\hat{p}^2}{2m} + V(x)$ then $\psi(t,x)$ also yields the probability that the energy of the quantum particle be the eigenvalue E_1, E_2, E_3, ... We reiterate that this is a direct consequence of the fact that each such probability is determined exclusively by the corresponding energy eigenstate $\psi_n(t,x)$ and $\psi(t,x)$ itself, as we shall now show.

Relation (3.V) makes a direct statement to the effect that the extent to which each eigenstate contributes to $\psi(t,x)$ is determined by the corresponding constant coefficient. This relation is really reminiscent of the elementary relation

$$\vec{v} = c_1\vec{v}_1 + c_2\vec{v}_2 + c_3\vec{v}_3$$

which yields an arbitrary vector $\vec{v}$ as a linear combination of three unit vectors $\vec{v}_1$, $\vec{v}_2$, $\vec{v}_3$ (each of them is of magnitude equal to 1: $|\vec{v}_1| = |\vec{v}_2| = |\vec{v}_3| = 1$) (Fig.5(a)). The only two differences between these two relations relate to the number of dimensions which the two relations respectively define and to the nature of the corresponding vectors. Whereas $\vec{v} \equiv (c_1, c_2, c_3)$ is a vector in the usual three-dimensional Euclidean space (whence, a linear combination of three vectors) with c_1, c_2, c_3 being real numbers, $\psi(t,x)$ in (3.V) is a function which is a linear combination of infinitely many eigenfunctions with $c_1, c_2, ...c_n, ...$ being complex numbers. For that matter, we formally define an abstract vector space whose vectors are functions. That is, we define an abstract infinite-dimensional functional *Hilbert space* for each operator of infinitely many eigenstates. In fact, as there are also operators of a finite number of eigenstates (for example, the spin-operator), we accordingly define a Hilbert space of finite or infinite dimensionality: *The Hilbert space of an operator is the collection of all functions $\psi(t,x)$ which emerge as linear combinations of the operator's eigenfunctions.* In particular, the Hilbert space of the energy operator $\hat{H}$ for a quantum particle which evolves against a static attractive potential is the collection of all functions $\psi(t,x)$ which vanish "adequately fast" at arbitrarily large distances (Fig.4).

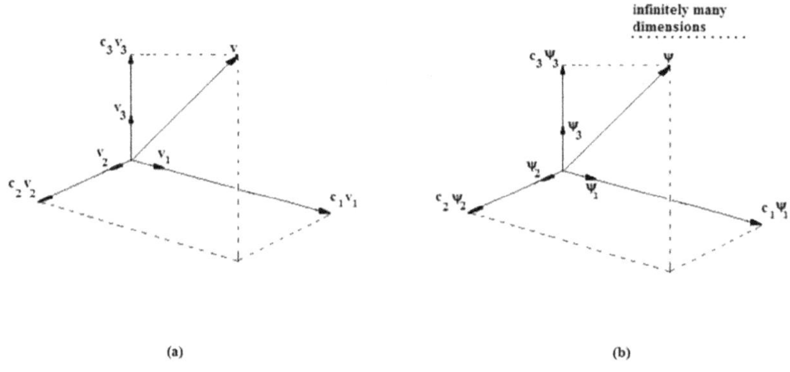

(a)

(b)

FIGURE 5. Each vector $\vec{v}$ in three-dimensional space is a linear combination of three independent unit vectors $\vec{v}_1, \vec{v}_2$ and $\vec{v}_3$ (none of these three is a linear combination of the other two) (a). Each wavefunction $\psi(t,x)$ in infinite-dimensional Hilbert space is a linear combination of infinitely many energy eigenfunctions (b).

The expression of the wavefunction in the abstract language of Hilbert spaces renders its character as a linear combination of eigenfunctions transparent. In order to appreciate this we first turn to certain elementary concepts in basic Euclidean geometry. A direct application of Pythagoras theorem yields the square of a vector's magnitude $|\vec{v}|$ as the sum of the squares of the corresponding three coefficients (Fig.5(a)):

$$|\vec{v}|^2 = |c_1|^2 + |c_2|^2 + |c_3|^2 \equiv c_1^2 + c_2^2 + c_3^2$$

A corresponding relation exists in all Hilbert spaces. We shall derive it at once in the Hilbert space of $\hat{H}$ and in the Hilbert space of $\hat{x}$ which are of immediate interest to us. To this end, we must first show that the energy eigenfunctions - which, in analogy to $\vec{v}_1, \vec{v}_2$ and $\vec{v}_3$, we consider orthogonal to each other (we shall see how) - are, each, of magnitude equal to 1. We exploit, in this respect, the simple yet fundamental property (already mentioned in the second section) which the energy eigenfunctions in static attractive potentials have. The total probability for position which each such energy eigenfunction yields throughout space is necessarily equal to 1. Although this property will be thoroughly analyzed in the sixth section, its essence should already be clear: Since it is certain that a quantum particle in an eigenstate of $\hat{H} = \frac{\hat{p}^2}{2m} + V(x)$ has a position at some point in infinite space, the total probability for position throughout space is, at any moment, equal to 1 (100%) (Fig.4). If we now recall that the probability for position in the immediate vicinity of each point x is $|\psi_n(t,x)|^2 dx$, it becomes clear that the continuous sum of all such probabilities yields:

$$\int_{-\infty}^{+\infty} dx |\psi_n(t,x)|^2 = 1 \; ; \; n = 1, 2, ... \qquad (3.VIII)$$

We observe now that this is very similar to the trivial relations $|\vec{v}_1|^2 = 1$, $|\vec{v}_2|^2 = 1$ and $|\vec{v}_3|^2 = 1$ which yield the magnitude of the three unit vectors, except that the continuous sum over all space is, to say the least, unusual. Of course, there is nothing surprising about this continuous summation. It is a direct consequense of (3.VI), of the statement that $\psi(x)$ is a vector in the infinite-dimensional Hilbert space of the position operator $\hat{x}$. Specifically, Pythagoras theorem implies that in this infinite-dimensional Hilbert space the square of the *magnitude of vector $\psi(x)$* is

$$|c_{a_1}|^2 + |c_{a_2}|^2 + ... + |c_{a_n}|^2 + ... \equiv \int_{-\infty}^{+\infty} da |c(a)|^2 \equiv \int_{-\infty}^{+\infty} dx |\psi(x)|^2 \qquad (3.IX)$$

From (3.VIII), (3.IX) and the fact that the left side of (3.VIII) (being equal to 1 at any moment) is time-independent, it follows that the latter is, indeed, the square of the magnitude of $\psi_n(x)$. Whence, each energy eigenfunction is, in the Hilbert space of $\hat{H}$, a vector of magnitude equal to 1.

Before we conclude the proof, we must stress the lesson inherent in (3.VIII) and (3.IX): *The square of the magnitude of <u>state vector</u> $\psi(x)$ constitutes the abstract geometric representation of the quantum state's total probability over all space.* As in a static attractive potential that total probability is invariably finite in view of the fact that it dwindles adequately fast over distance, the magnitude of a wavefunction which evolves against a static attractive potential is invariably finite and can be *normalized* (scaled) to the value 1. This is, in fact, the precise significance of the qualifying "adequately fast" which we used in order to characterize the behavior of wavefunctions in attractive potentials at arbitrarily large distances: They fall off fast enough to meet the *normalization condition*:

$$\int_{-\infty}^{+\infty} dx |\psi(x)|^2 = 1$$

Again, as the total probability over all space is independent of time, so is the magnitude of the corresponding wavefunction (it would not be much of a vector's magnitude if it were not!). The conservation of total probability for position in the course of time and the time-independence of the magnitude of the corresponding state vector are central aspects of quantum mechanics about which we shall say a lot more in the fifth chapter. Note, incidentally, that we have also defined the magnitude $|\psi(t,x)|$ of the complex number $\psi(t,x)$ at the specific point x and the specific moment t in (2.I). It should be clear that there is no relation between the two magnitudes. The magnitude in (2.I) is that of a number (for fixed x and t). The magnitude in (3.IX) is that of a state vector, of a vector in the Hilbert space of the position operator $\hat{x}$, that magnitude whose square yields the time-independent total (throughout space) probability for position.

Having thus established that the energy eigenfunctions are unit vectors, it is a trivial matter to deduce by analogy to $|\vec{v}|^2 = |c_1|^2 + |c_2|^2 + |c_3|^2$ (Fig.5(b)) that the square of the magnitude of $\psi(t,x)$ expressed through (3.V) as a vector in the Hilbert space of $\hat{H}$ is

$$\int_{-\infty}^{\infty} dx |\psi(t,x)|^2 = \int_{-\infty}^{\infty} dx |\psi(x)|^2 = |c_1|^2 + |c_2|^2 + ... + |c_n|^2 + ... \qquad (3.X)$$

which makes the same statement as (3.IX) albeit in terms of the discrete set of energy eigenfunctions in the Hilbert space of $\hat{H}$.

Since the left side expresses total probability over space in its entirety, each term on the right side also expresses probability. It is not difficult to see what that probability actually is. Since the coefficients c_1, c_2, ..., c_n, ... in (3.V) reveal the extent to which the corresponding energy eigenfunction $\psi_1(x)$, $\psi_2(x)$, ... $\psi_n(x)$, ... contributes to $\psi(x)$, the magnitudes of these coefficients in (3.X) necessarily express the probability that the quantum particle have the corresponding energy eigenvalue, the probability that the corresponding energy eigenvalue be the actual energy of the quantum particle. Specifically, the probability that in a measurement the energy of a quantum particle in a static attractive potential be the energy eigenvalue E_1, E_2, E_3, ... is respectively equal to $|c_1|^2, |c_2|^2, |c_3|^2, ...$. This result, which we derived here rather heuristically for $\hat{H}$, is valid for all operators, as we shall rigorously show soon. As is also the case with the probabilities for position in (3.IX) and with all probabilities, these probabilities are observable. Infinitely many identical experimental procedures, respectively administered on infinitely many identical quantum systems identically prepared, confirm that $|c_1|^2$, $|c_2|^2$, $|c_3|^2$, ... are indeed the probabilities for E_1, E_2, E_3,... The sum-total of all such probabilities is clearly finite and can be normalized to 1. Did we not expect this result? Of course we did from the very outset! The only possible values which the energy of a quantum particle in an attractive potential receives are the energy eigenvalues!

We reiterate that the square of the magnitude of $\psi(t,x)$ expressed through (3.VI) as a vector in the Hilbert space of $\hat{x}$ is given by (3.IX) with $|c(a)|^2 da \equiv |\psi(x)|^2 dx$ expressing the probability for position about each point in space. As stated just above, for a quantum particle in a static attractive potential the sum-total of all such probabilities throughout space is finite and can be normalized to 1, a result expected on the same grounds: The only possible positions which a quantum particle has, are the eigenvalues of $\hat{x}$.

We may now elucidate the statement to the effect that the eigenfunctions of $\hat{H}$, as well as the eigenfunctions of $\hat{x}$ and of all operators which represent physical attributes, are orthogonal to each other. In deriving (3.X) and (3.IX) I referred to Pythagoras theorem by way of emphasizing the fact that in Euclidean spaces the magnitude of a vector is in perfect conformity with the intuitive sense of length and can, for that matter, be thought of as "the length" of that vector. The square of the magnitude $|\vec{v}|^2$, this "squared length" as it were, is a special case of the *inner product* $\vec{v}.\vec{v}'$ between two arbitrary vectors $\vec{v} = (c_1, c_2, c_3)$ and $\vec{v}' = (c_1', c_2', c_3')$, that special case which emerges for $\vec{v}' = \vec{v}$. Thus, from

$$|\vec{v}|^2 \equiv \vec{v}.\vec{v} = c_1^2 + c_2^2 + c_3^2$$

it trivially follows that the general inner product between $\vec{v}$ and $\vec{v}'$ is

$$\vec{v}.\vec{v}' = c_1 c_1' + c_2 c_2' + c_3 c_3'$$

For that matter, it just as trivially follows that two vectors $\vec{v}$ and $\vec{v}'$ in the three-dimensional Euclidean space are orthogonal to each other if and only if their inner product is equal to zero:

$$\vec{v}.\vec{v}' = 0$$

For example, if $\vec{v}$ is in the "vertical direction" ($\vec{v} = (0, 0, c_3)$) and $\vec{v}'$ is in the "horizontal direction" ($\vec{v}' = c_1', 0, 0$) then $\vec{v}.\vec{v}' = 0c_1' + 0 + c_3 0 = 0$.

As was also the case with the square of the magnitude $|\vec{v}|^2$, it is a straightforward matter to extend the inner product from the three-dimensional Euclidean space of high-school geometry to the infinite-dimensional Hilbert spaces of quantum physics. From (3.V) it follows that the inner product between two vectors $\psi(x) = (c_1, c_2, ...c_n, ...)$ and $\psi'(x) = (c_1', c_2', ...c_n', ...)$ in the Hilbert space of $\hat{H}$ is:

$$(\psi, \psi') = c_1^* c_1' + c_2^* c_2' + ... + c_n^* c_n' + ... \tag{3.XI}$$

and clearly reduces to (3.X) for $\psi' \equiv \psi$:

$$(\psi, \psi) = |c_1|^2 + |c_2|^2 + ... + |c_n|^2 + ...$$

Accordingly, from (3.VI) it follows that the inner product between the same two wavefunctions, now expressed as vectors in the Hilbert space of $\hat{x}$, is:

$$(\psi, \psi') = \int_{-\infty}^{\infty} da\, c^*(a) c'(a) \equiv \int_{-\infty}^{\infty} dx\, \psi^*(x) \psi'(x) \tag{3.XII}$$

and clearly reduces to (3.IX) for $\psi' \equiv \psi$:

$$(\psi, \psi) = \int_{-\infty}^{+\infty} dx\, |\psi(x)|^2$$

This then is the significance of the preceding statement to the effect that the eigenfunctions of all operators which represent physical attributes are orthogonal to each other: If the eigenvalue spectrum of such an operator is discrete - as is the case with $\hat{H}$ - then in direct analogy to the mutually-orthogonal unit vectors $\vec{v}_1$, $\vec{v}_2$ and $\vec{v}_3$ in Fig.5(a), the eigenfunctions of that operator satisfy the orthogonality condition:

$$(\psi_m, \psi_n) = 0 \quad ; \quad m \neq n \tag{3.XIIIa}$$

If $m = n$, then (3.XI) and (3.X) imply that:

$$(\psi_n, \psi_n) = 1 \tag{3.XIIIb}$$

which reproduces (3.VIII).

We can succinctly convey (3.XIIIa) and (3.XIIIb) at once as the *orthonormality condition*:

$$(\psi_m, \psi_n) = \delta_{mn} \qquad (3.XIIIc)$$

with the *Kronecker-delta* δ_{mn} defined as:

$$\delta_{mn} = \begin{cases} 1 \; ; \; if \; m = n \\ \\ 0 \; ; \; if \; m \neq n \end{cases}$$

The case of operators which have a continuous spectrum of eigenvalues signifies a non-trivial extension of (3.XIIIc). For although the eigenfunctions of such operators are always orthogonal to each other, unlike the eigenfunctions of the discrete spectrum, they are not *normalizable*: They cannot be normalized to 1 (or to any other number) because their magnitude is infinite. Outstanding examples are the position eigenfunctions

$$\psi_a(x) = \delta(x - a) = \begin{cases} \infty \; ; \; if \; x = a \\ \\ 0 \; ; \; if \; x \neq a \end{cases}$$

and the momentum eigenfunctions in (3.IIb) which clearly yield infinite total probability throughout infinite space (they do not dwindle over distance). For that matter, the obvious extension of (3.XIIIc) to the continuous spectrum of eigenfunctions $\psi_a(t, x)$ is:

$$(\psi_a, \psi_{a'}) = \delta(a-a') \qquad (3.XIVa)$$

or, explicitly through (3.XII):

$$\int_{-\infty}^{+\infty} dx \psi_a^*(x) \psi_{a'}(x) = \delta(a-a') \qquad (3.XIVb)$$

where, of course:

$$\delta(a - a') = \begin{cases} \infty \; ; \; if \; a = a' \\ \\ 0 \; ; \; if \; a \neq a' \end{cases}$$

Let us confirm the validity of (3.XIV) in the Hilbert space of the position operator $\hat{x}$ and the momentum operator $\hat{p}$ respectively. Pursuant to (3.VII), in the former case it is:

$$\int_{-\infty}^{+\infty} dx \delta(x - a') \delta(x - a) = \delta(a - a')$$

and pursuant to (3.IIb), in the latter case it is:

$$\frac{1}{2\pi\hbar}\int_{-\infty}^{+\infty} dx\, e^{-\frac{i}{\hbar}px}e^{\frac{i}{\hbar}p'x} = \delta(p - p')$$

precisely as expected of the position eigenstates and the momentum eigenstates respectively[9].

Having established the orthonormality relation in (3.XIIIc) and its extension to the continuous spectrum in (3.XIV), it is necessary to make at least a passing reference to a technical but essential aspect of the latter and of the entire mathematical formalism of quantum mechanics. As $\delta(a - a')$ maps all values of a to zero except for the value $a = a'$ which it "maps" to infinity, the "δ-function" is not really a function. A function is a map of the numbers in a certain set, onto the numbers in another set (such that each number in the former maps onto only one number in the latter). Since infinity is not a number but a limit, the map of value $a = a'$ to infinity does not exist! The precise definition of $\delta(a - a')$ is that of the latter as a *functional* which maps function $c(a)$ to number $c(x)$ according to (3.VII) (again, x is arbitrary but fixed once chosen). We shall discuss the theory of functionals later in this chapter. On the understanding that the δ-function is the stated functional, we deem the definition which we have given to $\delta(a - a')$ as a function (0 if $a \neq a'$, ∞ if $a = a'$) adequately rigorous for our purposes and shall continue to use it[10].

We turn now to a remarkable implication inherent in the formalism of Hilbert spaces which we introduced in this section. In comparing the superposition of energy eigenfunctions in (3.V) to the superposition of momentum eigenfunctions in (2.II) we arrived at the conclusion that since a wavefunction can be expressed as a superposition of the energy eigenstates of distinct Hamiltonian operators, that wavefunction is essentially an abstract state vector differently expressed in different Hilbert spaces. This conclusion directly generalizes to the Hilbert spaces of all operators. Comparison between (3.XII) and (3.XI) as well as between (3.IX) and (3.X) directly reveals that the abstract *state vectors* ψ and ψ' are independently concretized in the Hilbert space of $\hat{x}$, in the Hilbert space of $\hat{H}$ and, accordingly, in the Hilbert space of any other operator which expresses a physical attribute. In order to stress this relation between abstract and concrete quantum states we advance a brief calculation which independently establishes that (3.XII) reduces indeed to (3.XI), this time by expressing $\psi(x)$ and $\psi'(x)$ in the former as superpositions of energy eigenfunctions through (3.V) rather than by expressing ψ and ψ' directly in a basis of the Hilbert space of $\hat{H}$ as we have already done. To this end, we introduce the shorthand $\Sigma_{n=1}^{+\infty}$ notation for the discrete sum and note that (3.XII) implies:

$$(\psi, \psi') = \int_{-\infty}^{+\infty} dx \left(\Sigma_{m=1}^{+\infty} c_m^* \psi_m^*(x)\right)\left(\Sigma_{n=1}^{+\infty} c_n' \psi_n(x)\right) =$$

---

[9] We have used again the integral representation of the δ-function mentioned in connection with (3.IIb).

[10] Historically, this is what Paul Dirac did. He introduced and treated $\delta(a - a')$ as a function. Only later was it rigorously defined as a functional by John von Neumann.

$$\Sigma_{m=1}^{+\infty} \Sigma_{n=1}^{+\infty} c_m^* c_n' \int_{-\infty}^{+\infty} dx \psi_m^*(x)\psi_n(x) = \Sigma_{k=1}^{+\infty} c_k^* c_k'$$

which is, indeed, (3.XI). In the last step, we used (3.XIIIc): For $m \neq n$, the continuous sum throughout space is equal to zero, for $m = n \equiv k$ the continuous sum throughout space, that square of the magnitude of corresponding eigenfunction $\psi_k(x)$, is equal to 1. Later in this chapter, we shall reproduce this result through use of the *Dirac notation*. We shall see, indeed, how the latter exploits the character of the wavefunction as an abstract state vector to spectacular effects!

Finally, let us return to the elementary relation $\vec{v} = c_1\vec{v}_1 + c_2\vec{v}_2 + c_3\vec{v}_3$ in order to settle the inherent issue of determining the coefficients c_1, c_2, c_3 for a given vector $\vec{v}$. Taking the inner product of both sides with $\vec{v}_1$ and using the fact that $\vec{v}_1$, $\vec{v}_2$ and $\vec{v}_3$ are unit vectors orthogonal to each other yields $c_1 = \vec{v}.\vec{v}_1$ (accordingly for c_2 and c_3). The extension of this result to infinite-dimensional Hilbert spaces is trivial. If the relevant operator has a discrete set of eigenvalues then from (3.V) and the *orthonormality* of the corresponding eigenfunctions (orthogonality of normalized eigenfunctions) it follows that

$$c_n = (\psi_n, \psi) \; : \; n = 1, 2, ... \tag{3.XV}$$

For convenience now we restate (3.X):

$$\int_{-\infty}^{+\infty} dy \psi^*(y)\psi(y) = c_1^* c_1 + c_2^* c_2 + ... + c_n^* c_n$$

and note that the expression of the coefficients c_i on the right side through (3.XV) and (3.XII) results in[11]

$$\int_{-\infty}^{+\infty} dy \psi^*(y)\psi(y) =$$

$$\int_{-\infty}^{\infty} dx \psi(x) \int_{-\infty}^{\infty} dx' \psi^*(x')[\psi_1^*(x)\psi_1(x') + \psi_2^*(x)\psi_2(x') + ... + \psi_n^*(x)\psi_n(x') + ...]$$

which is obviously valid only if

$$\psi_1^*(x)\psi_1(x') + \psi_2^*(x)\psi_2(x') + ... + \psi_n^*(x)\psi_n(x') + ... = \delta(x - x')$$

This *completeness relation* for the eigenfunctions of any operator which expresses a physical attribute is a fundamental aspect of quantum physics. We can state it in a more elegant manner as

$$\Sigma_{n=1}^{\infty} \psi_n^*(x)\psi_n(x') = \delta(x - x') \tag{3.XVI}$$

---

[11]The operation on the right side is really simple: With respect to the continuous summation over x', $\psi_i^*(x)$ is just a constant. Thus: $c_i^* c_i = \int_{-\infty}^{\infty} dx \psi(x)\psi_i^*(x) \int_{-\infty}^{\infty} dx' \psi^*(x')\psi_i(x') = \int_{-\infty}^{\infty} dx \psi(x) \int_{-\infty}^{\infty} dx' \psi^*(x')\psi_i^*(x)\psi_i(x') \; ; \; i = 1, 2, ..., n, ...$

I shall only add at this point that if the relevant operator has a continuous set of eigenfunctions - as is, for example, the case for $\hat{x}$ - the corresponding *completeness relation* is

$$\int_{-\infty}^{\infty} da\psi_a^*(x)\psi_a(x') = \delta(x-x') \tag{3.XVII}$$

as you can trivially confirm yourselves.

4. The Basic Statistical Concepts

I. The Expectation Value

Let us postulate a statistical quantity A the possible values of which form the discrete set $a_1, a_2, ..., a_n, ...$ If in a total number of measurements N, value a_1 appears N_1 times, value a_2 appears N_2 times and so on then the *average value*, or *mean value*, $\langle A \rangle$ of A is

$$\langle A \rangle = \frac{N_1 a_1 + N_2 a_2 + ... N_n a_n + ...}{N} = \frac{N_1}{N}a_1 + \frac{N_2}{N}a_2 + ... + \frac{N_n}{N}a_n + ... \equiv \Sigma_{n=1}^{\infty} a_n f_n$$

where $f_n = \frac{N_n}{N}$; $n = 1, 2, ...n, ...$ are respectively the *frequencies of occurence* of possible values $a_1, a_2, ..., a_n, ...$ At the limit $N \to \infty$, the frequencies f_n respectively tend to the probabilities of occurence P_n of the possible values, so that at $N \to \infty$ the average value - now acknowledged as the *expectation value* - of A is

$$\langle A \rangle = \Sigma_{n=1}^{\infty} a_n P_n \tag{4.I}$$

That is *the expectation value of a statistical quantity is the sum-total of the product of that quantity's possible values respectively multiplied by the corresponding probabilities.*

An arbitrary function $F(A)$ of A is itself a statistical quantity the possible values of which are $F(a_n)$, so that the expectation value of $F(A)$ is

$$\langle F(A) \rangle = \Sigma_{n=1}^{\infty} F(a_n) P_n \tag{4.II}$$

These results extend directly to the statistical quantities the possible values of which form a continuous set, usually extending from $-\infty$ to $+\infty$. This is the category in which the probability distribution for position and that for momentum fall. We have seen, of course, that in that case it is meaningless to refer to the probability of a specific value. The physically meaningful concept is the probability over a continuous range of possible values. This is expressed by the familiar *probability density* $P(a)$ defined by the demand that $P(a)da$ be the probability that a value between a given a and $a + da$ separated from a by an infinitesimal distance da appear in a measurement - that is, that a value appear in the immediate vicinity of the given a. In turn, the total probability that a value appear over a given interval is the continuous sum $\int P(a)da$ over that interval.

The expectation value $\langle A \rangle$ for a statistical quantity the possible values of which form a continuous set has the same form as that in (4.I) on the understanding that the discrete sum is now replaced by the continuous sum

$$\langle A \rangle = \int_{-\infty}^{\infty} aP(a)da \qquad (4.III)$$

Again, for any function $F(A)$ of a statistical quantity A which receives a continuous set of possible values the relation corresponding to (4.II) is

$$\langle F(A) \rangle = \int_{-\infty}^{\infty} F(a)P(a)da \qquad (4.IV)$$

Finally, let us reiterate that the sum-total - be it discrete or continuous - of probabilities for all possible values is equal to 100%. That is

$$\Sigma_{n=1}^{\infty} P_n = 1 \qquad (4.Va)$$

for the discrete case and

$$\int_{-\infty}^{\infty} P(a)da = 1 \qquad (4.Vb)$$

for the continuous case.

II. The Uncertainty

We proceed now to the definition of the *dispersion*, or *spread*, of a statistical distribution. This is the attribute which determines the extent to which the possible values are concentrated about the mean value. There are several notations for the dispersion of a statistical distribution. The notation which we shall use is ΔA, since this is the one used in quantum mechanics in which the dispersion is known by the, familiar from Fig.1, term *uncertainty of a probability distribution.*

The dispersion ΔA has the intended meaning only if it constitutes a measure of the average distance of possible values from the average value. A definition which conforms with this demand is

$$(\Delta A)^2 = \langle (A - \langle A \rangle)^2 \rangle \equiv \Sigma_{n=1}^{\infty} (a_n - \langle A \rangle)^2 P_n \qquad (4.VI)$$

That is, *the square of the dispersion is, by definition, equal to the average of the square of the deviation from the expectation value.* It should be obvious that the square of the deviation from the expectation value, as opposed to the deviation $A - \langle A \rangle$ itself is used because it is irrelevant whether a value a_n is to the "left" or to the "right" of $\langle A \rangle$, whether $A - \langle A \rangle$ is a positive, or a negative quantity.

At this point it is instructive to illustrate these concept with a brief example. Suppose that we wish to calculate the dispersion ΔA of a statistical quantity A which receives only the two possible values $a_1 = 4$ and $a_2 = 8$ with corresponding probabilities $P_1 = \frac{1}{4} \equiv 25\%$ and $P_2 = \frac{3}{4} \equiv 75\%$. To that end, we first calculate the expectation value $\langle A \rangle$:

$$\langle A \rangle = a_1 P_1 + a_2 P_2 = 4\frac{1}{4} + 8\frac{3}{4} = 7$$

so that

$$(\Delta A)^2 = (a_1 - \langle A \rangle)^2 P_1 + (a_2 - \langle A \rangle)^2 P_2 = (4 - 7)^2\frac{1}{4} + (8 - 7)^2\frac{3}{4} = 3$$

and thereby

$$\Delta A = \sqrt{3} \qquad \qquad \square$$

Expression (4.VI) for the dispersion may be recast into the equivalent and elegant form

$$(\Delta A)^2 = \langle A^2 \rangle - \langle A \rangle^2 \qquad (4.VII)$$

That is, *the square of the dispersion is equal to the expectation value of the square minus the square of the expectation value.* In order to prove (4.VII) we shall need the following, rather obvious, properties of the average value

$$\langle A + B + ... \rangle = \langle A \rangle + \langle B \rangle + ...$$

$$\langle cA \rangle = c\langle A \rangle$$

$$\langle c \rangle = c$$

In addition, since $\langle A \rangle$ is a given constant, we can merely set $\langle A \rangle \equiv \lambda$. Using now the elementary identity $(a - b)^2 = a^2 - 2ab + b^2$ we have

$$(A - \langle A \rangle)^2 = A^2 - 2\langle A \rangle A + \langle A \rangle^2 \equiv A^2 - 2\lambda A + \lambda^2$$

The expectation value of this relation and (4.VI) yield

$$(\Delta A)^2 = \langle (A^2 - 2\lambda A + \lambda^2) \rangle = \langle A^2 \rangle - 2\langle A \rangle\langle A \rangle + \langle A \rangle^2 = \langle A^2 \rangle - \langle A \rangle^2$$

which is the announced relation in (4.VII). Note, in this respect, that the expectation value of the square $\langle A^2 \rangle$ of the statistical quantity in (4.VII) is calculated according to (4.II) and (4.IV) respectively:

$$\langle A^2 \rangle = \Sigma_{n=1}^{\infty} a_n^2 P_n$$

for the discrete distribution and

$$\langle A^2 \rangle = \int a^2 P(a) da$$

for the continuous distribution.

It should be obvious now that (4.VII) directly implies the expected result: *If the dispersion - the quantum-mechanical uncertainty of a statistical distribution - vanishes, that distribution consists of only one possible outcome with probability 100%.* Equivalently, in this case, the statistical quantity is known with certainty (How could it not be! This is precisely the meaning of "vanishing uncertainty").

5. The Expectation Value in Quantum Mechanics

Suppose now that we wish to calculate the expectation value $\langle \hat{x} \rangle$ of the position of a quantum-mechanical particle which at a certain moment is in a given state described by wavefunction $\psi(x)$. According to (4.III), it is

$$\langle \hat{x} \rangle = \int_{-\infty}^{\infty} x P(x) dx = \int_{-\infty}^{\infty} x |\psi(x)|^2 dx \qquad (5.I)$$

Of course, the expectation value is not limited to $\langle \hat{x} \rangle$. Pursuant to (4.IV) we can calculate the expectation value $\langle \hat{F}(\hat{x}) \rangle$ of any function $\hat{F}(\hat{x})$:

$$\langle \hat{F}(\hat{x}) \rangle = \int_{-\infty}^{\infty} F(x) |\psi(x)|^2 dx \qquad (5.II)$$

In particular, through the expectation value $\langle \hat{x}^2 \rangle$ we obtain from (4.VII) the uncertainty (the dispersion) Δx of the possible values of position x about the expectation value $\langle \hat{x} \rangle$:

$$(\Delta x)^2 = \langle \hat{x}^2 \rangle - \langle \hat{x} \rangle^2 \qquad (5.III)$$

We reiterate here the premise associated with Fig.1. The uncertainty Δx yields a quantitative estimate of the range within which the probability for position is maximum. As the expectation value $\langle \hat{x} \rangle$ depends crucially on the wavefunction $\psi(x)$, so does the uncertainty Δx. Qualitatively, we may immediately determine whether Δx is large or small from the shape of the wavefunction. If $\psi(x)$ is very concentrated about a point then Δx is necessarily small. If, on the contrary, $\psi(x)$ extends over a wide range then Δx is necessarily large. It is just as obvious that the average position $\langle \hat{x} \rangle$ of the particle is necessarily close to that point x_0 where $\psi(x)$ - and, thus, also $P(x) = |\psi(x)|^2$ - are maximum. Clearly, if $\psi(x)$ is symmetric about x_0, it is $\langle \hat{x} \rangle = x_0$. Such is the case, for example, for the ground state in Fig.4.

Of course, the position is not the only observable attribute of a quantum-mechanical particle and, in most cases, it is not even the attribute of primary interest. The mathematical formalism of quantum mechanics is not, for that matter, complete until we have a general expression for the expectation value of any observable attribute. Since the attribute which allowed for the evaluation of $\langle \hat{x} \rangle$ is the probability distribution for position, we shall arrive at such an expression by generalizing the particular result in (5.I). We state:

$$\langle \hat{x} \rangle = \int_{-\infty}^{+\infty} x\psi(x)\psi^*(x)dx = \int_{-\infty}^{+\infty} \psi^*(x)x\psi(x)dx$$

At once, (3.VI) and (3.IV) imply that

$$\hat{x}\psi(x) = \int_{-\infty}^{+\infty} dac(a)\hat{x}\delta(x-a) = \int_{-\infty}^{+\infty} dac(a)a\delta(x-a) = xc(x) \equiv x\psi(x)$$

so that

$$\langle \hat{x} \rangle = \int_{-\infty}^{+\infty} \psi^*(x)(\hat{x}\psi(x))dx$$

with the bracket being a statement to the effect that $\hat{x}$ acts to the function on its right.

This expression directly suggests the general expression for the expectation value of an observable attribute expressed by operator $\hat{A}$:

$$\langle \hat{A} \rangle = \int_{-\infty}^{+\infty} \psi^*(x)(\hat{A}\psi(x))dx \qquad (5.IV)$$

We reiterate that the expectation value $\langle \hat{A} \rangle$ of operator $\hat{A}$ depends crucially on the wavefunction $\psi(x)$ which describes the state of the quantum-mechanical particle at a certain moment. This, in turn, has an immediate consequence in the event the wavefunction is an eigenfunction $\psi_n(x)$ of $\hat{A}$. Specifically, with a_n being the corresponding eigenvalue ($\hat{A}\psi_n(x) = a_n\psi_n(x)$), (5.IV) implies

$$\langle \hat{A} \rangle = \int_{-\infty}^{+\infty} \psi_n^*(x)(\hat{A}\psi_n(x))dx = a_n \int_{-\infty}^{+\infty} \psi_n^*(x)\psi_n(x)dx = a_n$$

where we have also used the normalization condition stated in (3.XIIIc)[12]. Moreover, since

$$\hat{A}^2\psi_n(x) = \hat{A}(\hat{A}\psi_n(x)) = \hat{A}(a\psi_n(x)) = a(\hat{A}\psi_n(x)) = a^2\psi_n(x)$$

it is:

$$\langle \hat{A}^2 \rangle = \int_{-\infty}^{+\infty} \psi_n^*(x)(\hat{A}^2\psi_n(x))dx = a_n^2 \int_{-\infty}^{+\infty} \psi_n^*(x)\psi_n(x)dx = a_n^2$$

so that the uncertainty of $\hat{A}$ is:

$$\Delta\hat{A} = \langle \hat{A}^2 \rangle - \langle \hat{A} \rangle^2 = a_n^2 - a_n^2 = 0$$

We have thus formally proved the statement whose validity we accepted on qualitative grounds in the preceding section:

If the wavefunction is an eigenfunction of physical attribute $\hat{A}$, the dispersion (uncertainty) of possible values vanishes. The only value which $\hat{A}$ receives and, for that matter,

---

[12]This result is also valid in the case of the continuous eigenvalue spectrum, with the "infinity" of the δ-function merely reflecting the fact that the total probability for position is infinite.

the only value which a measurement returns, is the eigenvalue a_n which corresponds to the stated eigenfunction with the direct implication that $\langle \hat{A} \rangle = a_n$.

The operators of immediate physical interest are $\hat{x}$, $\hat{p}$ and $\hat{H}$. We have just examined the expectation value $\langle \hat{x} \rangle$ of operator $\hat{x}$. In line with our conceptual approach, we defer the examination of $\hat{p}$ and $\hat{H}$ to the seventh section of the fifth chapter where we shall, to some extent, relax the restrictions which we have placed on the analysis of the relevant mathematical formalism. Now we state the central property of those operators which represent observable physical attributes, namely that they have real expectation values. So crucial is this property, that it serves, in fact, as a definition of the stated operators:

A necessary and sufficient condition that an operator represent an observable physical attribute is that its expectation value be invariably real.

The validity of this statement follows immediately from the fact that - in sharp contrast to the wavefunction and the operators - the expectation value $\langle \hat{A} \rangle$ of all operators which represent observable physical attributes is an observable quantity. Note that - in view of $\langle \hat{A} \rangle = a_n$ if the quantum system is in an eigenstate $\psi_n(x)$ of $\hat{A}$ - this statement immediately confirms the physical expectation to the effect that all operators which represent observable physical attributes have real eigenvalues: If $\langle \hat{A} \rangle$ is a real quantity, so is a_n.

Before we proceed, note that - pursuant to (3.XII) - (5.IV) may be recast in the concise form

$$\langle \hat{A} \rangle = (\psi, \hat{A}\psi) \qquad (5.V)$$

We shall now establish an independent and equivalent definition of the operators which represent observable physical attributes. Since the expectation value $\langle \hat{A} \rangle$ of such an operator is real, it is

$$\langle \hat{A} \rangle = \langle \hat{A} \rangle^* \qquad (5.VI)$$

in view of the fact that a real number is equal to its complex conjugate (refer to the Mathematical Preliminaries). For that matter, (5.IV) implies that

$$\langle \hat{A} \rangle^* = \left(\int_{-\infty}^{+\infty} \psi^*(x)(\hat{A}\psi(x))dx \right)^* = \int_{-\infty}^{+\infty} \psi(x)(\hat{A}\psi(x))^* dx \equiv \int_{-\infty}^{+\infty} (\hat{A}\psi(x))^* \psi(x)dx$$

which, in turn, recasts (5.VI) in the form

$$\int_{-\infty}^{+\infty} \psi^*(x)(\hat{A}\psi(x))dx = \int_{-\infty}^{+\infty} (\hat{A}\psi(x))^* \psi(x)dx \qquad (5.VII)$$

Since now (5.VII) is, by assumption, valid for an arbitrary wavefunction $\psi(x)$, it is a trivial matter to show (in fact, it can be seen by mere inspection) that for two arbitrary wavefunctions $\psi'(x)$ and $\psi(x)$ it is[13]

$$\int_{-\infty}^{+\infty} \psi'^*(x)(\hat{A}\psi(x))dx = \int_{-\infty}^{+\infty} (\hat{A}\psi'(x))^*\psi(x)dx \qquad (5.VIIIa)$$

or, in concise notation

$$(\psi', \hat{A}\psi) = (\hat{A}\psi', \psi) \qquad (5.VIIIb)$$

The stated independent and equivalent definition is, for that matter, inherent in (5.VIII):

A necessary and sufficient condition that an operator represent an observable physical attribute is that the transfer of its action from one function of the continuous sum to the other leave the continuous sum invariant.

Put in less formal and more simple language, the action of an operator $\hat{A}$ which represents an observable physical attribute can be transferred from one function of the continuous sum to the other.

Operators which express observable physical attributes are formally known as *self-adjoined operators*. The reason is inherent in the fundamental concept of *adjoined operator* $\hat{A}^\dagger$. In order to rigorously define $\hat{A}^\dagger$, merely consider the expression $\psi'^*(x)(a\psi(x))$, with a being a complex number. Then the simple relation

$$\psi'^*(x)(a\psi(x)) = (a^*\psi'(x))^*\psi(x)$$

implies that the action of complex number a can be transferred from function $\psi(x)$ to function $\psi'^*(x)$ on condition that a is replaced by its complex conjugate a^*. We may now view a complex number as a special case of an operator, as that operator whose action on a function is trivial - that is, as that operator whose action yields that function times that complex number. Conversely, operators may be viewed as an extension of the complex numbers. Like the latter they act on functions, although unlike the latter their action is non-trivial and, as we have seen, they do not commute. From this perspective, the fundamental relation

$$\int_{-\infty}^{+\infty} dx\psi'^*(x)(\hat{A}\psi(x)) = \int_{-\infty}^{+\infty} dx(\hat{A}^\dagger\psi'(x))^*\psi(x) \qquad (5.IXa)$$

is the obvious extension of the former relation with the continuous summation over all space justified by the fact that $\psi'(x)$ and $\hat{A}\psi(x)$ on the left side, $\hat{A}^\dagger\psi'(x)$ and $\psi(x)$ on the right side, are respectively vectors in the Hilbert space of the position operator $\hat{x}$, not merely numbers and must, for that matter, be multiplied as vectors by virtue of their inner product in (3.XII) ((5.IXa) clearly includes the trivial case $\hat{A} = a$). In fact, the concise notation for (5.IXa) is

$$(\psi', \hat{A}\psi) = (\hat{A}^\dagger\psi', \psi) \qquad (5.IXb)$$

---

[13]It is crucial to note that on the right side of (5.VIII), the complex conjugation is applied not on $\psi'(x)$ but on $\hat{A}\psi'(x)$.

Inherent in (5.IX) is the following fundamental implication: *The action of operator $\hat{A}$ can be transferred from one function of the continuous summation to the other on condition that $\hat{A}$ is replaced by its "complex conjugate" operator $\hat{A}^\dagger$.* Operator $\hat{A}^\dagger$ is the announced adjoint operator to $\hat{A}$. The stated self-adjoined character of all operators which represent observable physical attributes follows now from (5.IX) and (5.VIII).

We proceed now to the rigorous derivation of the heuristic result to which we arrived in the third section (immediately after (3.X)) concerning the probabilities which the possible values of self-adjoined operators have. We consider a self-adjoined operator $\hat{A}$ and a wavefunction $\psi(x)$ in the Hilbert space of that operator. Replacing (3.V) in (5.IV) yields

$$\langle \hat{A} \rangle = \int_{-\infty}^{+\infty} dx \left(\Sigma_{m=1}^{+\infty} (c_m^* \psi_m^*(x)) \right) \left(\hat{A} (\Sigma_{n=1}^{+\infty} c_n \psi_n(x)) \right) =$$

$$\Sigma_{m=1}^{+\infty} \Sigma_{n=1}^{+\infty} c_m^* c_n \int_{-\infty}^{+\infty} dx \psi_m^*(x) \left(\hat{A} \psi_n(x) \right)$$

which, in view of

$$\hat{A} \psi_n(x) = a_n \psi_n(x)$$

implies that

$$\langle \hat{A} \rangle = \Sigma_{m=1}^{+\infty} \Sigma_{n=1}^{+\infty} c_m^* c_n a_n \int_{-\infty}^{+\infty} dx \psi_m^*(x) \psi_n(x) = \Sigma_{m=1}^{+\infty} \Sigma_{n=1}^{+\infty} c_m^* c_n a_n \delta_{mn} = \Sigma_{n=1}^{+\infty} |c_n|^2 a_n$$

where, as in the third section, we used *the orthonormality relation* for the eigenfunctions of $\hat{A}$ this time expressed in the concise notation of the Kronecker-delta. Thus

$$\langle \hat{A} \rangle = \Sigma_{n=1}^{+\infty} |c_n|^2 a_n \qquad (5.X)$$

which is identical to (4.I). For that matter, we have rigorously proved that in the event that the state of the quantum system is, at a certain moment, described by wavefunction $\psi(x)$, the probability of eigenvalue a_n of physical observable $\hat{A}$ is $P_n = |c_n|^2$, where c_n is the coefficient of corresponding eigenfunction $\psi_n(x)$ in (3.V).

This concludes the introduction of the basic mathematical formalism which our conceptual analysis necessitates. We may now state the conclusions to which we have arrived in the last three sections.

The Six Fundamental Premises of Quantum Mechanics

(i) *To every physical system corresponds a wavefunction $\psi(t,x)$. This $\psi(t,x)$ contains all experimentally verifiable information about the physical system and its temporal development is determined by the Schrödinger equation.*

(ii) *All observable attributes of a quantum system are represented by self-adjoined operators each of which has its own Hilbert space. Each self-adjoined operator is constructed from the corresponding classical quantity through the replacement $x \to \hat{x}$; $p \to \hat{p}$. A* *wavefunction $\psi(t,x)$ which belongs to the Hilbert space of self-adjoined operator $\hat{A}$ is always expressed through (3.V) as a linear combination of that operator's eigenfunctions $\psi_n(x)$. The latter satisfy the eigenvalue equation*

$$\hat{A}\psi_n(x) = a_n\psi_n(x)$$

with a_n being the eigenvalue corresponding to eigenfunction $\psi_n(x)$. If at a certain moment the state of the quantum system is described by $\psi(x)$, the probability that eigenvalue a_n be realized in a measurement is equal to $|c_n|^2$, where c_n is the coefficient of eigenstate $\psi_n(x)$ in the linear combination (3.V).

(iii) *Two Hilbert spaces which correspond to two self-adjoined operators $\hat{A}$ and $\hat{B}$ coincide if the two attributes which $\hat{A}$ and $\hat{B}$ respectively represent are simultaneously observable with absolute certainty. Equivalently, two Hilbert spaces which correspond to two self-adjoined operators $\hat{A}$ and $\hat{B}$ coincide if the two operators commute:*

$$\hat{A}\hat{B} - \hat{B}\hat{A} = 0$$

that is, if their commutator vanishes:

$$[\hat{A}, \hat{B}] \equiv \hat{A}\hat{B} - \hat{B}\hat{A} = 0$$

(iv) *A self-adjoined operator has real eigenvalues. These eigenvalues are the only possible values which that observable physical attribute receives.*

This follows immediately from the fact that the eigenvalues of a self-adjoined operator are observable quantities. Observable quantities always receive real values.

(v) *The eigenfunctions of a self-adjoined operator form a complete set.*

This is the statement in (3.V): Each wavefunction $\psi(t,x)$ which belongs to the Hilbert space of a self-adjoined operator is a linear combination of that operator's eigenfunctions. This property is so crucially important that the entire quantum theory would be physically meaningless if it were not satisfied. For in that case there would exist wavefunctions which describe actual quantum systems, yet which would contradict (5.X), thereby precluding any prediction of the possible outcomes of an experimental procedure aiming at the observation of the physical attribute which the self-adjoined operator represents.

(vi) *The state of a physical system immediately after a measurement is the eigenstate of the eigenvalue which has just been measured.*

For the sake of conceptual consistency, we must add one more fundamental premise which concerns the internal degree of freedom of quantum particles, that of *spin* which we shall examine in the twelfth section of this chapter. In anticipation of this examination we revisit the *fundamental commutator*:

$$[\hat{x}, \hat{p}] = i\hbar \tag{5.XIa}$$

which expresses the fundamental incompatibility of position and momentum in a quantum-mechanical system - that is, the impossibility of simultaneous observation of a quantum-mechanical particle's position and momentum. In three dimensions where each position is specified by three coordinates

$$x_i = (x_1, x_2, x_3) \equiv (x, y, z)$$

and the momentum is a vector of three components

$$p_i = (p_1, p_2, p_3) \equiv (p_x, p_y, p_z)$$

the fundamental commutator in (5.XIa) applies to a position-coordinate and the *corresponding* component of momentum. That is:

$$[\hat{x}, \hat{p}_x] = i\hbar \; ; \; [\hat{y}, \hat{p}_y] = i\hbar \; ; \; [\hat{z}, \hat{p}_z] = i\hbar$$

On the contrary, commutators of the type $[\hat{x}, \hat{p}_y]$ (and so on) vanish, with the implication that x and p_y are simultaneously observable. In the fifth chapter it will become clear that (5.XIa) is an immediate consequence of the character of $\hat{p}$ as a differential operator and that $[\hat{x}, \hat{p}_y] = 0$ because x is a constant with respect to the y-differentiation of $\hat{p}_y$. Here, we express these results in a concise manner as:

$$[\hat{x}_i, \hat{p}_j] = i\hbar\delta_{ij} \tag{5.XIb}$$

Thus, we have accomplished far more than merely advance into the abstract mathematical language of quantum mechanics. We have developed the theoretical tools which reveal the essence of canonical quantization. As such, these tools will prove valuable to our exploration of quantum fields. Yet, we need to do more! In order to derive maximum insight into the essence of canonical quantization as applied to mechanical systems before we examine its application to fields we need to understand the dynamical behavior which emerges from it. We have already seen certain characteristic aspects of that behavior, such as the discrete energy spectrum in static attractive classical potentials and the penetration into the classically inaccessible regions of space. In the following sections of the present chapter we shall further examine the implications of such penetration and all qualitative aspects of quantum-mechanical behavior in attractive and in repulsive static potentials. First, however, we must give concrete meaning to certain concepts which we have only described heuristically.

6. Attractive Potentials and Quantum Particles

The static attractive potentials which we have hitherto considered are characterized by a finite value at each point in space. In order to advance beyond the qualitative results which we obtained in the third section in connection with Fig.4, we shall need the idealized case which involves infinite values for the potential. The relative simplicity of that potential will allow us to arrive qualitatively, without solving the Schrödinger equation, at general conclusions which encompass the dynamical behavior of quantum particles in all static attractive potentials.

The one-dimensional infinite square well in Fig.6 signifies a potential which is zero between $x = 0$ and $x = L$ and infinitely-high at any point out of this range. We have seen that the wavefunctions in static attractive potentials penetrate into the classically inaccessible regions of space and asymptotically vanish at infinite distance. The rate at which these wavefunctions fall off with increasing distance depends on the classical potential itself. Clearly, the steeper the attractive potential is, the more "difficult" it is for the quantum particle to penetrate into the classically inaccessible regions of space. Thus, the steeper the attractive potential is, the faster the energy eigenfunctions fall off as the distance increases. In the extreme case of Fig.6, all energy eigenfunctions vanish on the two walls at $x = 0$ and $x = L$ and no penetration into the classically inaccessible regions occurs. This underscores the intuitive value of the infinite square well. Whereas the dynamical behavior of a quantum particle which evolves against a general static attractive potential has general structural similarities to the dynamical behavior of the guitar chord both end-points of which are fixed, the dynamical behavior of a quantum particle which evolves against the infinite square well is in almost perfect conformity with the dynamical behavior of that classical system.

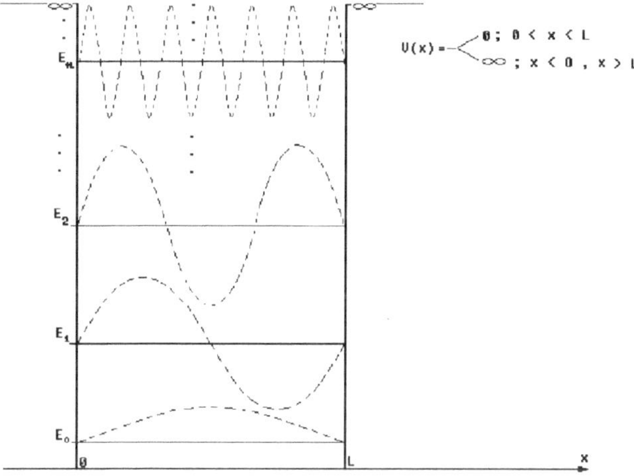

FIGURE 6. A quantum particle in a one-dimensional infinite square well. The particle can be in any quantum state out of infinitely many. All wavefunctions vanish strictly out of the well signifying strictly zero probability for the position of the particle past the two infinite walls.

In what sense "almost perfect"? In the sense that the analogy between the discrete energy spectrum of a quantum particle within an attractive potential and the discrete frequency spectrum of a guitar chord is strongest in the case of the infinite square well. Nevertheless, classical waves represent energy either propagating in space or confined within a certain volume as in the case of standing waves[14] on a guitar chord, whereas probability distributions in physical space represent the evolving probability density for position. This essential difference in physical content is reflected in the difference between the corresponding evolution patterns. In classical standing waves the amplitude

---

[14]A standing wave is a wave in which the location of its maximum and minimum amplitudes is fixed.

may, at certain moments, vanish throughout their extent without at all contradicting conservation laws and principles since at such moments all the potential energy of each molecule in the elastic medium is transformed into kinetic energy. Such a situation is physically impossible for probability distributions. At no moment can a wavefunction vanish at all points in space, as that would imply that the quantum particle has, at that moment, evolved out of existence! We recall in this respect that, unlike the classical waves, wavefunctions are complex functions of space and time. In particular, the energy eigenfunctions in static attractive potentials are of the form stated in (3.IIId). Clearly, only one of the two parts (real and imaginary) of the complex wavefunction may vanish at certain moments. Never both at once! This characteristic property of all wavefunctions is inherent in Fig.4 and Fig.6.

A property which characterizes the energy eigenstates in attractive potentials is the finite total probability for position. By their very nature, the probabilities for position of any quantum-mechanical system considered throughout space must invariably add up to unity (100%). The continuous sum of probabilities throughout space in (3.VIII), that total probability for position, must at any moment be equal to one since it is certain that at any moment the quantum particle has a position at some point in infinite space. All energy eigenstates in attractive potentials satisfy this physical demand because they are *bound states*, the energy eigenfunctions fall off in the classically inaccessible regions and vanish at infinite distances. Such a behavior is obvious in the attractive potential in Fig.4 in general and in the idealized potential in Fig.6 in particular. Nevertheless, this universal property of wavefunctions is obviously not satisfied by the plane waves of probability and the *scattering states*, the energy eigenstates in repulsive classical potentials, since such quantum states do not vanish at arbitrarily large distances. They yield, instead, infinite total probability for position over all space! This situation is not symptomatic of any underlying pathology in quantum physics. Physical space is neither empty, nor permeated by repulsive classical potentials. Plane waves of probability and energy eigenstates in repulsive potentials are tractable mathematical idealizations of actual spatially extended, yet essentially bounded, quantum states. A similar situation occurs in classical physics. The classical plane waves of energy extend throughout infinite space and are, for that matter, associated with infinite total energy (the classical analogue to the infinite total probability). Nevertheless, they are physically acceptable because they represent tractable mathematical idealizations of extended, yet bounded, classical waveforms.

Of particular interest are the static attractive potentials for which $V(-x) = V(x)$. These are potentials which are symmetric about a vertical axis, such as the vertical axis at point $x = x_0$ in Fig.4 and the vertical axis at point $x = \frac{L}{2}$ (the middle point between 0 and L) in Fig.6. The symmetry of such potentials introduces additional features in the dynamical behavior of quantum particles which evolve against them. The most significant of those features relates to the number of *nodes* which each energy eigenfunction has. That is, the number of points at which each energy eigenfunction permanently vanishes increases with the corresponding energy eigenvalue. This property which characterizes all energy eigenfunctions in symmetric static attractive potentials is almost obvious if you recall the analogy with the guitar chord. The energy eigenstates in symmetric static attractive potentials closely resemble the standing waves on the latter. In order to construct a standing wave on a guitar chord, a certain number of half-wavelengths (half the distance between two successive crests, or two successive troughs) must fit between the chord's two fixed end points. The standing wave of lowest frequency corresponds to one half-wavelength and has, for that matter, no nodes. Accordingly, the *ground state* of a quantum particle in a symmetric static attractive potential, that energy eigenstate which is characterized by energy eigenvalue E_0, has no nodes. The standing wave of the first frequency above the lowest corresponds to two half-wavelengths and has one node. Accordingly, the first excited state of a quantum particle in a symmetric static attractive potential, that energy eigenstate which is characterized by energy eigenvalue E_1, has one node. The second excited state is characterized by energy eigenvalue E_2 and has two nodes, and so on. Succinctly put, with $n = 1, 2, 3, \ldots$ the n-th energy eigenstate has n nodes. Energy eigenstates which correspond to high values of n are characterized by a high number of nodes. This attribute is very important to the emergence of the semi-classical limit in quantum mechanics and will be thoroughly discussed in the next section.

Another feature which deserves attention in its own merit concerns the relation between the ground-state energy eigenvalue E_0 and the region of space within which the attractive potential extends. The smaller that region is, the higher the ground-state energy E_0 is. For a qualitative justification of this statement we may invoke the uncertainty relation $\Delta x \Delta p \sim \hbar$. Since the region of space of highest probability for the quantum particle is the corresponding classically accessible region, the uncertainty Δx of the ground state is "more or less" equal to the width of that region. For the simplest case of the infinite square well in Fig.6 we may, for that matter, state that $\Delta x \sim L$. Thus, for the ground state it is $\Delta p \sim \frac{\hbar}{L}$. The intuitive value of the infinite square well becomes at this point obvious again. In general, no energy eigenvalue is separable into a kinetic and a potential term. However, within the infinite square well the energy of the quantum particle is purely kinetic. For that matter, it is consistent to state that the fixed ground-state energy eigenvalue E_0 is "more or less" equal to the average kinetic energy $\frac{(\Delta p)^2}{2m}$. That is,

$$E_0 \sim \frac{\hbar^2}{2mL^2} \tag{6.1}$$

The actual relation between the width of the potential and the ground-state energy which this mathematical statement expresses is, of course, valid only for the infinite square well. However, the inverse relation between these two quantities is clearly valid for all static attractive potentials: The smaller the width is, the higher the ground-state energy is!

The inverse relation between the ground-state energy E_0 and the mass m of the quantum particle, as well as the direct relation between E_0 and $\hbar$ are just as valid for all static attractive potentials. All three relations, that between E_0 and the width of the potential, that between E_0 and the mass m and that between E_0 and the Planck constant $\hbar$ relate to the semi-classical limit which we shall examine in the next section. However, the semi-classical implications which the first two relations have are already obvious. The larger the width is, the lower the ground-state energy is. If the potential is of infinite width $(L \to \infty)$ then the ground-state energy tends to vanish at the bottom of the classical potential $(E_0 \to 0)$, precisely in line with the lowest-energy state of classical particles. Thus, the smaller the width of the potential is, the more pronounced the quantum behavior of the quantum particle in it is! The bigger the width of the potential is, the more the quantum particle in it manifests a classical behavior! Accordingly, the smaller the mass of the quantum particle is, the higher E_0 is and the more pronounced the particle's quantum-mechanical behavior is! The bigger the mass of the quantum particle is, the lower E_0 is and tends to vanish at the bottom of the classical potential as the mass tends to infinity! The heavier a quantum particle is, the more it manifests a classical behavior!

We see, for that matter, that the usual statement which abounds in "popular science" to the effect that the quantum-mechanical behavior characterizes the microscopic particles is not altogether consistent. The physically consistent statement is that the quantum-mechanical behavior manifests itself in microscopic particles which are light and are constraint in a region of microscopic dimensions. When a microscopic and light particle, such as the electron, is free to move in a region of macroscopic dimensions (for example, in a cyclotron), its behavior is manifestly classical.

Let us now examine the implications which the relation between E_0 and L has for the distance between successive energy levels. At least for the first few energy levels, ΔE is of the same order of magnitude as the ground-state energy E_0. For example, as it turns out, in the case of the infinite square well it is $E_1 = 4E_0$, whereby $E_1 - E_0 = 3E_0$. Thus, the energy distance $\Delta E_n \equiv E_n - E_{n-1}$ between any two successive energy levels relates directly to the size of the quantum system through the ground-state energy E_0. In atoms, where the physically significant range of the classical electrostatic potential extends over an average radius of order[15] 10^{-8}cm, the distance ΔE between successive energy levels lies in the domain of electronvolts (ev). Any radiation - be that visible light, or X-rays - produced by excited atoms which make the transition to a lower state corresponds to that energy domain. By contrast, in the nuclei, where the range of the strong nuclear force extends over an average radius of order 10^{-13}cm, the distance ΔE between successive energy levels lies in the domain of mega electronvolts (Mev). If you now consider that one Mev is equal to one million ev $(1Mev = 10^6 ev)$, the reason for which the γ-rays - emitted when a nucleus makes a transition from an excited state to a lower (usually the ground) state of energy - are so energetic becomes immediately obvious! The crucial factor for such high energies is the minute size of the nucleus. It is also for that matter that only in exotic temperatures can thermal collisions cause an excitation of a nucleus. Indeed, if you consider that at room temperature $(T \approx 300^0 K)$ the energies involved are only of order $\frac{1}{40}$ eV, it becomes immediately clear that we must climb to temperatures 40×10^6 times higher than room temperature in order to excite a single nucleus! Hence, even in the highest temperatures of "every-day life" the nuclei in the atoms remain utterly "frozen" in their ground state.

Closely associated with the relation between ΔE and the size of a physical system is the fundamental fact that under no circumstances does a quantum particle fall to the bottom of a classical potential. A classical particle of minimum total energy within a

---

[15]It is $10^8 = 100000000$ and $10^{-8} = \frac{1}{100000000}$, just in case...

classical potential remains stationary at the potential's turning point at x_0 (Fig.3). A glance at Fig.4 and Fig.6 reveals that this can never be the case for a quantum particle in view of the fact that the minimum energy E_0 is invariably higher than the minimum total energy which a classical particle situated at x_0 would have (and which can safely be taken equal to zero). This fundamental feature of quantum theory can readily be understood. If a quantum particle "fell" to the bottom of the potential, its position would be the specific point x_0 with absolute certainty. That, however, is forbidden by the uncertainty principle as in that case the quantum particle could have any arbitrarily high value of momentum. In sharp contrast to a classical particle, not only would the "fall" to the bottom not minimize the quantum particle's energy, it would, instead, make it infinitely high! The uncertainty principle, for that matter, arms the quantum particles with an invincible capacity to resist fall! However strong a potential is, it shall never manage to capture a quantum particle at its bottom end. This is the reason that - contrary to the predictions of the classical theory - the electrons never fall on the atom's nucleus despite the powerful electromagnetic attraction to which they are subjected by the latter and despite the enormous pressure which collisions with other atoms exert (N3).

In line with the first and fourth premise laid out in the Introduction we have, up to this point, explored the manner in which quantum mechanics emerges from classical mechanics through the formal procedure of canonical quantization. We shall now move in the opposite direction in order to explore the content of the third premise, the manner in which classical mechanics emerges as a special aspect of quantum mechanics.

7. The semi-classical limit

The Planck constant $\hbar$ is a physical constant which lies at the foundation of quantum theory. It is a positive quantity much smaller than one, but not equal to zero! We saw in the second section that its presence in the mathematical description of a physical system is a statement of the fact that the physical system in consideration is a quantum system. As a matter of fact, $\hbar$ is never present in the mathematical description of classical systems. If the classical description emerges as a specific aspect of a quantum system's dynamical behavior then necessarily that aspect is characterized by the absence of $\hbar$. Formally then, the classical description emerges at that specific limit of a quantum system's dynamical behavior at which $\hbar \to 0$. This is the *semi-classical limit*, a term which underscores the fact that, although at that limit the behavior of a physical system is classical, the physical system itself is always a quantum system. The limit $\hbar \to 0$ itself has no physical content. It is a purely formal procedure which represents the emergence of the actual semi-classical behavior of a quantum system. For, clearly, as $\hbar$ is a constant, we have no right to treat it as a variable, let alone as a variable which tends to zero! The semi-classical behavior emerges in a variety of distinct physical situations as a specific limit of the quantum system's dynamical behavior. The physical content of that limit is exclusively determined by the specific physical situation itself, with $\hbar \to 0$ being the semi-classical limit's formal (abstract) representation, a representation which dispenses with the specific physical situation within which that limit occurred. In what follows, we shall examine this premise in specific physical contexts. We begin again with the one-dimensional infinite square well which best illustrates the essence of the matter in attractive potentials.

We have already seen the emergence of semi-classical behavior in the ground-state energy eigenvalue E_0. If either $L \to \infty$, or $m \to \infty$ then $E_0 \to 0$. It is clear from (6.I) that if instead of $L \to \infty$, or $m \to \infty$ the limit $\hbar \to 0$ is considered, the result is always $E_0 \to 0$ and that, for that matter, $\hbar \to 0$ is the formal representation of the actual semi-classical situation which emerges at either $L \to \infty$, or $m \to \infty$.

Let us turn now to the opposite end of the energy spectrum. The limit $n \to \infty$ signifies an independent semi-classical situation. The higher the energy eigenvalue E_n is, the more the characteristic aspects of classical motion emerge for the quantum particle in the corresponding energy eigenstate $\psi_n(t, x)$. The simple reason is inherent in the analysis in the preceding section. High energies are associated with small wavelengths. An energy eigenstate in Fig.6 corresponds to a standing wave and contains a certain number of half-wavelengths. This number increases as we ascend to eigenstates of higher n. Energy eigenstates which correspond to high values of n are characterized by a high number of nodes and by an accordingly high number of half-wavelengths. Within each such half-wavelength the quantum particle is "localized". The probability for position is maximum within the short range of each such half-wavelength and negligible between any two successive half-wavelengths. If you consider now that for enormously high values of n an enormously high number of half-wavelengths fits between the two infinite walls of the potential, it becomes obvious that at very high energies the quantum particle behaves like a classical particle which moves with constant speed, bouncing off the two perfectly elastic walls.

We have not yet arrived at the semi-classical limit. We must also demonstrate that at $n \to \infty$ the quantum particle satisfies the requirement of a continuous energy spectrum. Indeed, we would be very surprised to discover that at this limit the quantum particle behaves classically in terms of motion, but not in terms of energy! We have seen that in the discrete set of infinitely many energy levels in Fig.6, the distance $\Delta E_n = E_n - E_{n-1}$ between successive energy levels is of the same order of magnitude as the ground-state energy eigenvalue E_0 at least for the first few energy levels. As it turns out, for $n >> 1$, it is $\Delta E_n = 2nE_0$. Contrary to our expectations, the distance between successive energy levels increases! What has happened? Quite simply it is not true that the quantum discontinuities ΔE_n decrease on absolute terms. The quantities which have that property are the *relative discontinuities* $\frac{\Delta E_n}{E_n}$. In the infinite square well it is $E_n = (n+1)^2 E_0$. Thus, although the distance ΔE_n between successive energy levels increases with increasing n, the energy eigenvalues E_n increase much faster, causing the relative discontinuities to behave indeed as $\frac{\Delta E_n}{E_n} \to 0$ at the limit $n \to \infty$. It is in this respect that, as we ascend to progressively higher values of n, the spectrum of E_n becomes eventually so dense that it is practically indistinguishable from the continuous energy spectrum of the corresponding classical particle.

We have, thus, demonstrated that in a static attractive potential (fixed width) the semi-classical behavior of a quantum particle (fixed mass) emerges at the limit $n \to \infty$. We must now reveal the manner in which this semi-classical limit conforms with the expected absence of $\hbar$ so as to receive the formal representation $\hbar \to 0$. We have just seen that ΔE_n is proportional to the ground-state energy eigenvalue E_0 which features a $\hbar^2$ in the numerator. Since E_n is also proportional to E_0, the relative discontinuities $\frac{\Delta E_n}{E_n}$ do not feature $\hbar$, precisely as expected of the semi-classical limit! The fact that for very high values of n it is $\Delta E_n << E_n$ admits the physical interpretation that, for very high values of n, the energy of the quantum particle fluctuates about the value E_n by an amount equal to ΔE_n. The ratio $\frac{\Delta E_n}{E_n}$ expresses, for that matter, the *quantum fluctuation* which characterizes the energy value E_n for a certain $n >> 1$. The ensemble of these fluctuations for each $n >> 1$ actually signifies the quantum fluctuations of the particle's energy. Note, in this respect, that the "deep classical domain" corresponds to $n \sim 10^{27}$. For such values of n it is $\frac{\Delta E_n}{E_n} \sim 2 \times 10^{-27}$. In the deep classical domain the quantum discontinuities are 10^{-27} times smaller than the value of the energy E_n itself. It is clear that such discontinuities lie beyond the possibility of detection by any experimental procedure. Consequently, in terms of observation the energy spectrum at such values of n can be safely considered continuous and classical. We conclude that:

(i) *Vanishing quantum fluctuations signify the emergence of semi-classical behavior.*
(ii) *Vanishing quantum fluctuations are the exclusive physical content of $\hbar \to 0$.*

This conclusion has general significance. As our analysis advances we shall see the emergence of semi-classical behavior out of vanishing quantum fluctuations in several physical contexts in quantum mechanics and in quantum field theory.

In static attractive potentials, for that matter, the semi-classical behavior emerges at the limit $n \to \infty$ and is characterized by the absence of $\hbar$. In order to complete the derivation of this semi-classical limit, we must reveal the precise manner in which this absence is formally represented by $\hbar \to 0$. The latter is, in fact, the formal expression of emergence of semi-classical behavior in any physical context. Why is the electron and any subatomic particle a quantum object with manifestly quantum-mechanical behavior when constrained in a microscopic region of space, but a tennis ball (itself a quantum object like everything else...) a manifestly "classical object" under any circumstances? Why is the tennis ball not also a manifestly quantum object in view of the fact that it consist of electrons and subatomic particles? The very simple answer lies in the general relation $p = \hbar k = \hbar \frac{2\pi}{\lambda}$ introduced in the second section between the fixed momentum p and the wavelength λ of a plane wave of probability. The latter directly implies that $\lambda = \frac{h}{p}$; $(\hbar = \frac{h}{2\pi})$. In turn, the limit $h \to 0$ formally represents the physical limit of vanishing wavelengths $\lambda \to 0$ corresponding to infinite momenta $p \to \infty$. For very high momenta, the wavelength is so small that the quantum behavior lies beyond any possibility for detection. How high? Even a particle of mass $m = 1gr$ - let alone a tennis ball! - can be safely considered classical! For even if that particle's speed is as imperceptible as $v = 10^{-6}cm/sec$, the corresponding wavelength will be $\lambda \equiv 6 \times 10^{-21}cm$. If we recall from elementary wave theory that the characteristic aspects of wave behavior, such as interference, diffraction and refraction, manifest themselves only in the event that the wave is incident on slits or obstacles the size of which is comparable to the wavelength, it becomes obvious that even if the stated particle is incident on a crystal whose characteristic distance between adjacent atoms is of order $10^{-8}cm$, that particle will not reveal even a vestige of wave behavior!

Returning to the infinite square well, we observe that the enormously high number of half-wavelengths which fit between the two infinite walls for $n >> 1$ effectively renders the energy eigenstate a plane wave of probability of arbitrarily high momentum. In turn, the semi-classical limit $\lambda = \frac{h}{p} \to 0$ - and thus also $\frac{\Delta E_n}{E_n} \to 0$ - is formally represented by $\hbar \to 0$ with the direct implication that the semi-classical limit $n \to \infty$ is itself formally represented by $\hbar \to 0$.

We may now turn to the parabolic potential. As I remarked in the third section, it is difficult to overstate the physical importance of that potential if only because the simple quantum harmonic oscillator lies at the very foundation of quantum field theory! We shall study that oscillator in detail later, especially in the third chapter. For the present purposes, the ensuing qualitative analysis suffices.

8. The Simple Quantum Harmonic Oscillator

A classical particle within the parabolic potential in Fig.3 is a simple harmonic oscillator. In the third section we cited the simple examples of a particle which swings attached to the free end of a massless string (the pendulum) and of a particle which oscillates attached to the free end of a massless spring. The corresponding quantum-mechanical particle, that which evolves specifically against the parabolic potential in Fig.4, is a simple quantum harmonic oscillator. In addition to all the attributes which characterize quantum particles in static symmetric attractive classical potentials, the simple quantum harmonic oscillator has a particularity which sets it in a class of its own: The energy levels of the discrete energy eigenvalue spectrum are equally spaced. The physical justification of this particularity is relatively simple.

The incipient point of our analysis is the simple classical harmonic oscillator. It is basic knowledge that the frequency w of the latter is independent of the amplitude of oscillation (the maximum displacement from the point of equilibrium). It depends exclusively on the mass of the swinging particle and on how fast the restoring force grows with increasing displacement from the point of equilibrium. In the case of the ordinary pendulum, the restoring force is that of Newtonian gravity. In the case of the spring, the restoring force is that of the spring's tension (e.g. "loose" spring, "tight" spring). Whatever the case, the magnitude of the restoring force increases with progressively increasing displacement. It is obvious that the total energy of the classical oscillation depends directly on the amplitude of the latter and remains constant in the course of time as a direct consequence of time-translational invariance. Higher amplitude implies higher potential energy at the two points of maximum displacement from the point of equilibrium (zero kinetic energy) and higher kinetic energy at the point of equilibrium itself (zero potential energy). The frequency w is constant, independent of the amplitude and thus also of the total energy. The higher the displacement from the point of equilibrium is at a moment, the stronger the restoring force is at that moment and, thereby, the higher the acceleration toward the point of equilibrium. This can be seen directly in the parabolic potential in Fig.3. The parabolic potential determines the restoring force as a function of the displacement from the point of equilibrium. The frequency of oscillation w is independent of the maximum displacement from x_0 and depends exclusively on how steep the parabolic potential is.

Let us now postulate an electrically charged particle in the parabolic potential of Fig.3. A basic premise of classical electrodynamics is that accelerating charges emit electromagnetic radiation. Since the frequency of oscillation w is independent of the amplitude, the charged particle will emit an electromagnetic wave of a single frequency w determined exclusively by the "steepness" of the parabolic potential. In the quantum theory, the emission of electromagnetic radiation is the result of transition from a higher to a lower energy eigenstate[16]. At the semi-classical limit $n \gg 1$, the classical prediction for the frequency of the emitted electromagnetic wave must coincide with the prediction of the quantum theory. In turn, this enforces the following two conditions:

(i) The "transition rule" $\Delta n = 1$. That is, at least for $n \gg 1$, only transitions from an energy eigenstate to that immediately below are allowed.

(ii) At least for $n \gg 1$, the distance between successive energy levels is constant. The energy levels of the discrete energy eigenvalue spectrum are equally spaced.

Clearly, only if these two conditions are met is the classical prediction concerning the emission of a single-frequency electromagnetic wave valid. For these two conditions ensure the exclusive emission of photons whose frequency is determined by the constant distance $E_{n+1} - E_n$ between successive energy levels as $\frac{E_{n+1}-E_n}{\hbar}$ and which is, for that matter, itself constant. For $n \gg 1$, that constant frequency necessarily coincides with the amplitude- and energy-independent classical frequency of oscillation w:

$$w = \frac{E_{n+1} - E_n}{\hbar} \tag{8.I}$$

This establishes that, at least for $n \gg 1$, the distance between adjacent energy eigenvalues of the simple quantum harmonic oscillator is constant.

We see, for that matter, that the constant separation between adjacent energy eigenvalues of the simple quantum harmonic oscillator is inextricably linked to the fact that in the corresponding classical situation the frequency of oscillation is independent of the oscillation's amplitude and energy. We shall see in the fourth chapter that this characteristic property of the simple quantum harmonic oscillator has a profound consequence in quantum field theory.

The precise relation between the energy eigenvalues E_n and the frequency of classical oscillation w which emerges as a consequence of the solution to the Schrödinger equation for the parabolic potential is

$$E_n = \hbar w \left(n + \frac{1}{2}\right) \tag{8.II}$$

and can be seen to be in perfect conformity with (8.I) the validity of which we established only qualitatively and at least for $n \gg 1$ at that. Thus, (8.II) formally establishes our above-stated conclusion throughout the simple quantum harmonic oscillator's energy spectrum.

---

[16]This transcends the scope of quantum mechanics which is incompatible with the absorption and emission of photons, but it gets the point across!

We conclude by noting that (8.II) manifestly features the non-vanishing ground-state energy:

$$E_0 = \frac{1}{2}\hbar w \qquad (8.III)$$

which will be of crucial importance to our examination of the quantum vacuum.

9. Quantum Tunneling

We have yet to examine the premise stated in the second section to the effect that the probability distributions for position are observable quantities only in the classically accessible regions of space. In classical mechanics a particle has, in general, kinetic energy arising from its motion and potential energy arising from its interaction with a specific field. Both, the kinetic and the potential energy of a classical particle can be independently measured at each point in space. We have seen that whereas the motion of a free classical particle, or of a classical particle in a repulsive potential, is unrestricted, the motion of a classical particle in an attractive potential is restricted in the classically accessible region of space. By contrast, quantum-mechanical particles penetrate into the classically inaccessible regions of space yielding non-vanishing probabilities for position there. The fact that a quantum particle has a non-vanishing probability for position in a region in which classically the potential energy is bigger than the total energy (refer to (Fig.4)) could lead to the misconception that quantum mechanics is incompatible with energy conservation. The mistake in such a reasoning stems from an uncritical import into the quantum theory of the classical independence in the measurements of the potential and the kinetic energy. In quantum mechanics the total energy of the particle is a quantity which characterizes the quantum state in its entirety - that is, throughout space. It cannot, for that matter, be resolved into a kinetic term and a potential term at each point in space. The underlying reason is the fact that the potential energy $V(x)$ is exclusively a function of the position x, whereas the kinetic energy (being roughly equal to $\frac{(\Delta p)^2}{2m}$ if the position has been detected over a range Δx) is exclusively a function of the momentum p. Since precise simultaneous knowledge of x and p is impossible, just as impossible is the simultaneous knowledge of the particle's kinetic and potential energy. It is, for that matter, physically meaningless to state that at a point x the total energy E of the quantum-mechanical particle determined as an eigenvalue of $\hat{H} = \frac{\hat{p}^2}{2m} + V(x)$ is smaller than $V(x)$: Since

$$[\hat{H}, V(x)] \neq 0$$

precise simultaneous knowledge of E and $V(x)$ is impossible!

The implication which quantum penetration into the classically inaccessible regions of space has for observation is readily understood. In a classically accessible region of space an experimental procedure can always detect the position of a quantum particle. For since a classical particle has at each moment an observable position in the classically accessible regions of space, we expect a quantum particle to also extend the courtesy of an observable position in the same regions. It is, instead, in the classically inaccessible regions, in those regions in which no classical particle may exist, that no experimental procedure could ever detect the position of a quantum particle. Any experimental procedure which aims at the detection of a quantum particle's position in a classically inaccessible region of space inevitably disturbs the quantum particle's energy by an amount sufficient enough to "knock" the particle out of the potential well altogether[17]. The quantum-mechanical probability distributions for position are, indeed, observable quantities exclusively in the classically accessible regions of space.

The fact that the total energy of quantum particles admits no resolution in a kinetic and a potential term implies that quantum particles acknowledge no barriers! They penetrate into classically inaccessible regions of space with a non-vanishing probability for position. This fundamental property has general significance. As quantum particles can penetrate into classically inaccessible regions of space, so can they penetrate from a classically inaccessible region into a classically accessible region. This situation signifies the effect of *quantum tunneling*. Clearly, in the event that the wavefunction extends from one classically accessible region to another classically accessible region through a classically inaccessible region, the quantum particle emerges in the latter with smaller probability for position. This can clearly be seen in the scattering process (Fig.7) to which we turn now.

10. Scattering States

In the third section, we remarked that energy eigenfunctions in repulsive classical potentials are scattering states. Such states mirror the corresponding classical situation as much as bound states in attractive potentials do. The non-vanishing probabilities for position which a quantum particle evolving against a repulsive potential has at infinite distances mirror the transit of the corresponding classical particle to arbitrarily long distances. Accordingly, the continuous spectrum of energy eigenstates and eigenvalues in repulsive potentials mirrors the continuous spectrum of frequencies which characterizes unbounded classical waves such as those on an elastic chord which extends indefinitely in both directions.

A convenient physical context which illustrates the qualitative aspects of quantum-mechanical scattering is provided by the static repulsive "rectangular potential barrier" in Fig.7. The segment of one-dimensional space within which this potential extends is region II between 0 and L. The scattering process consists in sending in a set of quantum particles from a remote, formally infinite, distance in region I and determine how many of them get reflected by the repulsive potential and how many of them advance past region II to register their presence on a particle detector situated at a remote distance in region III. The scattering problem can, without loss of generality, be reduced to the evolution of the probability distribution of only one quantum particle throughout the one-dimensional space. We shall, for that matter, study the evolution of an incident plane wave of probability of momentum eigenvalue p and energy eigenvalue $E = \frac{p^2}{2m}$ in

---

[17]It should be clear that the reference here is not to the idealized potentials in Fig.3, Fig.4 and Fig.6, but to the realistic case in which no potential receives infinite values.

region I. As the geometry of the potential suggests, a distinction must be made between the situation in which the energy eigenvalue is $E < V_0$ and that in which the energy eigenvalue is $E > V_0$.

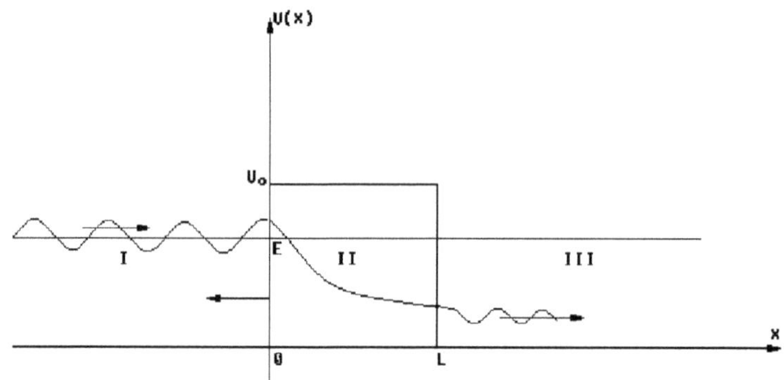

FIGURE 7. Scattering by a rectangular potential barrier. If $E < V_0$ then part of the incident plane wave is reflected back into region I by the repulsive potential (left-pointing arrow) and part of it tunnels through region II into region III. If $E > V_0$ then part of the incident wave is reflected and part of it advances in region III. Transmission to region III invariably occurs regardless of the value of E.

Only the situation in which $E < V_0$ is depicted in Fig.7. Part of the incident plane wave of probability in region I is reflected by the repulsive potential. The result is a plane wave of probability of the same p and E but of smaller probability amplitude. Part of the incident wave tunnels through the classically inaccessible region II to emerge in region III as a plane wave of the same p and E but of smaller probability amplitude. The scattering state is the superposition between the incident and the reflected plane wave in region I, the exponentially decreasing wavefunction in region II and the plane wave which emerges through tunnelling in region III.

The structural similarities with the classical waves and the correspondence with the most general properties of classical motion are obvious. A classical plane wave of energy would also be partially reflected by and partially transmitted through a region characterized by an abrupt change in the properties of the elastic medium. Both, the reflected and the transmitted segments, are plane waves of the same frequency. At once, the general properties of classical motion are preserved in the reflected plane wave of probability. A classical particle which bounces off a perfectly elastic wall reverses its direction of motion with the same kinetic energy and magnitude of momentum. The dynamical behavior of the incident quantum particle mirrors this classical property in the superposition between the incident and the reflected plane wave of probability in region I. Of course, the dynamical behavior of a quantum particle by far transcends all analogies with the corresponding classical situation. A classical particle incident upon the perfectly elastic wall in region I with energy $E < V_0$, would never emerge in region III. Transition through classically inaccessible regions ("classically forbidden" is just as valid) of space is an exclusively quantum-mechanical property[18]. In fact, the lighter a quantum particle is, the higher its ability to "pass through walls" is! We note in this respect that physical processes such as the formation of molecules, the chemical reactions, the formation of solid bodies with their characteristic structure and properties and eventually life itself, emerge as a consequence of quantum tunneling through classically inaccessible regions of space and are a testament to the penetrating ability and versatility which the electron has on account of its minute mass.

It is clear that region II is a classically accessible region for $E > V_0$. An incident classical particle of energy $E > V_0$ would advance from an arbitrarily high distance in region I to an arbitrarily high distance in region III. Such then is the property which the corresponding quantum particle also has. For $E > V_0$, no quantum tunnelling occurs. In this case, however, there is partial reflection[19] of the incident plane wave at both, $x = 0$ and $x = L$. A scattering state is the superposition of the incident and the reflected plane waves in region I, the superposition between the wave which advances toward increasing values of x and that which advances toward decreasing values of x in region II and the plane wave in region III. We also mention in passing that one of the most interesting features which scattering states manifest for $E > V_0$ is the fact that for certain values of the energy E the entire wavefunction propagates from region I to region III without reflection at $x = 0$ and $x = L$.

Before we advance our examination of quantum mechanics, it is important that we return to the abstract Hilbert spaces in order to study an aspect of the mathematical formalism which we shall henceforth extensively use, from advanced quantum mechanics to quantum field theory.

11. The Dirac Notation

An ingenious way of representing abstract quantum states is the *Dirac notation*, an aspect of quantum theory's mathematical formalism which is as relevant to quantum mechanics as it is to quantum field theory. This modest term suggests mere notational convenience, yet its scope extends well beyond that superficial level. For although there is nothing which the Dirac notation can accomplish that the conventional notation in the third, fourth and fifth section cannot, the Dirac notation renders the abstract geometrical character of the physical concepts involved immediately transparent without compromising the physical transparency of those concepts. The advantage of the Dirac notation is that it dispenses with the dependence on position, on momentum and on any

---

[18]From the fundamentally rigorous perspective of quantum field theory, there is no quantum tunnelling. The simple reason is that there are no classical potentials. There are only quantum fields in interaction with each other.

[19]For $E < V_0$ there is also partial reflection at $x = L$. As its effect is an eventual contribution to the reflected wave in region I, it is inconsequential to the preceding qualitative analysis.

physical attribute altogether. This is accomplished by exploiting the eigenvalue equation of some operator $\hat{A}$. The Dirac notation merely recasts the eigenvalue statement

$$\hat{A}\psi_a(t,x) = a\psi_a(t,x) \tag{11.I}$$

in the form

$$\hat{A}|a\rangle = a|a\rangle \tag{11.II}$$

and thus replaces the notation $\psi_a(t,x)$ for the eigenfunction of $\hat{A}$ by the notation $|a\rangle$ in which the same eigenfunction is merely labeled by the corresponding eigenvalue a. Both, (11.I) and (11.II) make the same mathematical statement. The only difference is that, unlike $\psi_a(t,x)$ in (11.I), $|a\rangle$ in (11.II) is an abstract quantum state. All dependence on position, or on momentum is absent! The abstract quantum state $|a\rangle$ is known as a *ket*. Formally then, we dispense with the expression of quantum states in the specific Hilbert space of position, or of momentum, or of any physical attribute and define instead an abstract Hilbert space of states. Those abstract states, the elements of that abstract Hilbert space, are the kets $|a\rangle$ and the abstract Hilbert space itself, the *Hilbert space of kets*. Clearly, the kets $|a\rangle$ are completely basis-independent. We concretize $|a\rangle$ only by specifying a basis - that is, the set of eigenfunctions of a self-adjoined operator - in which $|a\rangle$ is expressed. The following two examples illustrate the essence of the matter.

In terms of the Dirac notation, the eigenvalue statement

$$\hat{x}\delta(x-a) = a\delta(x-a) \tag{11.$IIIa$}$$

in (3.IV) is recast in the form

$$\hat{x}|a\rangle = a|a\rangle \tag{11.$IIIb$}$$

with $|a\rangle$ being the Dirac notation for position eigenfunction $\delta(x-a)$ on the understanding that abstract vector $|a\rangle$ concretizes to $\delta(x-a)$ only when that vector is expressed in the *basis* of position eigenstates. Accordingly, the eigenvalue statement

$$\hat{H}\psi_n(x,t) = E_n\psi_n(t,x) \tag{11.IVa}$$

in (3.I) is recast in the form

$$\hat{H}|n\rangle = E_n|n\rangle \tag{11.IVb}$$

with $|n\rangle$ being the Dirac notation for energy eigenfunction $\psi_n(t,x)$ in (3.IIId) on the understanding that abstract vector $|n\rangle$ concretizes to (3.IIId) only when that vector is expressed in the basis of position eigenstates.

Corresponding to the abstract Hilbert space of kets $|a\rangle$, there is the equally abstract *dual Hilbert space of bras*[20] $\langle a|$. The definition of the bra is somewhat more involved

---

[20]The Dirac notation is suffused with Dirac's ingenuity in terms of mathematical content and sense of humor in terms of terminology. The terms "bra" for $\langle\,|$ and "ket" for $|\,\rangle$ derive from the word "bracket".

than that of the ket. We shall give that formal definition very soon. For the moment, let it suffice to consider $\langle a|$ as the abstract version of $\psi_a^*(x)$. In order to understand what (11.I) and (11.II) imply for the relation of operator $\hat{A}$ to bra $\langle a|$, merely consider the fundamental relation (5.IX). We note first that, as mere inspection reveals, the corresponding eigenvalue statement to (11.I) is

$$\hat{A}^\dagger \psi_a^*(t,x) = a^* \psi_a^*(t,x) \tag{11.V}$$

The Dirac notation recasts (11.V) as

$$\langle a|\hat{A}^\dagger = a^*\langle a| \tag{11.VI}$$

thus replacing the, specific to space and time, wavefunction $\psi_a^*(t,x)$ by the abstract bra $\langle a|$.

For the purpose of illustration now, we respectively identify $\psi(x)$ and $\psi'(x)$ in (5.IX) with two eigenfunctions $\psi_m(x)$ and $\psi_n(x)$ of some self-adjoined operator (we shall see what happens if this is not the case) and note that if:

(i) $|m\rangle$ is the Dirac notation for $\psi_m(x) \equiv \psi(x)$ and $|n\rangle$ is the Dirac notation for $\psi_n(x) \equiv \psi'(x)$,

(ii) $\hat{A}$ acts exclusively on the set of kets in line with (11.II) and $\hat{A}^\dagger$ acts exclusively on the set of bras in line with (11.VI),

(iii) the bra-ket $\langle\ |\ \rangle$ implies the continuous summation $\int_{-\infty}^{\infty} dx$

then both sides receive the formal representation

$$\langle n|\hat{A}|m\rangle$$

since, clearly, the operation of $\hat{A}$ on $|m\rangle$ yields the left side $(\psi', \hat{A}\psi)$ of (5.IX) when translated to concrete wavefunctions and, accordingly, the operation of $\hat{A}^\dagger$ on $\langle n|$ yields the right side $(\hat{A}^\dagger\psi', \psi)$ of (5.IX). Inherent in this observation is the announced rigorous definition:

The bra $\langle n|$ is the map

$$\int_{-\infty}^{\infty} dx\, \psi_n^*(x)(\ )$$

of function $\psi_m(x)$ to complex number

$$\int_{-\infty}^{\infty} dx\, \psi_n^*(x)\psi_m(x) \qquad\qquad \square$$

That is, a given bra $\langle n|$ maps any ket $|m\rangle$ to complex number $\langle n|m\rangle$. In terms of concrete wavefunctions this translates to the statement that, given $\psi_n(x)$, the bra

$$\int_{-\infty}^{\infty} dx\, \psi_n^*(x)(\ )$$

maps whichever function $\psi_m(x)$ is replaced in ( ) to complex number

$$\int_{-\infty}^{\infty} dx\, \psi_n^*(x)\psi_m(x)$$

Before we proceed, we stress again this crucial aspect of the compact Dirac notation. In $\langle n|\hat{A}|m\rangle$, the operator acting to its right is $\hat{A}$, whereas the operator acting to its left is $\hat{A}^\dagger$. In terms of concrete wavefunctions, the result is $(\psi', \hat{A}\psi)$ in the former case and $(\hat{A}^\dagger\psi', \psi)$ in the latter case.

Note now the immediate implication which the abstract bra-ket procedure of the Dirac notation has for the (square of the) magnitude

$$\int_{-\infty}^{\infty} dx\, |\psi(x)|^2$$

of wavefunction $\psi(x)$. The latter is identically the map

$$\int_{-\infty}^{\infty} dx\, \psi^*(x)(\ )$$

of $\psi(x)$ to real number

$$\int_{-\infty}^{\infty} dx\, \psi^*(x)\psi(x)$$

For that matter

$$\int_{-\infty}^{\infty} dx\, |\psi_n(x)|^2 \equiv \langle n|n\rangle \qquad (11.VII)$$

and relation (3.VIII) for the energy eigenfunctions is recast as

$$\langle n|n\rangle = 1 \ ; \ n = 1, 2, \ldots$$

We can directly extend this to the *orthonormality relation* in (3.XIII) - that is, to the orthogonality of normalized eigenfunctions of self-adjoined operators which have a discrete spectrum:

$$\langle n|m\rangle = \delta_{mn} \qquad (11.VIII)$$

Clearly, the bra-ket expression of (3.XIV) which constitutes the extension of (11.VIII) to the continuous spectrum is:

$$\langle a'|a\rangle = \delta(a - a') \qquad (11.IX)$$

For example, $\langle x'|x\rangle = \delta(x - x')$ and $\langle p'|p\rangle = \delta(p - p')$ is (11.IX) specified in the set of position eigenstates and the set of momentum eigenstates respectively.

As another important demonstration of the flexibility and versatility which characterizes the Dirac notation, consider the statement in (3.VI). In terms of the Dirac notation, the right side is

$$\int_{-\infty}^{\infty} da c(a)|a\rangle$$

Since now the right side is abstract, the left side must also be abstract. For that matter, the left side is recast as $|\psi\rangle$, which features no dependence:

$$|\psi\rangle = \int_{-\infty}^{\infty} da c(a)|a\rangle \qquad (11.X)$$

Abstract ket $|\psi\rangle$ is the Dirac notation for $\psi(x)$. It is independent of any particular Hilbert space, basis and physical attribute (position, momentum, spin etc.). As such $|\psi\rangle$ is exemplary of the manner in which the Dirac notation produces abstract states which are not even labeled by eigenvalues.

By way of demonstrating the advantage of $|\psi\rangle$ we return to (5.IX), this time without the antecedent arbitrary identification $\psi(x) \equiv \psi_m(x)$ and $\psi'(x) \equiv \psi_n(x)$. In this general case, it is clear that both sides of (5.IX) receive the formal representation

$$\langle \psi'|\hat{A}|\psi\rangle$$

always on the understanding that the operator acting on $|\psi\rangle$ to its right is $\hat{A}$, whereas the operator acting on $\langle\psi'|$ to its left is $\hat{A}^\dagger$. In terms of concrete wavefunctions, the result is $(\psi', \hat{A}\psi)$ in the former case and $(\hat{A}^\dagger\psi', \psi)$ in the latter case.

These results allow us to demonstrate the power of the Dirac notation in the succinct derivation of the completeness relations to which we have already arrived through use of "conventional" mathematical techniques in the third section. The left side of the relation

$$\int_{-\infty}^{+\infty} dy\psi^*(y)\psi(y) = \sum_{n=1}^{\infty} c_n^* c_n$$

is now merely recast as $\langle\psi|\psi\rangle$. At once, (pursuant to (3.XV)) it is

$$c_n = \langle n|\psi\rangle \iff c_n^* = \langle\psi|n\rangle$$

Thus, in the Dirac notation the relation above is recast as

$$\sum_{n=1}^{\infty} \langle\psi|n\rangle\langle n|\psi\rangle = \langle\psi|\psi\rangle \qquad (11.XI)$$

which directly implies the completeness relation (3.XVI) in the simple form

$$\sum_{n=1}^{\infty} |n\rangle\langle n| = 1 \qquad (11.XII)$$

As a simple exercise you can show yourselves that in the Dirac notation, relation

$$(\psi, \psi') = \Sigma_{k=1}^{+\infty} c_k^* c_k'$$

in the third section is recast as

$$\langle\psi|\psi'\rangle = \sum_{n=1}^{\infty} \langle\psi|n\rangle\langle n|\psi'\rangle \tag{11.XIII}$$

The corresponding relation to (11.XII) for a continuous spectrum is

$$\int_{-\infty}^{\infty} da\,|a\rangle\langle a| = 1 \tag{11.XIV}$$

and can be seen to, directly, imply (3.XVII).

Starting now from (11.X), we may use (11.XIV) in order to retrace our steps and concretize the abstract ket $|\psi\rangle$ to any particular Hilbert space. Indeed, *inserting the complete set of eigenstates* of some self-adjoined operator $\hat{A}$ according to (11.XIV) in the left side of (11.X), yields:

$$\int_{-\infty}^{\infty} da\,|a\rangle\langle a|\psi\rangle = \int_{-\infty}^{\infty} da\,c(a)|a\rangle$$

which clearly reveals that

$$\langle a|\psi\rangle = c(a)$$

with the direct implication (recall that $c(x) \equiv \psi(x)$) that the familiar probability amplitude for position $\psi(x)$ is, in the Dirac notation, recast as $\langle x|\psi\rangle$, as the concrete expression of $|\psi\rangle$ in the complete set of position eigenstates:

$$\psi(x) = \langle x|\psi\rangle$$

As we have already seen in the third section, the above-stated relation

$$\psi(a) = \langle a|\psi\rangle$$

makes a statement of general significance, a statement to the effect that a wavefunction can be expressed as a vector in the Hilbert space of $\hat{x}$, or of $\hat{p}$, or of $\hat{H}$, or of any self-adjoined operator. The Dirac notation exploits this fact in order to elevate the wavefunction to an abstract state vector which can subsequently be concretized to any basis of our choice. It is a statement to the effect that $\langle a|\psi\rangle$ is the concrete expression of the abstract ket (or abstract vector, if you will) $|\psi\rangle$ in the specific basis (complete set of eigenstates) $|a\rangle$. We have just considered the case $a \equiv x$. You can consider the case $a \equiv p$ yourselves: Concretize $|\psi\rangle$ to the basis specified by the eigenstates of the momentum operator $\hat{p}$ and ensure that such a concrete expression is the wavefunction of momentum $\phi(p)$ in (2.II).

We conclude with three direct implications which the Dirac notation has for our analysis in the fifth section.

(i) The statement (5.IV) for the expectation value $\langle\hat{A}\rangle$ of operator $\hat{A}$ is, in the Dirac notation, recast as

$$\langle\hat{A}\rangle = \langle\psi|\hat{A}|\psi\rangle \tag{11.XV}$$

(ii) The statement (5.VIII) directly implies that a self-adjoined operator is that operator for which[21]

$$\hat{A}^\dagger = \hat{A} \qquad (11.XVI)$$

(iii) Relation (5.X) is recast as

$$\langle \hat{A} \rangle = \Sigma_{n=1}^{+\infty} \langle \psi | n \rangle \langle n | \psi \rangle a_n \qquad (11.XVII)$$

as you yourselves can easily confirm by using $c_n = \langle n | \psi \rangle$ and $c_n^* = \langle \psi | n \rangle$.

Up to this point we examined most of the essential aspects of quantum mechanics in the context of the canonical quantization using one-dimensional quantum systems as the vehicle of our analysis. Nevertheless, this simple approach does not encompass all essential aspects, since no physical system is one-dimesional. In order to study aspects of crucial importance which transcend the scope of the hitherto one-dimensional approach, aspects which are at once of central importance to quantum field theory, it is necessary that we examine quantum mechanics in the three-dimensional physical space. Although we shall introduce and explain certain new mathematical concepts, our approach in the next three sections will remain the same. The analysis will advance at the level of physical concepts with no mathematical derivations, or solutions to differential equations. The relevant new operators will be introduced on the grounds of physically plausible arguments and the only requirement for a thorough understanding of the subject matter will be committed thinking. The level of abstraction will be considerably higher than that at which our analysis has hitherto advanced. Readers who find it difficult to follow may skip the next three sections at first reading. Perhaps with the exception of the seventh section of the fourth chapter, such knowledge as the next three sections provide will not be required until the sixth chapter. However, the analysis in the next three sections is indispensible for quantum field theory. Thorough understanding of it is a requirement for all those who wish to advance past the fifth chapter.

12. Angular Momentum

Let us first make a brief reference to a few elementary mathematical concepts: A Cartesian coordinate system is a system of three axis perpendicular to each other with the vertical being the z-axis and the remaining two being respectively the x- and the y-axis. With respect to a Cartesian system, each point in space has three coordinates x, y, z and corresponds to vector $\vec{r}$ directed from the origin $(0,0,0)$ to that point: $\vec{r} = (x, y, z)$.

The *angular momentum* is a fundamental concept in classical mechanics. With respect to a Cartesian coordinate system, it is defined as the *cross product*:

$$\vec{L} = \vec{r} \times \vec{p} \qquad (12.I)$$

---

[21]Strictly speaking, (11.XVI) does not merely imply the "transfer" of $\hat{A}$ from ψ to ψ' in (5.VIII). It also implies the coincidence of the Hilbert space of $\hat{A}$ and that of $\hat{A}^\dagger$. This second requirement is not of immediate relevance to our conceptual approach (all physically important operators satisfy it).

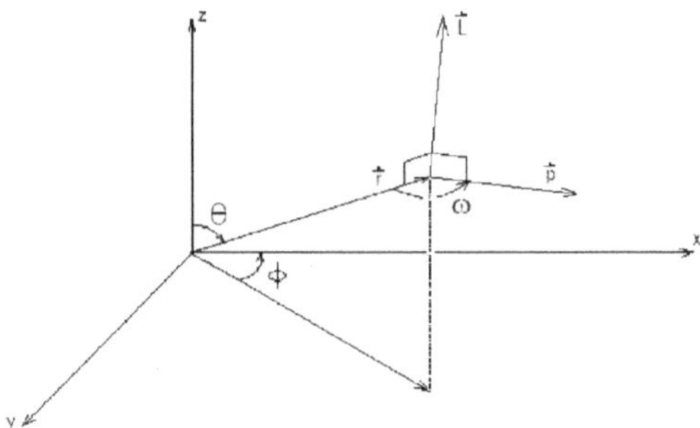

FIGURE 8. The cross product $\vec{r} \times \vec{p} = \vec{L}$ is a vector perpendicular to the plane defined by position vector $\vec{r}$ and momentum vector $\vec{p}$ whose magnitude is equal to $|\vec{L}| = |\vec{r}||\vec{p}|sinw$ with w being the angle between $\vec{r}$ and $\vec{p}$.

of a particle's position vector $\vec{r} = (x, y, z)$ and the particle's linear momentum $\vec{p} = (p_x, p_y, p_z)$. As, in general, these two vectors are not aligned, the angular momentum centrally involves the angle between them. As a consequence, the angular momentum transforms as a vector under all transformations except reflections. Under reflections, a *pseudovector* like $\vec{L}$ is opposite in direction to the reflection of a proper vector. For example, whereas a proper vector changes direction from left to right (and from right to left) under reflection, a *pseudovector* like $\vec{L}$ remains invariant. This particularity will not be essential to our analysis and, for that matter, we shall treat the angular momentum as a proper vector $\vec{L} = (L_x, L_y, L_z)$ in three-dimensional Euclidean space.

Since we shall not make a direct use of the definition in (12.I), it is not necessary that we elucidate its precise mathematical significance (see Fig.8). It is, instead, much more convenient to define $\vec{L}$ from the analytic expressions of its components which follow from (12.I):

$$L_x = yp_z - zp_y \tag{12.IIa}$$

$$L_y = zp_x - xp_z \tag{12.IIb}$$

$$L_z = xp_y - yp_x \tag{12.IIc}$$

Upon quantization the classical angular momentum in (12.I) is replaced by the angular-momentum operator:

$$\hat{\vec{L}} = \hat{\vec{r}} \times \hat{\vec{p}}$$

(12.*III*)

or, equivalently, by the three operators:

$$\hat{L}_x = \hat{y}\hat{p}_z - \hat{z}\hat{p}_y$$

(12.*IV a*)

$$\hat{L}_y = \hat{z}\hat{p}_x - \hat{x}\hat{p}_z$$

(12.*IV b*)

$$\hat{L}_z = \hat{x}\hat{p}_y - \hat{y}\hat{p}_x$$

(12.*IV c*)

Of course, it is hardly necessary to stress that since $\hat{\vec{L}}$ and, consequently, $\hat{L}_x, \hat{L}_y, \hat{L}_z$ represent observable attributes, they are self-adjoined operators. For that matter, they all have real eigenvalues and complete sets of eigenfunctions.

We have seen that the homogeneity of space implies momentum conservation and that time-translational invariance implies energy conservation. We note from (12.I) (and (12.III)) that since the angular momentum $\vec{L}$ involves the linear momentum $\vec{p}$ and the three directions of space in $\vec{r} = (x, y, z)$, a certain symmetry of space must imply the conservation of angular momentum. Such a symmetry must necessarily involve the angle between $\vec{r}$ and $\vec{p}$. Indeed, $\vec{L}$ is conserved if space is isotropic - that is, if its properties are the same in all directions. Such a spatial isotropy does not merely imply the empty space, but also the potentials which depend only on the magnitude of the position vector $\vec{r}$ - that is, the potentials of the form $V(r)$ with $r \equiv |\vec{r}|$. For, clearly, such potentials are isotropic about their centre $r = 0$. Of course, a classical potential implies a force at each position $\vec{r}$ and a classical potential of the form $V(r)$ necessarily implies a force which points in the direction of $\vec{r}$ (radially outward from $r = 0$) if the potential is repulsive, or opposite to the direction of $\vec{r}$ (radially inward from $r = 0$) if the potential is attractive. This is the reason that such potentials are known as *central*. With this elucidation we reiterate that *the isotropy of space implies angular-momentum conservation* and that this does not distinguish between classical and quantum behavior.

Although we have not yet stated the mathematical expression of the momentum operator $\hat{\vec{p}}$, a striking property of the angular-momentum operator is readily understood: The three components of $\hat{\vec{L}}$ do not commute with each other! This follows directly from (12.IV) in view of the fundamental non-commutativity between the position operator and the momentum operator: $xp_x \neq p_x x$, $yp_y \neq p_y y$, $zp_z \neq p_z z$. In fact, using the fundamental commutator in (5.XIb):

$$[\hat{x}, \hat{p}_x] = [\hat{y}, \hat{p}_y] = [\hat{z}, \hat{p}_z] = i\hbar$$

(12.*V*)

it is a matter of a trivial calculation to show that

$$[\hat{L}_x, \hat{L}_y] = i\hbar \hat{L}_z \qquad\qquad (12.VIa)$$

$$[\hat{L}_y, \hat{L}_z] = i\hbar \hat{L}_x \qquad\qquad (12.VIb)$$

$$[\hat{L}_z, \hat{L}_x] = i\hbar \hat{L}_y \qquad\qquad (12.VIc)$$

The implication of (12.VI) is that a measurement of a quantum particle's angular momentum can only yield one of the angular momentum's three components. Pursuant to the analysis in the third section, if more than one components were simultaneously observable with absolute certainty, (12.VI) would be violated. For example, if the L_z and the L_x components were simultaneously observable, (12.VIc) would be violated since in that case $[\hat{L}_z, \hat{L}_x] = 0$. Thus, contrary to popular misconceptions, neither do the electrons in atoms "orbit" the nucleus, nor do quantum particles in general follow well-defined paths in space.

The physical implication in (12.VI) is one of several counterintuitive aspects which the angular momentum has at quantum level. In the classical theory, $\vec{L}$ may have any orientation and, thereby, any projection along any of the three axis of a Cartesian coordinate system. That is, the possible values of the z-component (equivalently, of the x- or y-component) of a classical particle's angular momentum $\vec{L}$ form a continuous set. Not so in the quantum theory where observable attributes are represented by self-adjoined operators. In order that operator $\hat{L}_z$ which yields the observable z-component of a quantum system's angular momentum (again, this implies that the other two components are unobservable) be self-adjoined, its possible values must form the discrete set

$$\hbar(-l), \ \hbar(-l+1), \ ..., \ \hbar(l-1), \ \hbar l$$

As we shall soon see, this implies that the possible values of the angular-momentum magnitude also form a discrete set. The mathematical calculation which proves this statement is rather involved and not easy to conceptualize. There is, however, a simplified physical context which conveys the essence of the matter and reveals the manner in which this concept applies in a general physical context.

We consider a classical system of two particles respectively of masses m_1 and m_2 at fixed distance from each other. The system is constrained to rotational motion in the $x - y$-plane about the z-axis of a coordinate system situated at the system's centre of mass. In Fig.8, such an arrangement would correspond to m_1 at the fixed distance $r_1 = |\vec{r}_1|$ from the origin which the projection of $\vec{r}$ on the $x - y$-plane signifies ($\vec{r}$ itself is not relevant to this arrangement) and to m_2 at fixed distance r_2 from the origin in the direction opposite to $\vec{r}_1$ (not depicted). Clearly, in this context the angular momentum $\vec{L}$ is oriented along the positive z-axis. That is:

$$\vec{L} = \vec{L}_z$$

In order to state the energy operator $\hat{H}$ of the system which emerges from the quantization of this classical system, we reason by analogy to (3.IIIb): If in terms of the linear momentum of a free quantum-mechanical particle it is:

$$\hat{H} = \frac{\hat{p}^2}{2m}$$

then in terms of the angular momentum of the "diatomic molecule" in consideration it, necessarily, is:

$$\hat{H} = \frac{\hat{L}_z^2}{2I} \qquad (12.VIIa)$$

with the *moment of inertia* I defined in terms of the stated distances r_1 and r_2 of the two classical particles from the centre of mass by

$$I = m_1 r_1^2 + m_2 r_2^2$$

Of exclusive relevance to the present considerations is the character of I as the rotational equivalent to the mass m, a fact which brings (12.VIIa) in line with (3.IIIb). Besides that, $\frac{1}{2I}$ merely amounts to a multiplicative factor, rendering any discussion as to the precise physical significance of the moment of inertia unnecessary. We reiterate at this point that the classical statement $\vec{L} = \vec{L}_z$ does not imply that the angular momentum of the quantum system is oriented along the z-axis as such a statement would at once imply that the projections of the angular momentum along the x- and y-axis are respectively equal to zero in direct contradiction of (12.VI). It implies, instead, that in terms of absolute certainty no possible projection on the x- or y- axis is at once observable with a possible projection on the z- axis. By choice, only the eigenvalues of $\hat{L}_z$ are observable with absolute certainty. We note also that pursuant to the analysis in the third section, the obvious commutation relation

$$[\hat{L}_z, \hat{H}] = 0 \qquad (12.VIII)$$

is a statement to the effect that the z-component of the angular momentum is a conserved attribute of the quantum system.

Unlike $\hat{H}$ in (3.IIIb), $\hat{H}$ in (12.VIIa) has a discrete eigenvalue spectrum since the "diatomic molecule" is obviously a bound quantum state. It follows, for that matter, that $\hat{L}_z^2$ - and, thus, also $\hat{L}_z$ - has a discrete eigenvalue spectrum and that the set of its eigenstates is, at once, the set of the eigenstates of $\hat{H}$ in (12.VIIa). These results can be directly extended to the general case at which we arrive by relaxing the condition of rotational motion strictly in the $x - y$-plane. The "diatomic molecule" rotates now in any direction of space about its fixed centre of mass[22]. Thus, with

$$\vec{\hat{L}} = \hat{L}_x \hat{x}_0 + \hat{L}_y \hat{y}_0 + \hat{L}_z \hat{z}_0 \Longleftrightarrow \hat{L}^2 \equiv |\vec{\hat{L}}|^2 = \hat{L}_x^2 + \hat{L}_y^2 + \hat{L}_z^2$$

---

[22]As the issue here is the angular momentum of the quantum system, we disregard any translational motion of the centre of mass.

- $\hat{x}_0$, $\hat{y}_0$, $\hat{z}_0$ being the unit vectors along the x-, y- and z-axis respectively - (12.VIIa) generalizes to

$$\hat{H} = \frac{\hat{L}^2}{2I} \qquad (12.VIIb)$$

so that the commutation relation

$$[\hat{L}_z, \hat{L}^2] = 0 \qquad (12.IX)$$

whose validity follows from (12.VI), implies the direct extension of (12.VIII) to the general case - that is, with $\hat{H}$ stated in (12.VIIb). In turn, the discrete set of eigenstates of $\hat{H}$ in (12.VIIb) is, at once, the set of eigenstates of $\hat{L}_z$ and $\hat{L}^2$. Thus, $\hat{L}^2$ and $\hat{L}_z$ respectively receive a discrete set of eigenvalues. In order to determine the character of the discrete set of eigenvalues which $\hat{L}_z$ receives we further note that, always by analogy to (3.IIIb):

(i) Each eigenfunction of $\hat{L}_z$ depends exclusively on the angle ϕ which the position vector $\vec{r}_1$ on the $x - y$-plane forms with the x-axis, as each eigenfunction (3.IIb) of $\hat{p}$ depends exclusively on the position x.

(ii) In each such eigenfunction $\Phi_m(\phi)$ of $\hat{L}_z$, the corresponding eigenvalue m of $\hat{L}_z$ multiplies ϕ, as in $\psi_p(x)$ the eigenvalue p of the momentum operator multiplies x.

(iii) The eigenfunctions $\Phi_m(\phi)$ of $\hat{L}_z$ have the same form as the eigenfunctions $\psi_p(x)$ of $\hat{p}$.

We conclude, for that matter, that the solution to the eigenvalue equation

$$\hat{L}_z\Phi_m(\phi) = m\hbar\Phi_m(\phi) \qquad (12.Xa)$$

is of the form:

$$\Phi_m(\phi) = A_m e^{im\phi} \qquad (12.Xb)$$

with A_m being a constant[23].

From (12.Xb) now, it follows that each eigenfunction of $\hat{L}_z$ must be *single-valued*: It must receive the same value after each complete rotation $\phi = 2\pi$ of the position vector $\vec{r}_1$ about the z-axis. That is:

$$A_m e^{im(\phi+2\pi)} = A_m e^{im\phi} \qquad (12.XI)$$

This can only be possible if

$$e^{im2\pi} = e^0 = cos0 + isin0 = 1$$

which implies that:

---

[23]I stress that this derivation is only qualitative. The rigorous derivation can be found in any advanced book of quantum mechanics which I cite in the Bibliography.

$$m = 0,\ \pm 1,\ \pm 2,\ ... \qquad\qquad (12.XIIa)$$

and, thereby, the eigenvalue spectrum:

$$\hbar m = 0,\ \pm\hbar,\ \pm 2\hbar,\ ... \qquad\qquad (12.XIIb)$$

Although we have yet to determine the precise eigenvalue spectrum of $\hat{L}_z$, (12.XII) confirms our conclusion as to the discrete character of that spectrum and yields the distance between successive eigenvalues in it. We stress that (12.XII) characterizes the eigenvalue spectrum of $\hat{L}_z$ in the general case in which rotation is not restricted in the $x-y$-plane (this latter case is, of course, an aspect of the general case). The reason it is not yet the precise eigenvalue spectrum of $\hat{L}_z$, is the fact that we have derived it absent any information on the magnitude of the angular momentum itself. We shall see now that the relation between that magnitude and (12.XII) entails a peculiar twist.

The magnitude of the angular momentum relates, of course, to the eigenvalue spectrum of $\hat{\vec{L}}$. In order to establish that spectrum, we turn to the eigenvalue spectrum of $\hat{L}^2$. Our first observation, in this respect, is that $\hat{L}^2 = \hat{L}_x^2 + \hat{L}_y^2 + \hat{L}_z^2$ is a positive self-adjoined operator (again, all eigenvalues in (12.IV) are real):

$$\hat{L}^2 \geq 0 \qquad\qquad (12.XIII)$$

satisfying:

$$\hat{L}^2 \geq \hat{L}_z^2 \qquad\qquad (12.XIV)$$

From this point on rather than formally derive the eigenvalue spectrum of $\hat{L}^2$, we shall advance a conceptual argument which yields that spectrum and reveals the physical essence of the matter[24].

We would like to state the eigenvalues of $\hat{L}^2$ as $\hbar^2 l^2$, with l being a real number. Such a choice, however, is not particularly convenient for reasons associated with the mathematical operations which yield the solution to the eigenvalue equation of $\hat{L}^2$. For our purposes, it is sufficient to directly set l equal to the maximum value of m and state the eigenvalues of $\hat{L}^2$ as $\hbar^2 l(l+1)$. This choice is clearly consistent with (12.XIV) since

$$\hbar^2 l(l+1) \geq \hbar^2 m^2 \iff l(l+1) \geq m^2$$

From this, it follows that:

$$l^2 \geq m^2$$

and - since l and m^2 are non-negative integers - that:

---

[24]Again, I advance this conceptual argument in an attempt to render the theory as accessible to the reader as possible, yet to no detriment of the theory's physical rigor. The actual situation is substantially more involved (refer to the books of advanced quantum mechanics cited in the Bibliography).

$$l \geq |m| \iff -l \leq m \leq l \qquad\qquad (12.XV)$$

The statement in (12.XV) and the fact that pursuant to (12.XIIa) any two successive values of m differ by 1, have a crucial implication for m: Since a given maximum value of m

$$m = l > 0$$

and corresponding minimum value of m

$$m = -l < 0$$

in (12.XIIa), respectively signify the maximum and minimum projections on the z-axis, the expression

$$l - (-l) + 1 = 2l + 1$$

signifies the total number of possible projections

$$m = l,\ l-1,\ l-2,\ ...\ -(l-1),\ -l$$

yielding the complete eigenvalue spectrum

$$\hbar m = \hbar l,\ \hbar(l-1),\ \hbar(l-2),\ ...\ -\hbar(l-1),\ -\hbar l$$

of $\hat{L}_z$, precisely as promised at the outset! A few examples will illustrate the physical content of this result.

For $l = 0$ there is obviously only one eigenvalue of $\hat{L}_z$:

$$\hbar m = 0$$

For $l = 1$, there are three eigenvalues of $\hat{L}_z$ signifying a total of three projections:

$$\hbar m = \hbar(+1) = \hbar,\ \hbar m = \hbar(0) = 0,\ \hbar m = \hbar(-1) = -\hbar$$

For $l = 2$, there are five eigenvalues of $\hat{L}_z$ signifying a total of five projections:

$$\hbar m = \hbar(+2) = 2\hbar,\ \hbar m = \hbar(+1) = \hbar,\ \hbar m = \hbar(0) = 0,$$

$$\hbar m = \hbar(-1) = -\hbar,\ \hbar m = \hbar(-2) = -2\hbar$$

Likewise for $l = 3, 4, ...$

Yet, in addition to this physically transparent result, closer examination of our calculation points to an apparent peculiarity! For the constant separation between successive eigenvalues of $\hat{L}_z$ remains intact (equal to $\hbar$) also in the event that the maximum eigenvalue $\hbar l$ of $\hat{L}_z$ (the maximum projection of $\hat{\vec{L}}$ on the z-axis) receives the values

$$l = \frac{1}{2},\ \frac{3}{2},\ \frac{5}{2},\ ...$$

As in the case of integer l, the total number $2l + 1$ of possible z-projections is a positive integer number $n = 1, 2, 3, \ldots$. For example, for $l = \frac{1}{2}$ there are two eigenvalues of $\hat{L}_z$ signifying a total of two projections:

$$\hbar\frac{1}{2}, \ -\hbar\frac{1}{2}$$

For $l = \frac{3}{2}$ there are four eigenvalues of $\hat{L}_z$ signifying a total of four projections:

$$\hbar\frac{3}{2}, \ \hbar\frac{1}{2}, \ -\hbar\frac{1}{2}, \ -\hbar\frac{3}{2}$$

The inevitable conclusion, for that matter, is that the possible values of l are:

$$2l + 1 = n \Longleftrightarrow l = \frac{n-1}{2} \Longleftrightarrow$$

$$l = 0, \ \frac{1}{2}, \ 1, \ \frac{3}{2}, \ 2, \ \frac{5}{2}, \ 3, \ldots \qquad (12.XVI)$$

This result is in stark contrast to the eigenvalue spectrum of $\hat{L}_z$ which was the incepient point of this derivation. For now, in addition to the integer values in (12.XIIa), the maximum value l of m receives half-integer values. The resolution of this apparent contradiction can only be this: In addition to the *orbital angular momentum* which the operators in (12.III) and (12.IV) express[25], there is another kind of angular momentum which does not conform with $\hat{\vec{L}} = \hat{\vec{r}} \times \hat{\vec{p}}$, (12.X) and (12.XII). Let us examine this implication carefully.

We first establish the eigenvalue spectrum of $\hat{\vec{L}}$. The eigenvalue spectrum $\hbar^2 l(l+1)$ of $\hat{L}^2$ is determined by the *quantum number l of the angular momentum* in (12.XVI), as the eigenvalue spectrum $\hbar m$ of $\hat{L}_z$ is determined by the *quantum number m of the z-projection* in (12.XIIa). Thus, the eigenvalue spectrum of the angular-momentum operator $\hat{\vec{L}}$ is also determined by the quantum number l of the angular momentum as $\hbar\sqrt{l(l+1)}$. It would appear, for that matter, that the magnitude of the angular momentum receives the discrete set of values $\hbar\sqrt{l(l+1)}$, with l stated in (12.XVI). That this is not the case follows directly from (12.XV): For each value of quantum number l, quantum number m receives values from $-l$ to l with successive values of m differing by 1. The quantum number l is the maximum value of the quantum number m. From this, it follows that the magnitude of the angular momentum is not $\hbar\sqrt{l(l+1)}$, but $\hbar l$. On the contrary, $\hbar^2 l(l+1)$ is indeed that eigenvalue of $\hat{L}^2$ which corresponds to angular-momentum magnitude $\hbar l$ (on non-rigorous terms, the "square" of the physical $\hbar l$ is not the unphysical $\hbar^2 l^2$ but the physical $\hbar^2 l(l+1)$). The fact that the value of $\hat{L}^2$ is not $\hbar^2 l^2$, as would be expected from classical mechanics, but instead $\hbar^2 l(l+1)$, is a quantum-mechanical particularity. We note, nevertheless, that for very high values of the quantum number l: $l >> 1$, it is:

---

[25] I reiterate that there are no classical orbits. On non-rigorous terms, the *orbital angular momentum* signifies the evolution of the entire probability distribution for position about a centre.

$$\hbar^2 l(l+1) = \hbar^2 l^2 + \hbar^2 l \sim \hbar^2 l^2$$

a fact which implies that for $l >> 1$, a quantum-mechanical system is effectively identical to a classical system.

In sharp contrast to the eigenvalue spectrum of $\hat{L}_z$ in (12.XIIb), the eigenvalue spectrum of $\hat{L}_z$ inherent in (12.XVI) also includes half-integer values of m. This last premise is of utmost importance. Pursuant to the analysis which eventuated in (12.XII), if a quantum system is in a state which signifies revolution about a centre, the quantum number m for the possible projections of the associated angular momentum on the z-axis receives exclusively integer values. Consequently, the quantum number l of the angular momentum receives exclusively integer values. Rigorously put, if a quantum state is a state of revolution about a centre, the *quantum number of the orbital angular momentum is $l = 0, 1, 2, 3, ...$* The case of half-integer l cannot possibly represent orbital angular momentum as that would directly contradict (12.XII). It represents, instead, a new intrinsic - or, internal - quantum number, utterly unrelated to the orbital angular momentum: $l = \frac{n}{2}$; $n = 1, 2, 3, ...$ is the *quantum number of the intrinsic angular momentum*, commonly referred to as *the spin* of the quantum system. The spin is an altogether new degree of freedom specific and exclusive to quantum systems and as much a fundamental property of a quantum particle as the quantum particle's mass or charge. Before we examine this new degree of freedom, we stress the following two points:

(i) The converse is not true! The statement to the effect that a half-integer l represents spin does not imply that the spin is represented by a half-integer l. For whereas a half-integer l definitely represents spin, not every quantum particle's spin is represented by a half-integer l. Quantum mechanics refers almost exclusively to the electron, a particle whose spin is represented by $l = \frac{1}{2}$. As we shall see, however, no particle is really quantum-mechanical. Fundamentally, all particles express underlying quantum fields. Quantum fields fall into two major categories. That of integer quantum number l and that of half-integer quantum number l. In addition to particles, like the electron, whose spin is represented by a half-integer quantum number l, there are particles whose spin is represented by an integer quantum number l. The photon of quantum electrodynamics whose spin is represented by $l = 1$, the gluon of quantum chromodynamics whose spin is also represented by $l = 1$ and the graviton of quantum gravity whose spin is represented by $l = 2$, are outstanding examples.

(ii) The premise which follows (12.VI) is as relevant to the intrinsic angular momentum, as it is to the orbital angular momentum. Maximum projection on the z-axis ($m = l$) does not imply alignment! The magnitude of a quantum particle's angular momentum, be that orbital or intrinsic, is $\hbar l$. Clearly, any claim that the angular-momentum vector of this magnitude be aligned along the z-axis identically contradicts (12.VI) as it implies that the three eigenvalues of $\hat{L}_x, \hat{L}_y, \hat{L}_z$ - respectively equal to $0, 0, \hbar l$ - have been determined at once with absolute certainty.

The quantum number l determines the magnitude $\hbar l$ of a quantum system's angular momentum. Thus, we may refer to the angular momentum of a quantum system merely by l. For example, the electron is a spin-$\frac{1}{2}$ particle, the photon and the gluon are spin-1 particles and the graviton is a spin-2 particle. The distinct character which the intrinsic angular momentum has, necessitates a distinct notation. We reserve, for that matter, l for the orbital angular momentum and denote the intrinsic angular momentum (spin) by s. Thus, the magnitude of the orbital angular momentum is $\hbar l$, $l = 0, 1, 2, 3, \ldots$ and the magnitude of the spin angular momentum is $\hbar s$ with s receiving integer or half-integer values depending on the specific quantum system.

For the sake of consistency, it is necessary to reiterate that the derivation of the spectrum in (12.XVI) is qualitative. The rigorous approach involves "ladder operators" in order to derive (12.XII) and (12.XVI) at once (rather than derive the latter from the former) and is rather tedious and unsuitable for our conceptual approach. Nevertheless, it is essential to a thorough understanding of the concept of spin. Mathematically literate readers are strongly advised to study it in any of the books on quantum mechanics cited in the Bibliography.

13. Spin Angular Momentum

Classical particles do not have internal degrees of freedom. Knowledge of the position $\vec{r}$ and the momentum $\vec{p}$ of a classical particle suffices for that particle's complete description. The preceding analysis has revealed that knowledge of these two attributes alone does not determine the state of a quantum particle. In addition to the probability distribution over the eigenvalues of $\hat{\vec{r}}$ (or of $\hat{\vec{p}}$), the eigenvalues of the spin operator $\hat{\vec{s}}$ and the associated probabilities must be determined for a complete description of the quantum state. The wavefunction must at each moment be a function of the position (or of the momentum) and of the spin. For example, in addition to the momentum eigenvalue, the complete determination of a free quantum particle's state necessitates the determination of the eigenvalues of $\hat{\vec{s}}$. The spin as an internal degree of freedom is specific and exclusive to quantum particles and, fundamentally, to quantum fields. The theoretical considerations in the preceding section predict the existence of spin but do not, themselves, prove it. The ultimate proof is observation. Indeed, it was a decisive contribution to science the observation of Uehlenbeck and Goudsmit that the experimental data which in 1925 appeared to contradict the theoretical expectations founded exclusively on the orbital angular momentum can be completely explained by the assumption of an additional internal spin degree of freedom, utterly unrelated to the orbital angular momentum. In particular, their statement was that:

(i) The electron has an internal angular momentum $\hbar\hat{\vec{s}}$, known as the electron's spin, of magnitude $\hbar\frac{1}{2}$. This internal angular momentum is utterly independent of any orbital angular momentum which the electron may have.

(ii) The existence of the spin entails a magnetic moment additional to that which the orbital angular momentum entails.

This second premise effectively refers to the magnetic field which the orbiting electron generates and concerns as much the intrinsic angular momentum as it does the orbital angular momentum. We shall comment on it in the next section.

On the lines of our analysis in the preceding section, the operator $\hat{\vec{s}}$ which represents the internal spin degree of freedom satisfies the same commutation relations as those in (12.VI):

$$[\hat{s}_x, \hat{s}_y] = i\hbar\hat{s}_z \; ; \; [\hat{s}_y, \hat{s}_z] = i\hbar\hat{s}_x \; ; \; [\hat{s}_z, \hat{s}_x] = i\hbar\hat{s}_y \tag{13.I}$$

as well as:

$$[\hat{s}_z, \hat{s}^2] = 0 \tag{13.II}$$

which follows as a direct consequence of (12.IX). In addition - with an obvious notation regarding i and j - $\hat{\vec{s}}$ satisfies the following commutation relations:

$$[\hat{s}_i, \hat{L}_j] = [\hat{s}_i, \hat{r}_j] = [\hat{s}_i, \hat{p}_j] = 0 \quad ; \quad i,j = x, y, z \tag{13.III}$$

all of which are, clearly, a direct consequence of the fact that the intrinsic angular momentum is utterly unrelated to the orbital angular momentum. We reiterate that the magnitude of the electron's spin angular momentum (or merely, of the electron's spin) is $\hbar s = \hbar\frac{1}{2}$. In addition, pursuant to the analysis in connection with (12.X), (12.XII), (12.XV) and (12.XVI) in the preceding section, the quantum number s_z of the z-projection of the electron's spin (the spin-equivalent to m) satisfies the eigenvalue equation:

$$\hat{s}_z|\frac{1}{2}, s_z\rangle = \hbar s_z|\frac{1}{2}, s_z\rangle \tag{13.$IV$$a$}$$

with admissible values $s_z = \frac{1}{2}$ and $s_z = \frac{1}{2} - 1 = -\frac{1}{2}$. The only possible values of s_z are the two values $\pm\frac{1}{2}$. The corresponding two eigenvectors of $\hat{s}_z$ are $|\frac{1}{2}, \pm\frac{1}{2}\rangle$ and explicitly feature the quantum number of the intrinsic angular momentum $s = \frac{1}{2}$ which signifies spin $\frac{1}{2}$. Stated explicitly, for that matter, (13.IVa) is:

$$\hat{s}_z|\frac{1}{2}, \frac{1}{2}\rangle = \hbar\frac{1}{2}|\frac{1}{2}, \frac{1}{2}\rangle \tag{13.$IV$$b$}$$

and

$$\hat{s}_z|\frac{1}{2}, -\frac{1}{2}\rangle = -\hbar\frac{1}{2}|\frac{1}{2}, -\frac{1}{2}\rangle \tag{13.$IV$$c$}$$

These considerations reveal that the probability distribution for position which the wavefunction $\psi(\vec{r}, t)$ signifies does not completely determine the quantum state of the electron. The complete determination of a quantum state necessitates, in addition, the probability distribution over the eigenvalues of $\hat{s}_z$ - that is, the probability distribution over the possible values of the projection of the spin on the z-axis. Consequently, pursuant to the analysis in the eleventh section, a state of the electron is a vector in that Hilbert space which is spanned by the eigenvectors

$$|\vec{r}, s_z\rangle = |\vec{r}\rangle \, |s, s_z\rangle = |\vec{r}\rangle \, |\frac{1}{2}, s_z\rangle \tag{13.V}$$

of both, $\hat{\vec{r}}$ and $\hat{s}_z$. Thus, the complete wavefunction of the electron is

$$\psi(\vec{r}, s_z) = \langle \vec{r}, s_z | \psi \rangle \tag{13.VI}$$

and is the superposition

$$\psi(\vec{r}, s_z) \equiv \langle \vec{r}, s_z | \psi \rangle = \psi(\vec{r}, \tfrac{1}{2}) | \tfrac{1}{2}, \tfrac{1}{2} \rangle + \psi(\vec{r}, -\tfrac{1}{2}) | \tfrac{1}{2}, -\tfrac{1}{2} \rangle \tag{13.VII}$$

of the two $\hat{s}_z$-eigenstates $|\tfrac{1}{2}, \pm\tfrac{1}{2}\rangle$. This expansion is clearly of the form (3.V) in the third section:

$$\psi(\vec{r}, s_z) = c_1 | \tfrac{1}{2}, \tfrac{1}{2} \rangle + c_2 | \tfrac{1}{2}, -\tfrac{1}{2} \rangle$$

with c_1 and c_2 respectively identified as $\psi(\vec{r}, \pm\tfrac{1}{2})$. Inherent in the wavefunctions $\psi(\vec{r}, \pm\tfrac{1}{2})$ is, of course, the additional superposition (3.VI) of position eigenstates $\delta(\vec{r} - \vec{a})$ in the third section. The expansion in (13.VII) is, for that matter, a direct consequence of the fundamental premise stated in the third section: Pursuant to (13.I), we may only know the projection of the electron's spin on only one axis which we identify with the z-axis. If the actual eigenvalue $\hbar s_z$ which signifies that projection is unknown, the electron is in a superposition of $\hat{s}_z$-eigenstates.

It is now a straightforward matter to extend the concept of probability density which we examined in one dimension. Clearly, $|\psi(\vec{r}, s_z)|^2 d^3r$ *represents the probability that the position of the electron be within infinitesimal volume d^3r centred at point $\vec{r}$ with z-projection of the spin equal to $\hbar s_z$. The normalization condition is:*

$$\int d^3r \, |\psi(\vec{r}, \tfrac{1}{2})|^2 + \int d^3r \, |\psi(\vec{r}, -\tfrac{1}{2})|^2 = 1 \tag{13.VIII}$$

We turn now to the task of constructing the representation of spin-operator $\hat{\vec{s}}$ and the eigenstates of $\hat{s}_z$. In the previous section we gave a compelling argument as to the reason the solution to the eigenvalue equation (12.Xa) for $\hat{L}_z$ is (12.Xb). We refrained from studying the rigorous derivation of that solution which is founded upon the fact that $\hat{L}_z$ is a differential operator in space, an operator whose action on a function expresses the rate at which that function changes with respect to angle ϕ. This is not the case with $\hat{s}_z$ in view of the fact that the intrinsic angular momentum is precisely that, an internal degree of freedom utterly unrelated to continuous changes of coordinates in space. For that matter, all three operators $\hat{s}_x$, $\hat{s}_y$ and $\hat{s}_z$ which compose the spin operator $\hat{\vec{s}}$ must be constructed in such a manner as to satisfy the demands posed by the commutation relations and the entire preceding analysis in this section. The result is rather obvious. If, in the event that $s = \tfrac{1}{2}$ (as is the case for the electron), we represent the "spin-up" eigenstate of $\hat{s}_z$ as the two-entry column:

$$| \tfrac{1}{2}, \tfrac{1}{2} \rangle = \begin{bmatrix} 1 \\ 0 \end{bmatrix} \tag{13.IXa}$$

the "spin-down" eigenstate of $\hat{s}_z$ as the two entry column:

$$\left|\frac{1}{2}, -\frac{1}{2}\right\rangle = \begin{bmatrix} 0 \\ 1 \end{bmatrix} \tag{13.IXb}$$

then (13.IV) directly reveals that operator $\hat{s}_z$ receives the 2×2 matrix representation:

$$\hat{s}_z = \hbar \frac{1}{2} \begin{bmatrix} 1 & 0 \\ 0 & -1 \end{bmatrix} \tag{13.X}$$

In order to understand the reason this 2×2 matrix does indeed represent $\hat{s}_z$, we briefly digress to the relevant aspect of linear algebra.

A row is a horizontal array of n entries, a column is a vertical array of n entries and a $m \times n$ matrix is a vertical array of m rows each of which comprises n entries. Here the term "entry" refers to the number at the particular position in the array. The rows, columns and matrices relevant to the theory of spin angular momentum in quantum mechanics are characterized by $m = n = 2$. Multiplication of a matrix by a matrix is possible only if the number of columns of the left matrix $\mathbf{A}$ is equal to the number of rows of the right matrix $\mathbf{B}$. In that case we consider the product $a_{11}b_{11}$ between entry a_{11} of the first row, first column of $\mathbf{A}$ and entry b_{11} respectively of $\mathbf{B}$, accordingly the product $a_{12}b_{21}$ between entry a_{12} of the first row, second column of $\mathbf{A}$ and entry b_{21} of the second row, first column of $\mathbf{B}$, then the product $a_{13}b_{31}$ between entry a_{13} of the first row, third column of $\mathbf{A}$ and entry b_{31} of the third row, first column of $\mathbf{B}$ and so on. The sum-total $a_{11}b_{11} + a_{12}b_{21} + a_{13}b_{31} + ...$ of these products yields the first row, first column entry of $\mathbf{AB}$. Likewise the sum-total $a_{11}b_{12} + a_{12}b_{22} + a_{13}b_{32} + ...$ yields the first row, second column entry of $\mathbf{AB}$ and so on. For example, if $\mathbf{A}$ is the 2×3 matrix

$$\mathbf{A} = \begin{bmatrix} a_{11} & a_{12} & a_{13} \\ a_{21} & a_{22} & a_{23} \end{bmatrix}$$

and $\mathbf{B}$ is the 3×3 matrix

$$\mathbf{B} = \begin{bmatrix} b_{11} & b_{12} & b_{13} \\ b_{21} & b_{22} & b_{23} \\ b_{31} & b_{32} & b_{33} \end{bmatrix}$$

then the product $\mathbf{AB}$ is

$$\mathbf{AB} = \begin{bmatrix} a_{11} & a_{12} & a_{13} \\ a_{21} & a_{22} & a_{23} \end{bmatrix} \begin{bmatrix} b_{11} & b_{12} & b_{13} \\ b_{21} & b_{22} & b_{23} \\ b_{31} & b_{32} & b_{33} \end{bmatrix} =$$

$$\begin{bmatrix} a_{11}b_{11} + a_{12}b_{21} + a_{13}b_{31} & a_{11}b_{12} + a_{12}b_{22} + a_{13}b_{32} & a_{11}b_{13} + a_{12}b_{23} + a_{13}b_{33} \\ a_{21}b_{11} + a_{22}b_{21} + a_{23}b_{31} & a_{21}b_{12} + a_{22}b_{22} + a_{23}b_{32} & a_{21}b_{13} + a_{22}b_{23} + a_{23}b_{33} \end{bmatrix}$$

Rather than replace specific numbers for the entries a_{ij} of $\mathbf{A}$ and b_{kl} of $\mathbf{B}$ by way of concretizing this example, we directly consider the matrix of interest in (13.X). The multiplication rule applied to the square of this matrix yields:

$$\begin{bmatrix} 1 & 0 \\ 0 & -1 \end{bmatrix}\begin{bmatrix} 1 & 0 \\ 0 & -1 \end{bmatrix} = \begin{bmatrix} (1)(1)+(0)(0) & (1)(0)+(0)(-1) \\ (0)(1)+(-1)(0) & (0)(0)+(-1)(-1) \end{bmatrix} = \begin{bmatrix} 1 & 0 \\ 0 & 1 \end{bmatrix} \qquad (13.XI)$$

The matrix

$$\begin{bmatrix} 1 & 0 \\ 0 & 1 \end{bmatrix} \equiv \mathbf{1}$$

which appears in (13.XI) is the fundamental *unit matrix* of linear algebra.

The above-stated multiplication rule also applies to the multiplication between a matrix and a column and to the multiplication between a row and a matrix. As example we consider directly the case of interest:

$$\begin{bmatrix} 1 & 0 \\ 0 & -1 \end{bmatrix}\begin{bmatrix} 1 \\ 0 \end{bmatrix} = \begin{bmatrix} (1)(1)+(0)(0) \\ (0)(1)+(-1)(0) \end{bmatrix} = \begin{bmatrix} 1 \\ 0 \end{bmatrix}$$

Multiplication of both sides of this relation by $\frac{\hbar}{2}$ yields

$$\hbar\frac{1}{2}\begin{bmatrix} 1 & 0 \\ 0 & -1 \end{bmatrix}\begin{bmatrix} 1 \\ 0 \end{bmatrix} = \hbar\frac{1}{2}\begin{bmatrix} 1 \\ 0 \end{bmatrix} \qquad (13.XIIa)$$

The operation of multiplication between a matrix and a number yields the matrix with all its entries muliplied by that number. If, for example, λ and κ are two numbers then

$$\lambda\mathbf{A} = \begin{bmatrix} \lambda a_{11} & \lambda a_{12} & \lambda a_{13} \\ \lambda a_{21} & \lambda a_{22} & \lambda a_{23} \end{bmatrix} \; ; \; \kappa\mathbf{B} = \begin{bmatrix} \kappa b_{11} & \kappa b_{12} & \kappa b_{13} \\ \kappa b_{21} & \kappa b_{22} & \kappa b_{23} \\ \kappa b_{31} & \kappa b_{32} & \kappa b_{33} \end{bmatrix}$$

This rule also applies, of course, to the operation of multiplication between a number and a column and between a number and a row. For example,

$$(-1)\begin{bmatrix} 0 \\ 1 \end{bmatrix} = \begin{bmatrix} 0 \\ -1 \end{bmatrix}$$

from which it directly follows that

$$\hbar\frac{1}{2}\begin{bmatrix} 1 & 0 \\ 0 & -1 \end{bmatrix}\begin{bmatrix} 0 \\ 1 \end{bmatrix} = \hbar\frac{1}{2}\begin{bmatrix} 0 \\ -1 \end{bmatrix} = -\hbar\frac{1}{2}\begin{bmatrix} 0 \\ 1 \end{bmatrix} \qquad (13.XIIb)$$

In view of (13.X) and (13.IX), relations (13.XIIa) and (13.XIIb) are respectively identical to (13.IVb) and (13.IVc). As promised, we have arrived at the announced result: Operator $\hat{s}_z$ receives the matrix representation in (13.X) and the eigenstates of $\hat{s}_z$ receive the column representation in (13.IX). Of course, as stated, this by itself is not sufficient. Operator $\hat{s}_z$ must satisfy all commutation relations which we discussed thus far. We cite without proof, for that matter, the corresponding representation of the remaining two operators $\hat{s}_x$ and $\hat{s}_y$:

$$\hat{s}_x = \hbar\frac{1}{2}\begin{bmatrix} 0 & 1 \\ 1 & 0 \end{bmatrix} \qquad (13.XIIIa)$$

$$\hat{s}_y = \hbar \frac{1}{2} \begin{bmatrix} 0 & -i \\ i & 0 \end{bmatrix} \qquad (13.XIIIb)$$

and leave it as an exercise to the dedicated reader to confirm through a simple calculation that $\hat{s}_x$, $\hat{s}_y$ and $\hat{s}_z$ satisfy (13.I) and that, in addition, $\hat{s}_z$ satisfies (13.II). In fact, the validity of (13.II) is already obvious from (13.XI) and the fact that direct matrix multiplication reveals that the matrices in (13.XIII) also have the property which (13.XI) expresses:

$$\sigma_x^2 = \sigma_y^2 = \sigma_z^2 = 1 \qquad (13.XIVa)$$

with

$$\hat{s}_x = \frac{\hbar}{2}\sigma_x \; : \; \hat{s}_y = \frac{\hbar}{2}\sigma_y \; ; \; \hat{s}_z = \frac{\hbar}{2}\sigma_z \qquad (13.XIVb)$$

The matrices

$$\sigma_x = \begin{bmatrix} 0 & 1 \\ 1 & 0 \end{bmatrix} \; ; \; \sigma_y = \begin{bmatrix} 0 & -i \\ i & 0 \end{bmatrix} \; ; \; \sigma_z = \begin{bmatrix} 1 & 0 \\ 0 & -1 \end{bmatrix} \qquad (13.XV)$$

are known as *Pauli matrices* and are of fundamental importance in quantum mechanics and in quantum field theory. In turn, the eigenvectors of $\hat{s}_z$ in (13.IX) are known as *Pauli spinors*. It is obvious that in terms of Pauli matrices the spin operator is

$$\vec{\hat{s}} = (\hat{s}_x, \hat{s}_y, \hat{s}_z) = \hbar \frac{1}{2} \vec{\sigma} \qquad (13.XVI)$$

We have hitherto stated the basic operations of multiplication between matrices and between a matrix and a number. Additional aspects of the theory of matrices we shall examine according to the needs of our analysis. In particualr, we shall study, in the sixth chapter, the operation which yields the transpose and the adjoined matrix of a given matrix. At this stage, for the sake of completeness and for future use, we exemplify the above-stated operation of multiplication between a row and a matrix:

$$\begin{bmatrix} 1 & 0 \end{bmatrix} \begin{bmatrix} 1 & 0 \\ 0 & -1 \end{bmatrix} = \begin{bmatrix} (1)(1) + (0)(0) & (1)(0) + (0)(-1) \end{bmatrix} = \begin{bmatrix} 1 & 0 \end{bmatrix}$$

and remark that the addition of matrices is defined only if all terms in the operation are $m \times n$ matrices:

$$\begin{bmatrix} a_{11} & a_{12} & ...a_{1n} \\ a_{21} & a_{22} & ...a_{2n} \\ ... & ... & ... \\ a_{m1} & a_{m2} & ...a_{mn} \end{bmatrix} + \begin{bmatrix} b_{11} & b_{12} & ...b_{1n} \\ b_{21} & b_{22} & ...b_{2n} \\ ... & ... & ... \\ b_{m1} & b_{m2} & ...b_{mn} \end{bmatrix} = \begin{bmatrix} a_{11}+b_{11} & a_{12}+b_{12} & ...a_{1n}+b_{1n} \\ a_{21}+b_{21} & a_{22}+b_{22} & ...a_{2n}+b_{2n} \\ ... & ... & ... \\ a_{m1}+b_{m1} & a_{m2}+b_{m2} & ...a_{mn}+b_{mn} \end{bmatrix}$$

Moreover, we note that, unlike numbers (real or complex), multiplied matrices do not commute. That is, in general, for two matrices $\mathbf{A}$ and $\mathbf{B}$, it is

AB $\neq$ BA

This can be directly confirmed in the set of the three Pauli matrices, for example. In fact, we have already confirmed it through direct replacement of (13.XIVb) in (13.I):

$$[\sigma_x, \sigma_y] = 2i\sigma_z \qquad\qquad (13.XVIIa)$$

$$[\sigma_y, \sigma_z] = 2i\sigma_x \qquad\qquad (13.XVIIb)$$

$$[\sigma_z, \sigma_x] = 2i\sigma_y \qquad\qquad (13.XVIIc)$$

In addition to expressing a fundamental property of angular momentum, (13.I) (and all non-vanishing commutators) is at once a statement to the effect that operators behave like matrices: In general, they do not commute. It is, for that matter, no wonder that all linear operators can be represented as matrices.

Although in line with (13.I) the Pauli matrices do not commute with each other, they have a remarkable property: As can be confirmed through direct multiplication, they *anti-commute*:

$$\sigma_x\sigma_y + \sigma_y\sigma_x = 0 \ ; \ \sigma_x\sigma_z + \sigma_z\sigma_x = 0 \ ; \ \sigma_y\sigma_z + \sigma_z\sigma_y = 0$$

In order to state these relations in a concise manner, we introduce the *anti-commutator* $\{\,,\,\}$:

$$\{\sigma_x, \sigma_y\} = 0 \qquad\qquad (13.XVIIIa)$$

$$\{\sigma_x, \sigma_z\} = 0 \qquad\qquad (13.XVIIIb)$$

$$\{\sigma_y, \sigma_z\} = 0 \qquad\qquad (13.XVIIIc)$$

Finally, we note one more important property which the three Pauli matrices have. Clearly, all three of them have a vanishing *trace* - that is, the sum of the diagonal entries (from up left to low right) in each of them is equal to zero:

$$Tr\sigma_x = 0 \ ; \ Tr\sigma_y = 0 \ ; \ Tr\sigma_z = 0 \qquad\qquad (13.XIX)$$

Let us now examine the implications which the Pauli spinors and matrices have for (13.VII). Replacing (13.IX) in the latter yields

$$\psi(\vec{r}, s_z) = \psi(\vec{r}, \tfrac{1}{2}) \begin{bmatrix} 1 \\ 0 \end{bmatrix} + \psi(\vec{r}, -\tfrac{1}{2}) \begin{bmatrix} 0 \\ 1 \end{bmatrix} = \begin{bmatrix} \psi(\vec{r}, \tfrac{1}{2}) \\ \psi(\vec{r}, -\tfrac{1}{2}) \end{bmatrix} \qquad (13.XX)$$

This is a statement of the wavefunction as a two-entry column, or more rigorously as a *spinor* in the Hilbert space spanned by the two Pauli spinors in (13.IX). The upper entry represents the probability amplitude that the position of the quantum particle be within infinitesimal volume d^3r centred at point $\vec{r}$ with z-projection of the spin equal to $\hbar s_z = \hbar\frac{1}{2}$, whereas the lower entry likewise with $\hbar s_z = -\hbar\frac{1}{2}$. The operators which act in the Hilbert space of spin are 2×2 matrices. On the contrary, the operators which act in the Hilbert space of position are differential operators. As we shall see, these are operators the action of which yields a rate of change in space.

Thus, we have accomplished a comprehensive examination of the orbital and the intrinsic angular momentum. In order to gain a solid understanding of the latter (spin), we consider it in two important physical contexts.

14. Spin in the Absence and in the Presence of a Magnetic Field

We first consider the case of a free electron. A plane wave of probability in one dimension is given by (3.IIb) which we reproduce here:

$$\psi_p(x) = \frac{1}{\sqrt{2\pi\hbar}} e^{\frac{i}{\hbar}px} \tag{14.Ia}$$

Clearly, since no restrictions have been imposed on p, this expression involves as much the propagation with momentum $-p$, as it involves the propagation with momentum $+p$. However, as in the procedure which follows it will be necessary to distinguish between the corresponding two quantum states, we reserve (14.Ia) for a plane wave which propagates with $+p$ and likewise express the one-dimensional plane wave of probability which propagates in the opposite direction with momentum $-p$ as:

$$\psi_{-p}(x) = \frac{1}{\sqrt{2\pi\hbar}} e^{-\frac{i}{\hbar}px} \tag{14.Ib}$$

These two states are now common eigenstates of both, the momentum operator $\hat{p}$ with momentum eigenvalues $+p$ and $-p$ respectively and the energy operator

$$\hat{H} = \frac{\hat{p}^2}{2m} \tag{14.IIa}$$

in (3.IIIb) with energy eigenvalue equal to

$$E_p = \frac{p^2}{2m} \tag{14.IIb}$$

In the language of quantum theory, the latter situation signifies a *degeneracy* of degree equal to two for the energy eigenvalue E_p: Two distinct energy eigenstates having the same energy eigenvalue E_p.

In physical three-dimensional space the internal spin degree of freedom drastically changes this one-dimensional energy spectrum. Since the electron has spin equal to $\hbar s = \hbar\frac{1}{2}$, the eigenfunctions of the energy operator are

$$\hat{H} = \frac{|\hat{\vec{p}}|^2}{2m} \tag{14.III}$$

$$\psi_{\vec{p}}^{\uparrow}(\vec{r}) = A_p e^{\frac{i}{\hbar}\vec{p}.\vec{r}} \begin{bmatrix} 1 \\ 0 \end{bmatrix} \qquad s_z = \frac{1}{2} \tag{14.$IV\,a$}$$

$$\psi_{\vec{p}}^{\downarrow}(\vec{r}) = A_p e^{\frac{i}{\hbar}\vec{p}.\vec{r}} \begin{bmatrix} 0 \\ 1 \end{bmatrix} \qquad s_z = -\frac{1}{2} \tag{14.$IV\,b$}$$

$$\psi_{-\vec{p}}^{\uparrow}(\vec{r}) = A_p e^{-\frac{i}{\hbar}\vec{p}.\vec{r}} \begin{bmatrix} 1 \\ 0 \end{bmatrix} \qquad s_z = \frac{1}{2} \tag{14.$IV\,c$}$$

$$\psi_{-\vec{p}}^{\downarrow}(\vec{r}) = A_p e^{-\frac{i}{\hbar}\vec{p}.\vec{r}} \begin{bmatrix} 0 \\ 1 \end{bmatrix} \qquad s_z = -\frac{1}{2} \tag{14.$IV\,d$}$$

and clearly signify a degeneracy of order four for each energy eigenvalue

$$E_p = \frac{|\vec{p}|^2}{2m}$$

In (14.IV), the multiplicative constant A_p is the three-dimensional "extension" of the constant in (14.I):

$$A_p = \left(\frac{1}{\sqrt{2\pi\hbar}}\right)^3 = \frac{1}{(2\pi\hbar)^{\frac{3}{2}}} \tag{14.$IV\,e$}$$

We also remind that $|\vec{p}|$ is the magnitude of momentum $\vec{p} = (p_x, p_y, p_z)$ and $\vec{r} = (x, y, z)$ is the vector which signifies point (x, y, z) in three-dimensional space. The inner product $\vec{p}.\vec{r} = xp_x + yp_y + zp_z$ is already familiar to us from the preceding third section.

The reason these eigenfunctions are products of momentum and spin eigenfunctions is because the three operators $\hat{s}_x$, $\hat{s}_y$ and $\hat{s}_z$ commute with $\hat{H}$ in (14.III) as (13.III) immediately reveals, a fact which implies that the eigenfunctions common to both operators can only be the products in (14.IV). For, only then is it

$$\hat{H} A_p e^{\frac{i}{\hbar}\vec{p}.\vec{r}} \begin{bmatrix} 1 \\ 0 \end{bmatrix} = \begin{bmatrix} 1 \\ 0 \end{bmatrix} \hat{H} A_p e^{\frac{i}{\hbar}\vec{p}.\vec{r}} = \begin{bmatrix} 1 \\ 0 \end{bmatrix} \frac{|\vec{p}|^2}{2m} A_p e^{\frac{i}{\hbar}\vec{p}.\vec{r}} = \frac{|\vec{p}|^2}{2m} A_p e^{\frac{i}{\hbar}\vec{p}.\vec{r}} \begin{bmatrix} 1 \\ 0 \end{bmatrix}$$

and

$$\hat{s}_z A_p e^{\frac{i}{\hbar}\vec{p}.\vec{r}} \begin{bmatrix} 1 \\ 0 \end{bmatrix} = A_p e^{\frac{i}{\hbar}\vec{p}.\vec{r}} \hat{s}_z \begin{bmatrix} 1 \\ 0 \end{bmatrix} = A_p e^{\frac{i}{\hbar}\vec{p}.\vec{r}} \frac{\hbar}{2} \begin{bmatrix} 1 \\ 0 \end{bmatrix} = \frac{\hbar}{2} A_p e^{\frac{i}{\hbar}\vec{p}.\vec{r}} \begin{bmatrix} 1 \\ 0 \end{bmatrix}$$

The eigenfunctions in (14.IV) are a particular case of (13.XX). They arise in the event that the z-projection of the <u>free</u> electron's spin is known with absolute certainty. If this is not the case then the wavefunction is

104

$$\psi_p(\vec{r}, s_z) = c_1 A_p e^{\frac{i}{\hbar}\vec{p}\cdot\vec{r}} \begin{bmatrix} 1 \\ 0 \end{bmatrix} + c_2 A_p e^{\frac{i}{\hbar}\vec{p}\cdot\vec{r}} \begin{bmatrix} 0 \\ 1 \end{bmatrix} \equiv$$

$$\psi_p(\vec{r}, \tfrac{1}{2}) \begin{bmatrix} 1 \\ 0 \end{bmatrix} + \psi_p(\vec{r}, -\tfrac{1}{2}) \begin{bmatrix} 0 \\ 1 \end{bmatrix} = \begin{bmatrix} \psi_p(\vec{r}, \tfrac{1}{2}) \\ \psi_p(\vec{r}, -\tfrac{1}{2}) \end{bmatrix} = A_p e^{\frac{i}{\hbar}\vec{p}\cdot\vec{r}} \begin{bmatrix} c_1 \\ c_2 \end{bmatrix} \qquad (14.V)$$

(accordingly with momentum eigenvalue: $-\vec{p}$).

In general, $\hat{H}$ has a spin-dependent part as a consequence of which none of $\hat{s}_i$; $i = x, y, z$ commutes with $\hat{H}$. We shall not advance the analysis in that direction. We shall, instead, study a non-trivial interaction between the spin and a classical field. We consider the spin of the electron in the presence of a magnetic field.

The magnetic moment of the electron which we stated in the opening paragraph of the preceding section is essentially a reference to the magnetic field which the orbital and internal angular motion generate in line with the classical premise to the effect that loops of electric currents generate magnetic fields. The orbital angular momentum $\vec{L}$ generates the magnetic moment

$$\vec{M}_L = \frac{e\hbar}{2\mu c}\vec{L} \qquad (14.VIa)$$

and the intrinsic angular momentum $\vec{s}$ accordingly generates the additional magnetic moment

$$\vec{m}_s = g_e \frac{e\hbar}{2\mu c}\vec{s} \qquad (14.VIb)$$

where here we denote the mass of the electron by μ in order to distinguish it from the magnetic moment m_s, e is the charge of the electron, c is the speed of light in vacuum and, always for the electron, $g_e = 2$.

We consider now a magnetic field of constant strength $\vec{B}$ oriented along the z-axis. From (3.IIIc), (14.VIb) and (13.XIVb), the Hamiltonian (energy) operator for an electron evolving against that field is

$$\hat{H} = \frac{|\hat{\vec{p}}|^2}{2\mu} + \frac{\mu_e}{2}B\sigma_z \qquad (14.VIIa)$$

with

$$\mu_e = \frac{|e|\hbar}{\mu c}$$

The solution to the Schrödinger equation which describes the electron in the magnetic field at some moment $t = 0$ is $\psi(0, \vec{r}, s_z)$ in (13.XX). Thus, the probability that the position of the electron be within infinitesimal volume d^3r centred at point r with z-projection of spin equal to $\hbar s_z = \hbar\tfrac{1}{2}$ - equivalently, the probability that a measurement yield that value about point r - is $|\psi(0, \vec{r}, \tfrac{1}{2})|^2$ and, accordingly, the probability for $\hbar s_z = -\hbar\tfrac{1}{2}$ within d^3r is $|\psi(0, \vec{r}, -\tfrac{1}{2})|^2$.

In what follows we shall not concern ourselves with the dependence of the wavefunction on the coordinates $\vec{r}$. For that matter, we keep only the potential term in (14.VIIa) and state:

$$\hat{H} = \frac{\mu_e}{2} B \sigma_z \qquad (14.VIIb)$$

From (13.XII) and (14.VIIb) now we obtain

$$\hat{H} \left| \frac{1}{2}, \frac{1}{2} \right\rangle = E_0 \left| \frac{1}{2}, \frac{1}{2} \right\rangle \qquad (14.VIIIa)$$

and

$$\hat{H} \left| \frac{1}{2}, -\frac{1}{2} \right\rangle = -E_0 \left| \frac{1}{2}, -\frac{1}{2} \right\rangle \qquad (14.VIIIb)$$

where the energy eigenvalue E_0 is

$$E_0 = \frac{\mu_e B}{2}$$

Moreover, by directly applying the evolution operator $\hat{U}(t)$:

$$e^{-\frac{i}{\hbar}\hat{H}t} \left| \frac{1}{2}, \frac{1}{2} \right\rangle = e^{-\frac{i}{\hbar}E_0 t} \left| \frac{1}{2}, \frac{1}{2} \right\rangle \; ; \; e^{-\frac{i}{\hbar}\hat{H}t} \left| \frac{1}{2}, -\frac{1}{2} \right\rangle = e^{\frac{i}{\hbar}E_0 t} \left| \frac{1}{2}, -\frac{1}{2} \right\rangle$$

we obtain the evolved energy eigenstates:

$$e^{-\frac{i}{\hbar}E_0 t} \left| \frac{1}{2}, \frac{1}{2} \right\rangle = e^{-\frac{i}{\hbar}E_0 t} \begin{bmatrix} 1 \\ 0 \end{bmatrix} \; ; \; e^{\frac{i}{\hbar}E_0 t} \left| \frac{1}{2}, -\frac{1}{2} \right\rangle = e^{\frac{i}{\hbar}E_0 t} \begin{bmatrix} 0 \\ 1 \end{bmatrix}$$

Thus, from (14.VIII) it follows that (disregarding the spatial dependence in $\psi(0, \vec{r}, \pm\frac{1}{2})$) the wavefunction of the electron at $t = 0$ is

$$\psi(0, s_z) = c_1 \begin{bmatrix} 1 \\ 0 \end{bmatrix} + c_2 \begin{bmatrix} 0 \\ 1 \end{bmatrix} = \begin{bmatrix} c_1 \\ c_2 \end{bmatrix} \qquad (14.IXa)$$

and from the above two expressions for the evolved energy eigenstates, that the time-dependent wavefunction of the electron is

$$\psi(t, s_z) = e^{-\frac{i}{\hbar}\hat{H}t} \psi(0, s_z) = e^{-\frac{i}{\hbar}E_0 t} c_1 \begin{bmatrix} 1 \\ 0 \end{bmatrix} + e^{\frac{i}{\hbar}E_0 t} c_2 \begin{bmatrix} 0 \\ 1 \end{bmatrix} = \begin{bmatrix} e^{-\frac{i}{\hbar}E_0 t} c_1 \\ e^{\frac{i}{\hbar}E_0 t} c_2 \end{bmatrix} \qquad (14.IXb)$$

We conclude that (14.VIII) yields the two energy eigenvalues and (14.IXb) the temporal evolution of the quantum system.

We reiterate that in (14.IXa) the probability that the z-projection of the spin be $\hbar\frac{1}{2}$ is equal to $|c_1|^2$ and the probability that the z-projection of the spin be $-\hbar\frac{1}{2}$ is equal to $|c_2|^2$. The two constants c_1 and c_2 we may calculate from the given initial conditions. Let us assume, for example, that $\psi(t = 0, s_z)$ represents an electron with spin $+\frac{1}{2}$ along the x-axis. This implies:

$$\sigma_x \psi(0, s_z) = 1\psi(0, s_z)$$

or, from (13.XV) and (14.IXa):

$$\begin{bmatrix} 0 & 1 \\ 1 & 0 \end{bmatrix} \begin{bmatrix} c_1 \\ c_2 \end{bmatrix} = \begin{bmatrix} c_1 \\ c_2 \end{bmatrix}$$

which, in turn, yields:

$$\begin{bmatrix} c_2 \\ c_1 \end{bmatrix} = \begin{bmatrix} c_1 \\ c_2 \end{bmatrix}$$

with the implication that

$$c_1 = c_2$$

Yet, the renormalization condition in (13.VIII) implies that

$$|c_1|^2 + |c_2|^2 = 1$$

which, in view of $c_1 = c_2$ yields:

$$c_1 = c_2 = \frac{1}{\sqrt{2}}$$

Replacing this in (14.IXb) we obtain the final result:

$$\psi(t, s_z) = \frac{1}{\sqrt{2}} \begin{bmatrix} e^{-\frac{i}{\hbar} E_0 t} \\ e^{\frac{i}{\hbar} E_0 t} \end{bmatrix} \tag{14.X}$$

This is the wavefunction of the electron in a magnetic field given the stated initial conditions.

We conclude with an important observation concerning the expectation value $\langle \psi | \hat{\vec{s}} | \psi \rangle$. It is[26]:

$$\langle \psi | \sigma_z | \psi \rangle = \frac{1}{2} \begin{bmatrix} e^{\frac{i}{\hbar} E_0 t} & e^{-\frac{i}{\hbar} E_0 t} \end{bmatrix} \begin{bmatrix} 1 & 0 \\ 0 & -1 \end{bmatrix} \begin{bmatrix} e^{-\frac{i}{\hbar} E_0 t} \\ e^{\frac{i}{\hbar} E_0 t} \end{bmatrix} = \frac{1}{2}(1-1) = 0 \tag{14.XIa}$$

accordingly:

$$\langle \psi | \sigma_y | \psi \rangle = sin(\frac{2E_0}{\hbar} t) \tag{14.XIb}$$

and

$$\langle \psi | \sigma_x | \psi \rangle = cos(\frac{2E_0}{\hbar} t) \tag{14.XIc}$$

---

[26]Notice that the entries in the left row are complex conjugate to the entries in the right column. At this stage, this suffices. Again, we shall study the operation which yields the former from the latter later.

Thus:

$$\langle\psi|\hat{\vec{s}}|\psi\rangle = \frac{\hbar}{2}\left(cos(\frac{2E_0}{\hbar}t)\hat{x}+sin(\frac{2E_0}{\hbar}t)\hat{y}\right) \qquad (14.XII)$$

with $\hat{x}$ and $\hat{y}$ respectively being the unit vectors in the x- and y-direction. (14.XII) directly reveals that in the course of time the expectation value of the spin revolves around the z-axis. On account of the initial conditions, this expectation value is a vector oriented along the x-axis at $t = 0$. Its projection on the z-axis remains constant equal to zero, as (14.XIa) reveals. As can be directly deduced from (14.VII), the underlying reason for this constancy is that

$$[\hat{s}_z, \hat{H}] = 0$$

Pursuant to the third section, this implies that the z-projection of the spin's expectation value is a conserved quantity. We have repeatedly stressed that a physical quantity which is conserved in the classical problem will also be conserved in the quantum-mechanical problem. As expected, for that matter, the motion of expectation value $\langle\psi|\hat{\vec{s}}|\psi\rangle$ is realized according to the classical equations of motion (interested readers may study the classical effect of *Larmor precession*).

15. Concluding Remarks

Quantum mechanics predicts the existence of spin but does not explain it! For quantum mechanics this internal degree of freedom is an externally given fact, a fact which quantum mechanics explains away with the statement: "That's how it is"! The true origin of spin does not merely lie in quantization, but in the consistent combination of quantization and relativistic covariance. That is, the spin is rigorously explained only by the relativistic quantum field theory! We shall see, in fact, that the Dirac field, a prominent example of which is the electron quantum field, comprises four degrees of freedom at each point in space. Two of the four degrees of freedom which that quantum field comprises are indeed respectively the spin-up $\hbar\frac{1}{2}$ and spin-down $-\hbar\frac{1}{2}$ states of the electron and the other two degrees of freedom are the spin-up and spin down states of the electron's antiparticle, the positively charged positron. As observable antiparticles are equivalent to unobservable particles propagating backward in time, the electron has - on qualitative terms - an orientation in space ($s_z = \pm\frac{1}{2}$) and an orientation in time (electron - positron), in perfect conformity with the relativistic covariance of the electron (Dirac) field.

Thus, the spin of quantum particles points to the relativistic quantum field theory as the rigorous theory of Nature and, for that matter, to the limited physical significance of non-relativistic quantum mechanics. In the fourth and the fifth chapter we shall see how any attempt at formulating a theory of relativistic quantum mechanics by quantizing relativistic classical particles in the same manner we quantized non-relativistic classical particles yields unphysical negative energies and unphysical negative probabilities, not to mention other disasters! We shall see how the demand that quantum systems conform with relativistic covariance necessarily yields the infinitely many degrees of freedom of quantum field theory[27], rendering quantum mechanics an, at best, approximate theory of Nature, a theory which signifies an approximation physically acceptable only at non-relativistic speeds and very low energies. We reiterate that quantum mechanics is, indeed, a complete physical theory in the sense that it completely describes quantum systems characterized by a finite number of degrees of freedom. The actual discrepancy with physical reality lies in the physically inconsistent assumption that Nature accommodates systems characterized by a finite number of degrees of freedom, not in the mathematical and conceptual foundations of quantum mechanics. It should be mentioned in this respect that, starting with Albert Einstein and his associates in 1934, numerous attempts have been made to demonstrate that the conceptual foundations of quantum mechanics, its replacement of classical functions by operators and the concomitant probability distributions, are in contradiction with physical reality on account of an alleged prediction of "instantaneous action at a distance" and that, for that matter, quantum mechanics is a conceptually incomplete theory which conceals hidden physical variables at a more fundamental level. The claim was made that a consistent approach to hidden variables would replace quantum mechanics by a (non-linear) classical theory. However, numerous experiments appear to have conclusively ruled this eventuality out.

The issue of causality which lies at the core of such claims as the "instantaneous action at a distance" is, is yet another issue which points to the limited physical significance of quantum mechanics as an, at best, approximate theory and to quantum field theory as the consistent and rigorous theory of Nature. We shall see that although a quantum particle may, indeed, propagate between two points in space in arbitrarily short time, quantum field theory safeguards causality through a spectacular cancellation between the transition amplitude of that particle and the amplitude of the inevitable transition of its antiparticle between the same two points, yet in reverse temporal order between the event of particle emergence and the event of particle annihilation! Quantum mechanics, on the contrary, having no inherent capacity for antiparticles, is utterly unable to confront the issue of causality.

Despite such serious limitations, quantum mechanics cannot be dismissed as a physically irrelevant theory if only because, as we shall see, at the non-relativistic limit of very low energies, quantum fields assume the semblance (but never the reality) of quantum-mechanical behavior. This fact alone, makes it necessary that we examine the epistemological implications of quantum mechanics, the interpretation of the quantum-mechanical wavefunction, before we turn to the advanced path-integral formulation of quantum mechanics and, thereby, to quantum field theory.

A century after the emergence of quantum mechanics, the debate over the epistemological issue as to "what it all means" has yet to be settled. Contrary to the prevailing view which dismisses such a question as philosophical and scientifically irrelevant, the interpretation of the inherent probabilities is, in fact, an essential aspect not only of

---

[27]Quantum field theory is one approach. There is another approach which also meets this demand while retaining the relativistic quantum particle. That is the approach of string theory. I shall only mention in this respect that in no way does string theory contradict quantum field theory. On the contrary, the latter emerges as the low energy limit of the former.

quantum theory, but of all Physics. For it is clearly meaningless to strive for knowledge if objective reality is inaccessible to the human mind, or - far worse - an illusion! It is for this reason that I reject convention and treat the interpretation of the wavefunction as an essential stage to the development of quantum field theory. There are several conflicting interpretations all of which are experimentally indistinguishable, yet not all of which are rational! A thorough analysis on this issue would be lengthy and beyond the scope of this book. For the purpose of my conceptual approach, it is necessary that I expose the irrational character of the dominant *Copenhagen Interpretation* and confront it with the rational approach to quantum theory.

CHAPTER 2.

PHYSICS AND METAPHYSICS

> *"Matter disappears" means that the limit*
> *within which we have hitherto known*
> *matter disappears and that our*
> *knowledge is penetrating deeper.*
>
> *V. I. Lenin*
> *Materialism and Empirio-criticism,* 1908

In both, classical and quantum physics, the measurement is a process of interaction between a detector and a physical system. The interaction modifies the state of the detector by, for example, shifting the pointer to the position corresponding to the value of the measured quantity. At once, the detector affects the physical system by reducing it to a different state from that at which the physical system was prior to the measurement.

Yet, the measurement of a quantum system radically differs from the measurement of a classical system. The two basic characteristic properties of the classical measurement are:

(i) *We are invariably at liberty to reduce at will the influence of the detector on the physical system.*

When, for example, we measure the temperature, we are at liberty to arbitrarily minimize the thermometer-effected change by using a thermometer of very small size in comparison to the size of the body whose temperature is being measured.

(ii) *The influence of the detector on the physical system may be retroactively subtracted from the result of the measurement.*

This second property is all the more important as such a subtraction allows for the recovery of the initial state of the physical system. For example, using the (final) temperature of the body registered on the thermometer, the heat capacity of that body, the heat capacity of the thermometer and the initial temperature of the latter, we can calculate the initial temperature of the body. It is very important to note the implication inherent in these two properties: The classical measurement allows for simultaneous precise knowledge of all attributes. No measurement performed on an attribute of a classical system compromises the simultaneous measurement on another of that system's attributes.

Neither of these two properties is relevant to quantum physics. In the classical electromagnetic theory, a microscopic object will be discerned only if light of accordingly small wavelength is directed at it. According to the preceding two premises, we can simultaneously minimize the influence of the incident light on the position and the momentum of the microscopic object by reducing the incident light's luminosity as well as, independently, subtract that influence from the result of the observation. Not so in quantum physics! Light is not a classical electromagnetic wave but a collection of photons each of which is a quantum particle of energy $E = \hbar w$. This makes it impossible to obtain precise knowledge of a unique position and a unique momentum of a quantum particle at a given moment. In order to obtain precise knowledge of a unique position, we must direct a photon at the quantum particle which is in a given initial state. Such a procedure must eventuate in a binary outcome: Either point x is at a given moment the position, or it is not. Clearly, the shorter the wavelength of the incident photon is - on rigorous terms, the shorter the wavelength of the plane wave of probability which expresses the photon is - the more accurate the measurement of the position is. At once, however, the shorter that wavelength is, the higher the corresponding energy E is. A "hard" photon will yield accurate information as to the actual position but completely disrupt the probability distribution for position which the quantum particle has just prior to the incidence of the photon, causing in fact the instantaneous collapse of the corresponding wavefunction to a position eigenstate (Fig.2(a)). Any attempt to minimize the influence of the incident photon on the particle's probability distribution for position by decreasing the photon's energy would drastically compromise the accuracy of the measurement on the position, as such a decrease in energy identically implies an increase in wavelength. In stark contrast to the classical measurement, not all attributes of a quantum system can be simultaneously observed. As I shall amply demonstrate in this chapter, this is in fact the exclusive consequence of the quantum particle as a specific dialectical unity and utterly irrelevant to the act of observation itself. In conformity with the analysis associated with Fig.2, if the quantum particle is at a moment in a position eigenstate then at that moment the quantum particle allows for precise knowledge of its actual position and precludes all knowledge of its actual momentum merely because the latter is irrelevant to the former. Conversely, if the quantum particle is in a momentum eigenstate (a plane wave of probability) then it allows for precise knowledge of its actual momentum and precludes all knowledge of its actual position merely because the latter is irrelevant to the former. For that matter:

(i) *It is impossible to reduce the influence of the detector on the quantum system without irreversibly affecting the accuracy of the measurement.*

This first property of quantum measurement is inextricably linked to the second property to the effect that

(ii) *It is impossible to retroactively subtract the influence of the detector from the result of the measurement.*

This second property follows directly from the fact that there is no correspondence between the final interaction-effected state and the initial quantum state. The former is independent of the latter. In our example, the position which a high-energy photon reveals could be associated with any of infinitely many distinct initial quantum states, even if the initial quantum state on which the measurement has been performed were given. Formally, this is inherent in the general fundamental relation (3.V) stated in the preceding chapter since any wavefunction $\psi(t, x)$ can be expressed as a linear combination of eigenfunctions $\delta(x - a)$ of the position operator (see (3.VI)) with the implication that the quantum state $\delta(x - a)$ which emerges at the moment of observation could be associated with any $\psi(t, x)$.

With these preliminary remarks on the theory of measurement in mind we turn to the interpretation of quantum mechanics. In what follows, I present the basic premises of - what I perceive as - the consistent materialist approach and advance a critique of the Copenhagen Interpretation from the standpoint of Marxism.

Any rational interpretation of the wavefunction and the probability distribution which it signifies can only proceed from the following logically-consistent premise: Since the position of the electron - which we consider here as representative of all quantum particles - is an observable attribute, the electron has at each moment a position which is either "here" or "there" but never "here and there". The dominant *Copenhagen Interpretation of quantum mechanics,* developed by Niels Bohr and Werner Heisenberg between 1924 and 1927, would have us believe that one can only make a physically meaningful statement as to the existence of the electron when an observation of the electron is realized. According to the Copenhagen Interpretation, in the event of such an observation the point particle is indeed either at point A or at point B. However, in the absence of observation any statement as to the existence of the electron at point A or at point B is metaphysical - that is, utterly speculative and void of logical content. In the absence of observation all we know - insists the Copenhagen Interpretation - is a probability distribution for the position with the direct implication that the electron may indeed be "here and there" at once or, far worse, nowhere!

Such views call into question the very existence of the physical reality and render physics a meaningless endeavor. They range from the positivist standpoint to the effect that objective existence is confirmed only through observation and that as long as we do not observe something no claim can be made as to whether *it* exists or not, to the idealist standpoint that only the *Idea,* the mathematical concept of probability, exists with the direct implication that particles themselves are artifacts of observation and, thereby, constructs of the mind! The irony of this state of affairs is that although the Copenhagen Interpretation questions, if not altogether denies, the existence of the very reality which theoretical physics strives to explain, the majority of physicists has uncritically accepted the Copenhagen Interpretation and dismissed all objections to it as irrelevant to physics altogether. The consolidated view is that the probabilities which the wavefunction signifies are the only physical reality and that since quantum mechanics allows for testable predictions of these probabilities, quantum mechanics is a sound physical theory. Any inquiry into the physical content of the wavefunction and the probabilities, the issue of "what it all means" is physically irrelevant and best left to the philosophers! The essence of the matter is, of course, identically the opposite. Not only is the issue of "what it all means" centrally scientific and not at all philosophical, the majority of physicists never turned a moment's thought to the fact that their cherished Copenhagen Interpretation, as openly speculative and metaphysical as it is, emanates from the unscientific philosophy of positivism and eventuates in the absurd philosophy of idealism!

In their landmark work *The German Ideology* Karl Marx and Friedrich Engels specified the difference between philosophy and science with the statement that, whereas philosophy merely speculates on the facts, science actually deals with them! ([10], p.48):

Where speculation ends - in real life - there real, positive science begins: the representation of the practical activity, of the practical process of development of men. Empty talk about consciousness ceases and real knowledge has to take its place. When reality is depicted, philosophy as an independent branch of knowledge loses its medium of existence.

What then is the real knowledge which must replace the positivist speculation that in the absence of observation the probabilities are all we know and the empty idealist talk about particles being products of the mind? How does the logically consistent premise to the effect that the position of the electron is, at a given moment, either "here" or "there" conform with the fact that each point in space is associated with a specific probability for being the position?

The incipient point of our analysis is the empirically concrete fact that the electron is an objective entity. It exists independently from the mind, regardless of whether it is observed or not. We can be reasonably certain of this, for if that were not the case we ourselves would not be here to discuss this matter! The probabilities for position which describe the electron in a certain quantum state are precisely that: Probabilities for position! They are the predictable probabilities that each point in space be the position of the electron. Thus, the first criticism which must be levelled against the Copenhagen Interpretation is this: *Absence of observation is not observation of absence!* It is absurd to state that in the absence of observation such probabilities describe a non-existent electron! The position of the electron is a certain point x at a given moment. Pursuant to (3.VI), absence of knowledge as to that specific point $x = a$ implies a superposition $\psi(x)$ of eigenfunctions $\delta(x-a)$ and the concomitant predictable probability $|\psi(x)|^2 dx$ for position about each point x in space. The set of all such predictable probabilities evolving in the course of time is indeed the electron described by $\psi(t,x)$. This specific quantum state, this specific electron, is itself observable. In sharp contrast to the probabilities associated with classical systems, the probability distribution associated with $\psi(t,x)$ is a central aspect of the physical system itself with an observable wave-like behavior (prominently manifest in the *two-slit experiment*). In turn, this observable probability distribution, this electron which is observed as a wave of probability, implies the electron as a particle: as that physical system whose position is, at a given moment, a certain point x. We are dealing here with the dialectical unity of two aspects of the physical reality. The electron is a global entity which implies the local entity, as the local entity implies the global entity. From the perspective of classical physics these two aspects are mutually exclusive since a physical system cannot at once be a global and a local entity. Quantum physics, on the contrary, refutes this formal opposition and reveals either opposite as a *Moment* in the development of the other. In order to grasp the essence of this dialectical unity, the following three points must be understood:

(i) No point x in space is in any respect associated with a value of momentum p. This is an immediate consequence of (2.II) which reveals that all dependence of $\psi(t,x)$ on p has been "summed out". Conversely, no value of momentum p in $\phi(t,p)$ is in any respect associated with a point x. Clearly, the electron is not a classical particle and does not, for that matter, have a certain momentum associated with its position at a certain moment. In fact, the popular statement "the more accurately you try to measure the position of the particle, the less accurately you can measure its speed, and vice versa"[28] expresses just this erroneous standpoint, as though the position and the momentum receive certain values simultaneously at each point but because of some whim of Nature would not be simultaneously observed. As a probability distribution for position expressed by $\psi(t,x)$, the electron is at each moment a superposition of position eigenstates $\delta(x-a)$. Equivalently, it is a superposition of momentum eigenstates $\psi_p(x)$ according to (2.II), itself a direct consequence of (3.V). As $\psi(t,x)$ implies an actual unknown position at any given moment, so $\phi(t,p)$ independently implies an actual unknown momentum at the

---

[28]See for example Stephen Hawking's *A Brief History of Time* p.55.

same moment. These two descriptions are equivalent and independent. The quantum particle is described either by $\psi(t,x)$ or by $\phi(t,p)$, not by $\psi(t,x)$ and $\phi(t,p)$. It is erroneous, for that matter, to attribute to the quantum particle an actual position and an actual momentum at once. The information which $\psi(t,x)$ reveals as to the actual position through Δx, is inversely proportional to the information which $\phi(t,p)$ reveals as to the actual momentum through Δp (Fig.1). As however $\psi(t,x)$ and $\phi(t,p)$ provide two independent and equivalent descriptions of the same quantum state, these descriptions are independently complete. If the electron is at a moment in the position eigenstate $\delta(x-a)$ then at that moment the complete absence of momentum-related information inherent in $\Delta p \to \infty$ signifies as complete a description of the electron as the precise knowledge of the position $x = a$ inherent in $\Delta x = 0$ (Fig.2(a)) signifies.

(ii) Any observed attribute which a measurement returns - be that position, momentum, spin, etc. - constitutes the *material expression of the electron*, a term justified by the fact that we have explicitly revealed the electron's objective existence through interaction. For clearly, the electron does not have an objective position (or an objective momentum, spin, etc.) unless the electron itself is an objective entity. It follows that each observable attribute constitutes a material expression of the electron regardless of whether observation has occurred, or not. The probability distributions for all observable attributes of the electron (position, momentum, spin, etc.) in the absence of observation constitute the *potentialities of the electron* - that is, the possibilities which determine its dynamical behavior as expressed in the evolution of the wavefunction. Central to the rational interpretation of the wavefunction is the premise that the electron is no less the set of its possibilities and potentialities than it is its material expression. The immediate consequence is - as we have already seen - that the former is no less observable than the latter. The possibilities and potentialities of the electron are directly observable in the probability distribution for position and indirectly observable in the probability distribution for momentum, since the latter relates to the former through a mere mathematical transformation.

(iii) Although, in general, the quantum state of the electron prior to measurement does not determine the outcome of that measurement, it does determine the probability for each possible outcome of that measurement. Contrary to the claim of the Copenhagen Interpretation, this renders quantum mechanics a *determinate theory* since the evolution of the wavefunction and of the probabilities which it signifies for all possible results is determined by its character as the solution to the Schrödinger equation and is clearly deterministic. Again, the objection of the Copenhagen Interpretation is predicated on a formal conception of determinism. The fact of the matter is this: If the fact that a single cause yields many possible results signifies the negation of determinism then, logically, the predictable probabilities of these possible results signify the negation of the negation!

The metaphysical claims of the Copenhagen Interpretation emanate from an arbitrary and uncritical identification of the electron with its position, an identification which reduces the electron to an entity occupying a specific point x at a given moment, to "the particle at point $x = a$". From such an arbitrary and uncritical identification the Copenhagen Interpretation proceeds to confer the status of objective existence exclusively upon the "particle" observed at $x = a$ and (since particles have no other existence than that of an entity occupying a point at a moment) refuse to acknowledge the objective existence of the quantum particle in the observable probability distributions. The source of such metaphysical claims and of all the confusion which has enmeshed the Copenhagen Interpretation lies in the latter's preoccupation with the concepts of classical physics and the categories of formal logic. Neither the former, nor the latter are at all relevant to the reality of the quantum systems. Indeed, the electron is a process, not a classical point particle which can be observed at rest. The physical essence of the electron is revealed only in the course of the electron's development. The identification of the actual quantum particle with its observed position at a single moment is a formal abstraction void of physical content. At the moment of observation the position $x = a$ is only the material expression of the quantum particle. The quantum particle itself is the development of the position eigenstate $\delta(x - a)$. For, it is $\delta(x - a)$ which, as a *boundary condition*, uniquely determines $\psi(t, x)$ as a solution to the Schrödinger equation. This already reveals a fundamental aspect of the quantum particle: The dialectical unity!

As the observed position determines the future probability distribution for position and thereby expresses the potentialities of the electron, so a probability distribution for position expresses the unobservable actual position which the Copenhagen Interpretation identifies with the "would-be point particle". This is precisely the physical content of the predictable probabilities for position inherent in the corresponding wavefunction: At any given moment each point in infinite space qualifies for that actual position with a specific probability. These probabilities express the fact that the quantum particle has a unique actual position irrelevant to momentum and the corresponding probabilities for momentum the fact that, at the same moment, the quantum particle has a unique actual momentum irrelevant to position. The position as the material expression of the quantum particle is only an attribute of the quantum particle, not the quantum particle itself! The other attribute are the potentialities. The quantum particle's potentialities considered in isolation from its material expression - a prominent consideration of the Copenhagen Interpretation - is yet another formal abstraction from the actual quantum particle, an abstraction void not only of physical but also of logical content. The quantum particle is the dialectical unity of these two opposite aspects, the global and the local, the probability distribution for position and the actual position itself. The unity is dialectical in the sense that either of these two aspects is an expression of the other and either of these is a *Moment* in the development of the other (N1). As the observed actual position expresses the potentialities of the electron, so the unobservable actual position expresses the infinitely many potentialities and objective probabilities which the quantum particle has for its position and no less at that than the global probability distribution for position expresses the unobservable actual position. In determining and expressing at the moment of its observation the future probabilities for position, the position $x = a$ is a *Moment* in the development of its opposite, of the probability distribution for position[29] [30]. As the wavefunction $\psi(t, x)$ which emerges from the observed position $x = a$ is

---

[29]The preceding conditional statement "If the electron is at a moment in a position eigenstate" must really be understood as "If the electron emanates from a position eigenstate".

[30]Contrary to the Copenhagen Interpretation, the complete absence of knowledge regarding the momentum at the moment at which $x = a$ is the exclusive consequence of the stated dialectical unity and, thus, utterly irrelevant to the interaction with the detector.

already inherent in $\delta(x - a)$ at the moment of observation, the probability distribution for position associated with $\psi(t, x)$ is a *Moment* in the development of $\delta(x - a)$ and thereby in the development of the observed actual position $x = a$.

We see, for that matter, that the electron is a relation and a process. It is a relation between the opposite aspects of global wave-like probabilities and local particle-like position in dialectical unity. It is a process because this relation is dynamical, either aspect expresses its opposite in the course of time. It is as erroneous to identify the quantum particle with the unique actual position, observed or unobservable, as it is erroneous to identify the quantum particle with an abstract probability distribution for position, with a distribution which does not express an actual position and is for that matter void of physical and logical content. In sharp contrast to the Copenhagen Interpretation, the dialectical method commences from the whole, from the dialectical unity, and proceeds inward to the parts, to the actual position and its potentialities, which it conceives as processes in relations of mutual dependence. The development of the dialectical unity which defines the quantum particle emerges as a result of the inherent contradiction between the local actual position and its potentialities realized in the global probability distribution for position. By way of concretizing this abstract statement it is imperative to stress the following point.

The unobservable actual position does not evolve to a trajectory. In the course of time there is only an incoherent succession of unobservable actual positions. The simple reason is that a trajectory is characterized by a specific momentum at each specific position no less than it is characterized by a specific position at each specific moment. The irrelevance of the unobservable actual momentum to the unobservable actual position at any given moment precludes the latter from evolving to a coherent trajectory in the course of time. The incoherent succession of unobservable actual positions may be counterintuitive, but it is logically consistent. For, in dealing with the position, we are dealing with only one attribute of the quantum particle, not with the quantum particle itself. The evolution of the unobservable actual position is a physically meaningless concept. The physically meaningful concept is the evolution of the quantum particle itself - that is, the development of the dialectical unity of the unobservable actual position and the observable objective probability distribution for position. The development of that dialectical unity, the evolution of the probability distribution for position which expresses the actual position, is completely deterministic. We shall see in the next chapter that the evolution of the wavefunction for position is indeed equivalent to the contribution of infinitely many unobservable trajectories which intercept two given points at two given moments precisely because that infinity encompasses all possibilities and potentialities of the quantum particle. As to the counterintuitive character of the incoherent succession of unobservable actual positions in the course of time, note that such a succession is no more counterintuitive than the instantaneous "collapse of the wavefunction" at the moment of observation. Such a collapse from a given probability distribution to a $\delta(x - a)$ does not at all follow form the mathematical formalism of quantum mechanics. Yet, despite its strongly counterintuitive character, it is readily accepted because it conforms with sensory perception: The position $x = a$ is observed instantaneously, not as a result of evolution of the wavefunction in the context of its interaction with the "classical" detector.

The deterministic character and continuity which characterize the evolution of the wavefunction may appear to advocate a certain similarity between the dialectical unity which I advance and the position which Schrödinger advanced in his well-known papers ([21], [22]). The man who introduced the wavefunction was indeed a strong opponent of the Copenhagen Interpretation (a fact clearly evident in the pejorative connotations with which his term "quantum jumps" is laden). Nevertheless, despite his strong opposition, his position is irrelevant to mine. For, Schrödinger commits the same methodological error as the Copenhagen Interpretation. He confers the status of objective existence exclusively upon the observable and denies the objective existence of the unobservable. Whereas the Copenhagen Interpretation acknowledges objective existence exclusively for the "particle" observed at a point and denies the objective existence of the particle in the observable probability distribution for position, Schrödinger acknowledges objective existence for the probability distribution for position and denies the objective existence of the particle as a point entity. *"The wave acts simultaneously throughout the whole region it covers, not either here or there"*, states Schrödinger with respect to the characteristic interference patterns produced by probability distributions in specific experimental contexts. Thus, in his interpretation the observable objective probability distribution for position does not express an unobservable actual position and the dialectical unity between the former and the latter is lost. Therein - from my Marxist perspective - lie the limitations of his epistemological approach, clearly evident in its inability to explain the observed discontinuous collapse of the probability distribution to a $\delta(x - a)$.

Returning to my critique of the Copenhagen Interpretation, it is necessary to stress a point which I have already made. The dialectical method commences from the whole and proceeds inward to the parts. The dialectical unity between the material expression, the actual position observed or unobservable, and the observable objective probability distribution for position stands in direct opposition to Bohr's *principle of complementarity* which envisages a purely formal contradiction between the particle and the probability distribution in the sense that the quantum particle expresses itself as a point entity in certain experimental conditions and as a wave of probability in certain others, with the implication that it is either a point entity or a wave of probability, but not both at once. It is impossible to overstate this fundamental premise: Absence of observation is not observation of absence! It is absurd to confer absolute reality upon the probabilities for position (and momentum) and deny all knowledge as to the existence of position (and momentum) itself - far worse, deny the existence of position (and momentum) itself - unless an observation is made! It is absurd to confer absolute reality upon the Idea (the abstract mathematical probabilities) and deny the reality of the empirically concrete particle (the object of human experience through the observation of "this" or "that" attribute) and the objective reality altogether!

Such is the absurdity of the Copenhagen Interpretation! The following famous statement quoted in Aage Petersen's *The Philosophy of Niels Bohr*, in *Bulletin of the Atomic Scientists 1963, XIX, 7, p.12* is most revealing in this respect:

There is no quantum world. There is only an abstract quantum physical description. It is wrong to think that the task of physics is to find out how Nature is. Physics concerns what we can say about Nature.

In other words, trying to understand Nature and reach its essence is a futile exercise! All we can do is construct theories which allow us to predict the outcomes of measurements without any hope of ever understanding what these outcomes mean. This is the essence of the Copenhagen Interpretation. The human mind is impotent! We can never obtain actual knowledge as to the objective world and shall never obtain such knowledge. And this uncritical positivism which directly contradicts the scientific endeavor and the objective of physics, this pusillanimous negation of the human mind's powerful ability to theoretically conquer the objective world, has been passed off as science for almost a century! It may be argued that in the above quotation Bohr was merely expressing his own views, that somehow this statement is an exception which does not accurately convey the spirit of his Copenhagen Interpretation, or even that this statement, personally made by Bohr to his long-term assistant, was inaccurately reproduced by the latter. Such objections are irrelevant. The fact of the matter is that not only was Bohr identically expressing the spirit of the Copenhagen Interpretation, he was expressing the Copenhagen Interpretation itself! Despite the carefully constructed language which permeates his public statements, his philosophy never detracted from that expressed in the above quotation. Compare, for example, the latter to his conclusion in 1927 [23]:

Accordingly, an independent reality in the ordinary physical sense can neither be ascribed to the phenomena nor to the agencies of observation.

and to his statement ([11], p.18):

In our description of nature the purpose is not to disclose the real essence of the phenomena but only to track down, so far as it is possible, relations between the manifold aspects of our experience.

This is the Copenhagen Interpretation! This is the interpretation which proceeds from the positivist premise "As long as I do not observe *it*, I do not know whether *it* exists or not!" in order to arrive at the stated article of faith to the effect that the particle exists only when an observation of an attribute (especially the position) is made. In the absence of observation, all physics can do is determine the probabilities for all possible outcomes. These probability distributions are all we can say about Nature. It is instructive, in this respect, to let a contemporary famous physicist declare the explicitly positivist character of the Copenhagen Interpretation. In his highly-publicised debate with Roger Penrose which took place at Cambridge University in 1995, Stephen Hawking stated ([12], p.121) (N2):

He [R. Penrose] is a Platonist and I am a positivist. He is worried that Schrödinger's cat is in a quantum state, where it is half alive and half dead. He feels that cannot correspond to reality. But that doesn't bother me. I don't demand that a theory correspond to reality because I don't know what it is. Reality is not a quality you can test with litmus paper. All I am concerned with is that the theory should predict the results of measurements. Quantum theory does this very successfully. It predicts that the result of an observation is either that the cat is alive or that it is dead. It is like you can't be slightly pregnant: you either are or you aren't.

In terms of content (although certainly not in terms of language), the identity with Bohr's preceding statements is striking! Reality is not a quality which you can test with scientific theories. You only test it when you observe it, when you observe the "point particle" and the cat inside the box! In the absence of such an observation all you can do is predict abstract probabilities void of physical content. To the careful observer, the conceptual ambiguity and uncritical positivism which characterize this standpoint, the standpoint of the Copenhagen Interpretation, are obvious. For Marxism, conflicts, contradictions and paradoxes which arise in the field of thought are essentially the result of an uncritical abstraction from reality, of an analytical process which confers absolute significance upon boundaries which only have social and cultural significance:

All boundaries in Nature are conditional, relative, movable and express the gradual approximation of our mind toward knowledge of matter.

states V. I. Lenin in his *Materialism and Empirio-criticism* ([13], p.289 - compare this also to the quotation which opens this chapter [13], p.267). In the case of the Copenhagen Interpretation the arbitrary boundary is the concept of point-particle, the only concept which conformed with the limited human experience up to the first world war. Particles whose dynamical evolution is determined by potentialities and possibilities which contradict the concept of point-particle do not exist! Such possibilities had to be external to the electron and, thereby, hypostatized into independent existence. The Copenhagen Interpretation arbitrarily separates the potentialities and possibilities of the quantum particle from the quantum particle's position (its material expression), elevates the former to an abstract mathematical Idea by dismissing the objective existence of the latter as physically irrelevant and then just as arbitrarily reintroduces the latter when observation is made. For, it is not the "collapse of the wavefunction" which in the event of observation confers objective existence to the electron (such a collapse occurs in any interaction and does not necessitate an observer), but the observation of the electron as a "point-particle" itself. Yet, since the Copenhagen Interpretation has failed to construct a bridge leading from the abstract mathematical Idea of probability to the concrete material existence which observation reveals (and in all eternity would never construct such a bridge), all the Copenhagen Interpretation can really do is smuggle the empirical world in again, in an underhand fashion.

Although Bohr and Heisenberg imagined that the metaphysical claims of their Copenhagen Interpretation signify a new trend in physics[31], their claims were certainly nothing new in the field of thought. In fact, in a glaring demonstration of self-contradiction Heisenberg himself stated [24]:

The epistemological analysis of quantum theory, especially in the form which Bohr gave it, contains many features which resemble the methods of the Hegelian philosophy.

This is indeed the case! We are dealing here with the famous *subject-predicate inversion* in philosophy. The following two excerpts from the critique which young Karl Marx levels at the philosophy of Hegel are most revealing as to the positivist and idealist origins of

---

[31]N. Bohr was enamored with the "paradox" and the negation of all reason. His guiding principle was "I believe it because it is absurd!". If nothing else, his statement: *"We are all agreed that your theory is crazy. The question which divides us is whether it is crazy enough to be correct."* and exclamation: *"How wonderful that we have met with a paradox! Now we have some hope of making progress."* speak volumes!

the Copenhagen Interpretation. The first from his famous *Economic and Philosophical Manuscripts* ([14], p.20 and p.385)[32]:

Hegel's philosophy suffers from the double defect of being at once "uncritical positivism" and "equally uncritical idealism". It is uncritical idealism because Hegel denies the empirical, sensible world and acknowledges true reality only in abstraction, in the Idea. And it is uncritical positivism because Hegel cannot help in the end restoring the empirical object-world originally denied - the Idea has no other possible earthly incarnation or meaning.

The second from his *Critique of Hegel's Doctrine of the State* ([14], p.21 and p.69):

But since he [Hegel] has failed to construct a bridge leading from the general idea of the organism to the particular idea of the organism of the state or the political constitution (and in all eternity would never construct such a bridge), all Hegel can really do is smuggle the empirical world in again, in underhand fashion.

Formally, the Copenhagen Interpretation does not deny the empirical, sensible world and acknowledge true reality only in the Idea of abstract mathematical probabilities. Instead, in the absence of observation it dismisses the objective existence of the world as physically-irrelevant speculation. In essence, however, this is tantamount to idealism. Idealism is implicit in the Copenhagen Interpretation and it only takes one small step in order to arrive at it. For, if quantum particles exist only as point-particles and if no claim as to their objective existence can be made unless they are detected as such then there are only two possibilities. Either quantum particles objectively exist and their objective existence precedes observation, or they do not objectively exist but is instead the act of observation and, by extension, the mind of the observer which creates the quantum particles out of a world of abstract mathematical probabilities. Since the Copenhagen Interpretation vehemently denies the first possibility, it implicitly embraces the second. Whence, the uncritical idealism which acknowledges true reality only in the Idea and the uncritical positivism which surreptitiously restores the objective electron at the moment of observation! The remark which N. Bohr returned to A. Einstein "Prove that the moon is there when you don't look at it!" is indicative of Bohr's true idealist inclinations. It is, nevertheless, hardly necessary to speculate on Bohr's conscious or unconscious psychological inclinations and symbols when W. Heisenberg brought out the inherent idealism of the Copenhagen Interpretation with "transparent clarity"! According to Heisenberg, the transition of physics from classical to quantum must be treated not as a penetration of the human mind into the essence of matter and of objective reality but, instead, as ([15], p.28):

[...] the dissolution of the objectively real world in the transparent clarity of mathematics, the laws of which govern the possible and not the actual.

In his famous book *Physics and Philosophy* we read that ([16], p.95):

The ontology of materialism rested upon the illusion that the kind of existence, the direct "actuality" of the world around us, can be extrapolated into the atomic range. This extrapolation is impossible, however.

that ([16], p.128):

---

[32]This is, in fact, an excerpt from the Introduction in [14] (p.20) which conveys the essence of the extended premise in the text (p.385).

In the experiments about atomic events we have to do with things and facts, with phenomena that are just as real as any phenomena in daily life. But the atoms or the elementary particles themselves are not as real; they form a world of potentialities or possibilities rather than one of things or facts.

and that ([16], p.82 - p.83):

The Copenhagen interpretation of quantum theory has led the physicists far away from the simple materialistic views that prevailed in the natural science of the nineteenth century. [...] However, all the opponents of the Copenhagen interpretation do agree on one point. [...] They would prefer to come back to the idea of an objective real world whose smallest parts exist objectively in the same sense as stones or trees exist, independently of whether or not we observe them. This, however, is impossible...

just to present a few of the bluntest affirmations[33] of the idealist absurdity which pervades the Copenhagen Interpretation. As though the act of measurement is not an interaction between physical systems just like any other interaction, as though the human mind is not itself a physical system, the Copenhagen Interpretation, implicitly yet concretely (Heisenberg and many others, explicitly), arrives at the absurd conclusion that it is "the mind" which "incarnates" and gives substance to an otherwise abstract world of mathematical probabilities! Heisenberg's *Physics and Philosophy* is, in fact, a desperate attempt to discredit not only the ontology of materialism but also those who adhere to it. Let us, for example, look at the critique which he levels at Soviet physicists who defend the view that the probability distribution expresses the objective state of the quantum particle ([16], p.89):

In his presentation Alexandrov overlooks the fact that the formalism of quantum theory does not allow the same degree of objectivation as that of classical physics. For instance, if the interaction of a system with the measuring apparatus is treated as a whole according to quantum mechanics and if both are regarded as cut off from the rest of the world, then the formalism of quantum theory does not as a rule lead to a definite result; it will not lead, e.g., to the blackening of a photographic plate at a given point.

Heisenberg's argument is hardly of physical relevance, of course. For, while it is true that the formalism of quantum theory does not as a rule lead to a definite result, it is certainly not true that this signifies the "dissolution of the electron in the transparent clarity of mathematics". If the interaction with the measuring apparatus does not reveal the position then the electron is a probability distribution for position which expresses the actual, albeit unobservable, position itself. The electron is, in fact, the dialectical unity of these two attributes. In this respect, Alexandrov is right. The probability distribution for position expresses indeed the objective state of the electron with observable wave-like effects such as interference. According to Heisenberg, the absence of a black spot on the photographic plate signifies the absence of objective actual position for the electron and thereby the absence of the electron itself. As I have stressed, however, it is fundamentally erroneous to identify the quantum particle as a point particle and thereby deny its objective existence when it is not observed as such. It is fundamentally

---

[33]*Heisenberg provides here some of the bluntest affirmations of the denial of the objective reality of the external world that I have seen,* gleefully observes idealist and religious propagandist Paul Davies in the Introduction to Heisenberg's book.

erroneous to consider its possibilities as something abstract and external to it. The quantum particle is the dialectical unity of the observable objective probability distribution for position and unobservable actual position itself. The position $x = a$ observed as a "point particle" in position eigenstate $\delta(x - a)$ is only a Moment in the development of the actual quantum particle itself. The absence of a black spot on the photographic plate signifies the presence of the interference pattern with which the probability distribution for position, the electron itself, registers its objective existence.

Heisenberg never lacked in poetic imagination. To him views which conform with dialectical materialism are nothing but *"distressing attempts which mislead us into continually occupying ourselves with the inevitable cracks in the old bottles instead of rejoicing over the new wine"*! Of course, when poetry falls short of matching the objective, there are always less eloquent means of attack available: *"Here we are dealing not with science but with a confession of faith, with adherence to a certain creed"*! And this provocative denunciation of dialectical materialism as "faith" and "religious creed" comes from Heisenberg, one of the two foremost representatives of that philosophy which categorically denounces any attempt to advance past it as unscientific and futile on the openly metaphysical and patently absurd claim that physics will never find out how Nature is, that *the best we can ever do* - as Bohr put it - is predict abstract probabilities! This is the very same Heisenberg who, in a demonstration of blatant intellectual dishonesty, dismisses the dialectical materialism of 1908 as an "old cracking bottle" while joyfully pouring his "new wine" of idealism into the Hegelian bottles of 1805 and - far worse - into the Platonist bottles of 2.400 years ago! For ultimately, his views amount to nothing more than Plato's objective idealism expressed in the language of modern theoretical physics.

Rather than continue with the refutation of the openly metaphysical and patently absurd claims of the Copenhagen Interpretation, it would be more instructive to let Karl Marx himself ridicule them with his penetrating analysis and scathing sense of humor. Here follows a famous excerpt from his equally famous work *The Poverty of Philosophy* of 1846 [17]:

If we abstract thus from every subject all the alleged accidents, animate or inanimate, men or things, we are right in saying that in the final abstraction, the only substance left is the logical categories. Thus the metaphysicians who, in making these abstractions, think they are making analyses, and who, the more they detach themselves from things, the more they imagine themselves to be getting all the nearer to the point of penetrating to their core - these metaphysicians in turn are right in saying that things here below are embroideries of which the logical categories constitute the canvas. This is what distinguishes the philosopher from the Christian. The Christian, in spite of logic, has only one incarnation of the 'Logos'; the philosopher has yet to finish with incarnations.

I need hardly stress that my Marxist critique is directed against the Copenhagen Interpretation of quantum mechanics, not against quantum mechanics itself, the conceptual and mathematical consistency of which I have stressed. The scientific merit of N. Bohr and W. Heisenberg is not into question. It is their openly metaphysical philosophical views with which they polluted physics which I reject. Although this standpoint may appear contradictory, it is not. For, Bohr, Heisenberg and all those physicists who introduced and embraced mysticism in modern physics, including those who shrugged it off as "physically irrelevant", merely view reality from the prism of the social class to which they belong. The unqualified attack which Heisenberg directed against dialectical materialism and those who adhere to it is merely a reflection of his character and utterly irrelevant to his ideological inclinations. It is naive to ascribe intensions of deliberate propaganda to them and thereby denigrate them as physicists, as certain critics of the Copenhagen Interpretation do[34]. According to Marxism, their philosophical views are a clear expression of *false consciousness*: If the Copenhagen Interpretation is deception, it is also self-deception! Lenin had no qualms giving due credit to prominent physicists whose philosophical outlook he attacked, while making a clear distinction between their personal intensions and the objectively reactionary character of their philosophy ([13], p.355):

Not a single one of these professors, who are capable of making very valuable contributions in the special fields of chemistry, history, or physics, can be trusted one iota when it comes to philosophy. [...] Taken as a whole, the professors of economics are nothing but learned salesmen of the capitalist class, while the professors of philosophy are learned salesmen of the theologians.

This, however, does not extend to our time. Since the emergence of quantum mechanics, the Copenhagen Interpretation has come a long way. It and the equally absurd "Many-Worlds interpretation" (denounced as metaphysical and abandoned even by its own authors) transcended the mathematical and conceptual complexity which separates the "scientific community" from the "public" and imbued social consciousness. A spate of books, "authoritative web-sites" and TV-documentaries inculcated the idea of the illusion of an objective world, of "the matter myth", of the ridiculous "link between quantum physics and eastern mysticism", of the equally ridiculous parallel universes in which Nazi-Germany has won the Second World War and so on. A typical such example of "popular physics" which advances the absurdity of philosophical idealism is John Gribbin's *In Search of Schrödinger's Cat*. The basic theme of that book is summed up in its lines ([18], p.162):

If we can not say what a particle does when we are not looking at it, neither can we say if it exists when we are not looking at it, then it is reasonable to claim that nuclei and positrons did not exist prior to the twentieth century, because nobody before 1900 ever saw one. In the quantum world, what you see is what you get, and nothing is real; the best you can hope for is a set of delusions that agree with one another.

The first sentence is merely a crude rendition of the inherently idealist Copenhagen Interpretation in general and the views of W. Heisenberg in particular. However, it is the last sentence which betrays Gribbin's intensions. For, not only does he renounce the objective reality, he renounces all knowledge of the abstract mathematical possibilities and their "transparent clarity", of the Idea itself, as a delusion. The exclusive purpose of deliberate propaganda is all too obvious in this logically-incoherent and meaningless

---

[34]Nor are Heisenberg's Nazi affiliations a relevant consideration. Authors who stress this point, only reveal their naivety. The Copenhagen Interpretation is false consciousness, not moral choice: Outstanding Soviet physicist Lev Landau, richly decorated by the socialist state as "Hero of Socialist Labour", also adhered to the Copenhagen Interpretation, as we shall see later in this chapter.

jumble of words! Behind such attempts at mass-brainwashing stand, of course, powerful corporate interests, their political establishment and their mass media, class interests which have every reason to promote the idea of an illusory world, of a world which we can never understand, let alone change!

The deliberate propaganda of the political establishment does not merit serious consideration by any rational individual. The essential question concerns the false consciousness itself, the reason the class position of the physicists dictated an interpretation of modern physics in the direction of idealism and mysticism. I shall not advance into social theory. The discriminating reader can deduce the answer from my critical analysis in this chapter and my relevant remarks in the Introduction: Every advance in theoretical physics is inextricably linked to social change. Metaphysical, mystical and absurd interpretations of rigorous scientific theories are a reflection in the field of thought of stagnated social practice, of social relations which severely impede further development of the production forces and historical advance. I have stated the rigorous Marxist distinction between philosophy (which merely speculates on the facts) and science (which actually deals with the facts). Marxism rejects philosophy and embraces science. Marxism itself is science. The pollution of science by philosophy is a particular expression of social crisis. The immobility of scientific thought is conditioned by the contradictions which split society. It is the practical activity of men which directly produces all material wealth and the practical activity of men which indirectly produces all intellectual wealth. For that matter, it is only a qualitative leap of that practical activity which is now fettered by the particular production and social relations of capitalism, the advance to a higher historical stage, which will set captive thought free of all mysticism and permit it to attain its full development. So striking is the expression of this fact in K. Marx's famous *Eighth Thesis on Feuerbach* ([14], p.423):

All social life is essentially practical. All mysteries which lead theory to mysticism find their rational solution in human practice and in the comprehension of this practice.

$$* * * *$$

Before I conclude, it is necessary to make at least a few general remarks regarding certain issues inherent in the preceding analysis. The Marxist approach to the interpretation of quantum mechanics which I have advanced is dialectical and materialist. Is this then the interpretation of dialectical materialism? What is dialectical materialism anyway? Is the Marxist interpretation which I have advanced in identity with the position which the Soviet physicists and all proponents of dialectical materialism held?

Friedrich Engels developed a philosophical theory in works which appeared in the later years of Marx's own life, most notably in his *Anti-Dühring* (1878). First Georgi Plekhanov and subsequently Vladimir I. Lenin stressed the general character of Engels' theory as *the* theory of development of Nature, history and thought. As it purported to apply the dialectical method in the materialist outlook, it was labelled "dialectical materialism" and viewed as a necessary preliminary to the more 'particular' theory of historical materialism. Yet, there is no evidence that Engels' philosophical positions reflected Marx's thought. In fact, Marx himself, the principal founder of historical materialism, never made a reference to Engels' "dialectical materialism". The reason, in my opinion, must be sought in Marx's characteristic discipline in scientific formulation. Although the dialectical method is as relevant to Nature as it is to history and to political economy, its application in the fields of physics, geology and biology must be demonstrated in each specific physical, geological and biological context if dialectical materialism is to be a truly scientific method, as opposed to mere philosophical speculation. The reason works like the *German Ideology* and the *Capital* have a solid scientific foundation is precisely because they rigorously demonstrate the dialectical method in history and in political economy respectively.

Engels' formulation of dialectical materialism, on the contrary, is founded on "universal laws of development of Nature, history and thought, equally valid in the animal and plant kingdoms, in geology in mathematics, in history and in philosophy" (*Anti-Dühring*). The extent to which such "universal laws" are consistent with the dialectical method, the supreme expression of which is Marx's *Capital* [19], is not the issue here. The issue is that from Engels' philosophical standpoint, the universal flexibility of concepts reflecting the eternal development of the objectively real world is not transparent and the dialectical method appears oversimplified and rigid. As a result, not only is the high dexterity required in applying the dialectical method in a specific context not emphasized, it is instead downplayed. The immediate danger inherent in this is voluntarism: Dialectical materialism no longer serves as a method for analyzing a physical system's development. Rather, it is the physical system itself which provides a "striking confirmation" of dialectical materialism.

Of course, Engels' and Lenin's intensions lay in the opposite direction. In stressing the general character of Engels' theory, Lenin was, in fact, stressing the direct relevance of the dialectical method to Nature and to the field of thought. His premise to the effect that a central aspect of Marxism is a *"specific analysis of a specific situation"* is nothing short of the demand that Marxist dialectics be demonstrated in each specific context. This is particularly evident in his consistent application of that method in *Materialism and Empirio-criticism*. Yet, not all adherents to "dialectical materialism" adhered to Lenin's premise. The book which - to my knowledge - more than any other purports to advance the standpoint of "dialectical materialism" in modern physics, especially in its critique of the Copenhagen Interpretation, is that by Soviet professor M. E. Omelyanovsky entitled *Dialectics in Modern Physics* [20]. Despite its references to Lenin's *Materialism and Empirio-criticism*, this book has no soul. What impresses the knowledgable reader from the outset is its conciliatory orientation toward the Copenhagen school. We read, for example, that *Niels Bohr made a definite advance towards a materialist approach to quantum mechanics* ([20], p.50) and specifically ([20], p.47):

Thus, one can find explicit expression in Bohr's essay "Quantum Physics and Philosophy" of a position that is basically materialist and dialectical. By linking the mathematical apparatus of quantum mechanics with visualisable notions of classical concepts he reveals, as a philosopher would say, the antithetical character of the corpuscular and wave conceptions. Contrasting these ideas as a certain antinomy always played a decisive role in Bohr's notion of complementarity. In his early work on quantum mechanics, however, this antithesis was masked by the ida of 'uncontrollable interaction'; in "Quantum Physics and Philosophy", however, this drawback is eliminated.

"Uncontrollable interaction" is a term which conveys the impossibility of cognising Nature in the sense that the very interaction which aims at measuring "this" or "that" attribute of a physical system will irretrievably alter that system. What Omelyanovsky states here is basically that Bohr refined his language. Niels Bohr, on the evidence a man of uneasy conscience, constantly refined the language in terms of which he stated his principle of complementarity and the Copenhagen Interpretation in an attempt to mask - primarily from his own self[35] - the contradiction of defending as a scientist a thoroughly unscientific position. The fact is, however, that Bohr never detracted from his explicitly positivist and inherently idealist position. Let it suffice to quote Omelyanovsky himself on this point ([20], p.56):

And yet Bohr's complementarity principle does not finally [i.e.: in "Quantum Physics and Philosophy" (the elucidation is mine)] *resolve the problem of particle-wave dualism. According to his ideas, the contradiction between the particle and wave properties of atomic objects is supposedly frozen in the form of an opposition of two classes of mutually exclusive experimental situations with which the 'complementary phenomena' are associated. The true solution of the 'antinomy of complementarity', however, consists in considering the particle and wave properties of an atomic object as a unity of opposites. That is why quantum theory concepts reflecting the dual nature of atomic objects must differ qualitatively from classical ones.*

That is, the abstract probabilities, utterly void of objective physical content, are in certain "experimental situations" the only reality. So much for Niels Bohr who later in his life saw the light of "dialectical materialism"! In fact, not only does Omelyanovski extend such a conciliatory gesture to Bohr, he is inclined to extend it also to Heisenberg! ([20], p.44):

Hence it would seem to follow that Heisenberg's argumentation deserves the closest attention and support from the standpoint of dialectical materialism.

Omelyanovski was not expressing only his personal views. Such views as he expressed in his book were the views which the entire USSR Academy of Sciences held at the time. At first sight this certainly appears to be contradictory. Why would the bastion of "dialectical materialism" come to such a compromise with a most reactionary ideology? The answer lies in the solution which Omelyanovski and his colleagues gave to the centrally important problem of developing a rational interpretation of the wavefunction, to the problem of applying the dialectical method in quantum physics. That solution explains at once the perspective from which the Soviet thinkers approached all other interpretations of quantum physics. As is obvious from the preceding quotation, they followed Engels' guidelines and considered a quantum particle as a "unity of opposites" which they identified as "the unity of the particle and wave properties of an atomic object", for it is this "unity of opposites" which at first sight impresses as the obvious

---

[35]Omelyanovski himself implicitly comes to the same conclusion ([20], p.52): *Bohr needed the concept of uncontrollability in principle in order to hide the logical inconsistency resulting from the concepts of classical mechanics being used outside their field of applicability.*

and straightforward solution. The highly non-trivial actual solution, of course, is predicated upon the method which reveals the objective existence of the quantum particle as a specific dialectical unity and thereby refutes the reactionary positivist and idealist claims of the Copenhagen school. In this respect, I have stressed that it is erroneous to identify the quantum particle either with the observed position of that particle at a given moment, or with the unobservable unique and objective position of that particle in the absence of such observation and that in the latter case the quantum particle is the dialectical unity of the unobservable objective position and the observable probability distribution for position which expresses both, the former and the potentialities of the quantum particle. In acknowledging the dialectical unity not between the unobservable objective position and the observable probability distribution for position but between "the particle and the wave", the first error which Omelyanovski and his colleagues commit is to identify the particle with its position, precisely as the Copenhagen school does. This, in turn, leaves "dialectical materialism" with only one possibility: Although the probability distribution - "the wave property", as Omelyanovski crudely calls it[36] - is indeed objective, it does not express an unobservable objective position. Such a position would express a "particle" which cannot exist at once with "the wave" due to the "relativity with respect to the means of observation", as Bohr himself put it. Unless "the particle" is observed at a point, "the wave property" is itself the "atomic object". Whence, the logical contradiction! The probabilities for position exist independently of the mind, yet without expressing a position which itself exists independently of the mind! In the absence of observation any claim as to an objective position is logically meaningless, only "the wave of probability" objectively (!) exists. This kind of "dialectical unity of opposites" does not constitute a "striking confirmation of dialectical materialism", as Omelyanovski claims, but a "striking confirmation" of Bohr's principle of complementarity! Is it then any wonder that the defenders of "dialectical materialism" felt obliged to "shake hands" with the proponents of the most unscientific and reactionary ideology which ever emerged in theoretical physics?

What a "comedown" for revolutionary Marxism! The responsibility does not fall on Engels, of course. Engels, himself a man of principle, resented dogmatism. He believed that dialectical materialism - as rigid and oversimplified as it appears in his formulation - alters its form with every new advance of science, an allusion to the fact that the splitting of a single whole and the cognition of its contradictory parts is not at all as trivial as Omelyanovski and his colleagues in the USSR Academy of Sciences imagined. That their views were not at all limited to the department of philosophy is evident in the position which outstanding theoretical physicist Lev Landau and his student Evgeni Lifshitz express in their famous text *Quantum Mechanics*, first published in 1958. In it, they state ([3], p.3):

The more exact the measurement, the stronger the effect exerted by it [...] This property is logically related to the fact that the dynamical characteristics of the electron appear only as a result of the measurement itself.

---

[36]The wave is essentially a field, merely labelled by each point in space. On the contrary, the wavelike pobability distribution expresses probability for the position at a point as a dynamical variable: In quantum mechanics, the dichotomy between particles and fields unconditionally stands. Only in quantum field theory is this dichotomy refuted!

What then is the material basis of such a dismal compromise with the reactionary Copenhagen school? In response, I shall only state the following. Those who have studied the evolution of socialism in the Soviet Union should be able to see the stagnation of social practise at the time such views became prevalent in the theoretical superstructure of the Soviet society and the first serious signs which foreshadowed the momentous events of the Soviet Union's final stage almost two decades later. As social practise goes, so does the theory! A number of metaphysical positions became accepted as an expression of "dialectical materialism" in physics, in philosophy and in political economy. It is, for that matter, that I refrain from characterizing my epistemological approach to quantum physics by the term "dialectical materialism". Although it is materialist and dialectical, I characterize it as "the Marxist approach" in order to preclude association with the lifeless "dialectical materialism" which was officially promoted in the USSR of that time. Indeed, those gentlemen who purported to defend "dialectical materialism" did a lot to deface its once glorious edifice by reducing it to a set of ready-made universally applicable recipes instead of realizing it as the specific analysis of the specific situation, by arriving at the correct conclusion of "striking confirmation" through use of incorrect and devious means. It is to them that young Karl Marx speaks[37]:

The truth includes not only the result but also the path to it. The investigation of truth must itself be true!

$$* * * *$$

I conclude with a relevant point which does not receive due consideration in debates on such matters. Quantum particles are - in the rational sense analyzed in this chapter - probability distributions and as such, global entities. In quantum field theory in which particles are rigorously interpreted as an expression of underlying quantum fields, this global character renders them nebulous, all the more so in curved space-times. The particle concept itself loses its fundamental physical significance. Observers in a certain state of motion may register particles when others in a different state of motion will register none. In the spirit of *general covariance*, each observer is correct in his claim that he either registers particles or that he does not. This fundamental aspect of quantum field theory exceeds the scope of this book. We shall only deal with quantum fields in flat Minkowski space-time as registered by non-accelerating inertial frames all of which agree on the existence of particles. This primary aspect of quantum field theory is a prerequisite for understanding quantum fields in the presence of gravity. Nevertheless, I do mention the relative physical significance which particles have in order to stress the absolute physical significance of quantum fields: Quantum fields exist regardless of who observes them and of whether he observes them. They are the objective and ultimate physical reality!

---

[37]*Comments on the latest Prussian Censorship Instruction* (1842).

CHAPTER 3.

PATH-INTEGRAL QUANTIZATION

1. The Action Functional in Quantum Mechanics

Not only is the concept which we are going to introduce and discuss in this section a central aspect of path-integral quantization, it is central to theoretical physics in its entirety. It is probably necessary, for that matter, to first elucidate certain basic mathematical concepts for the benefit of those readers who are not altogether familiar with them.

Consider a function $f(t)$ and form the ratio

$$\frac{f(t) - f(t_0)}{t - t_0}$$

where t_0 is a given fixed value of independent variable t. If t is considered arbitrarily close to the given t_0 - rigorously put: if t *tends to* t_0 $(t \to t_0)$ - then, depending on what $f(t)$ actually is, the limit of the ratio itself may or may not exist. If the limit of the ratio as $t \to t_0$ exists then little reflection is required in order to see that it expresses, in fact, the *rate of change of $f(t)$ at $t = t_0$*. This rate is formally the *first derivative of $f(t)$ at $t = t_0$*:

$$\frac{df(t)}{dt}\Big|_{t_0} \equiv \frac{d}{dt}\Big|_{t_0} f(t) = \lim_{t \to t_0} \frac{f(t) - f(t_0)}{t - t_0} \tag{1.I}$$

The first derivative of $f(t)$ at $t = t_0$ directly implies the *first derivative of $f(t)$*: $\frac{df(t)}{dt} \equiv \frac{d}{dt} f(t)$. The latter is the first derivative of $f(t)$ at t_0 with t_0 now being any value of the independent variable t in a subset of the function's domain of definition. If that subset is the entire domain of definition of $f(t)$ then $f(t)$ is a *differentiable function*. A *stationary point* of a differentiable function $f(t)$ is any ordered pair $(t_0, f(t_0))$ in the entire set of ordered pairs $(t, f(t))$ at which $\frac{df}{dt}\big|_{t_0} = 0$. On the graph of $f(t)$, the stationary point is a point at which the slope of the tangent line vanishes, a fact which implies that the tangent line is horizontal (parallel to the x-axis).

We shall also need the *second derivative of $f(t)$* the definition of which trivially follows now from the first derivative of $f(t)$. The second derivative of $f(t)$ is merely the first derivative of the first derivative:

$$\frac{d^2}{dt^2} f(t) = \frac{d}{dt}\left(\frac{d}{dt} f(t)\right) \tag{1.II}$$

Higher-order derivatives are accordingly defined.

It is common knowledge that in non-relativistic classical mechanics the motion of material bodies is determined by Newton's laws. It is, nevertheless, possible to develop theoretical mechanics entirely from a new and independent principle, *Hamilton's principle*, from which Newton's laws are themselves derived. The main advantage of Hamilton's principle is that - in contrast to Newton's laws - it is just as relevant to fields. In addition, it allows for a unified description of several theorems in classical mechanics. Let us, for that matter, revisit the opening paragraph of the first chapter in which we defined the concept of degrees of freedom for a system of particles and gave concrete examples.

Consider on such lines a system of particles which comprises n degrees of freedom described in terms of n *generalized coordinates* $q_1(t), q_2(t), ..., q_n(t)$ (such as, for example, the six translational, rotational and vibrational degrees of freedom of the diatomic gas molecule). Hamilton's principle amounts essentially to the following statement:

In the set of infinitely many possible trajectories available to a dynamical system which advances from specified position $P_1 \equiv q_1(t_1), q_2(t_1), ..., q_n(t_1)$ at moment t_1 to specified position $P_2 \equiv q_1(t_2), q_2(t_2), ..., q_n(t_2)$ at moment t_2, the one which is realized is the trajectory which renders the integral[38]

$$S[q_1, q_2, ..., q_n] = \int_{t_1}^{t_2} L\big(q_1(t), q_2(t), ..., q_n(t), \dot{q}_1(t), \dot{q}_2(t), ..., \dot{q}_n(t), t\big)dt \qquad (1.III)$$

stationary.

In this statement: (i) *the Lagrange function L is a function of the generalized coordinates* $q_i(t)$; $i = 1, 2, ..., n$, of the corresponding generalized velocities $\dot{q}_i(t) \equiv \frac{dq_i(t)}{dt}$ and of time t. For a system of <u>non-relativistic</u> particles, it is defined as the difference between the total kinetic energy T of the dynamical system (e.g. the sum-total of translational, rotational and vibrational kinetic energies) and the total potential energy V (for example, the potential energy due to vibration plus that due to any external field). That is

$$L = T - V \qquad (1.IV)$$

(ii) As the entire set of generalized coordinates $q_i(t)$; $i = 1, 2, ...n$ describes positions P_1 and P_2 respectively specified at $t = t_1$ and $t = t_2$, so each choice of generalized coordinates $q_i(t)$ at each moment t defines a single trajectory in an abstract *configuration space* (in the same sense that a choice of Cartesian coordinates $x_1(t), x_2(t)$ and $x_3(t)$ defines a particle-trajectory in physical Euclidean space). There are infinitely many possible such choices, whence infinitely many possible trajectories available to a dynamical system which advances from $t = t_1$ to $t = t_2$. To each such trajectory $q(t) \equiv q_1(t), q_2(t), ..., q_n(t)$ in configuration space, corresponds the value $S[q] \equiv S[q_1, q_2, ..., q_n]$ (note that t has been "summed out"). The set of all such values is the *action functional*[39] $S[q]$.

A *stationary point of the action functional is a point at which the functional derivative vanishes*:

$$\frac{\delta S}{\delta q}\bigg|_{q=q_0} = 0$$

as is the case, for example, with point (q_0, S_0) in Fig.9. Since, according to Hamilton's principle, the trajectory $q_0(t)$ is the trajectory which is realized, the trajectory $q_0(t)$ is clearly the classical trajectory, the solution to the classical equations of motion. The

---

[38] "Integral" is, of course, the formal term for the continuous sum (here defined between the two values t_1 and t_2) which we extensively used in the first chapter.

[39] In contrast to functions (which are maps of numbers onto numbers), functionals are maps of functions onto numbers. The action functional which maps functions representing trajectories onto numbers is but one example of a functional. The bra which we introduced in the first chapter is yet another.

derivation of the classical *Euler-Lagrange equations* from Hamilton's principle proceeds in the context of the *calculus of variations*, an aspect of advanced Mathematics. Nevertheless, on condition that differential calculus has been understood, the essence of that derivation is not difficult to grasp. For that matter, in the derivation of the classical Euler-Lagrange equations of motion which follows, I provide for the benefit of those readers who are not as confident in differential calculus as much verbal explanation as the subject matter of the present chapter possibly allows. Readers who still find it difficult to follow this derivation may disregard it at this stage. However, as the importance of the Euler-Lagrange equations cannot be overstated, they should constantly return to this derivation with every advance they make in the course of studying this book.

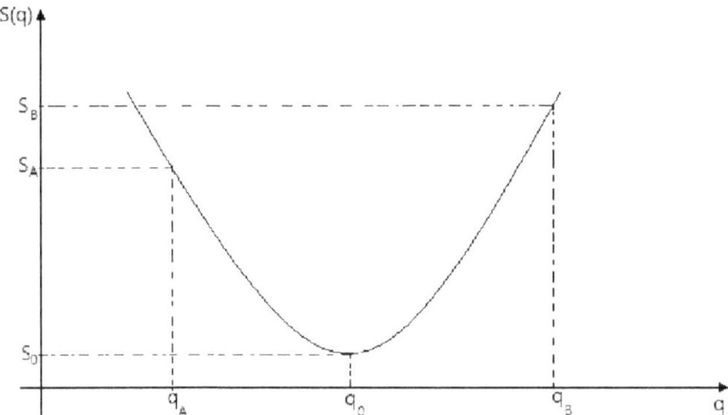

FIGURE 9. To each trajectory in configuration space corresponds an action!: The action functional. Each point on the q-axis represents a trajectory in an n-dimensional configuration space determined by the n degrees of freedom of the dynamical system. The minimum action S_0 corresponds to the classical trajectory $q_0(t)$.

We consider a dynamical system described by one generalized coordinate $q(t)$ such that $q(t_1) \equiv P_1$ and $q(t_2) \equiv P_2$. In the set of infinitely many possible trajectories which satisfy these two *boundary conditions* in configuration space, we wish to determine that $q(t) = q_0(t)$ which *extremizes* the action functional $S[q]$ (i.e. which renders it stationary). Let us, for that matter, also consider the one-parameter family of possible trajectories $q(t)$, subset of the above-stated set, defined by

$$q(t, \epsilon) = q_0(t) + \epsilon \eta(t) \implies \dot{q}(t, \epsilon) = \dot{q}_0(t) + \epsilon \dot{\eta}(t)$$

where $\eta(t)$ is an arbitrary, but fixed once chosen, differentiable function satisfying

$$\eta(t_1) = 0 \; ; \; \eta(t_2) = 0$$

and the "small" parameter ϵ receives values between 0 and some number $\epsilon_0 << 1$ (so that each fixed value of ϵ determines a unique trajectory $q(t, \epsilon)$).

In order to determine $q_0(t)$, it suffices that we consider this one-parameter family of possible trajectories about $q_0(t)$. As a direct consequence, the action functional receives the expression

$$S[q, \epsilon] = \int_{t_1}^{t_2} L\big(q(t, \epsilon), \dot{q}(t, \epsilon), t\big) dt$$

which clearly renders it a function of ϵ. Since it becomes stationary at $\epsilon = 0$, it satisfies

$$\frac{dS}{d\epsilon}_{|\epsilon=0} = 0$$

Note now that since $S[q, \epsilon]$ is differentiated with respect to ϵ whereas $S[q, \epsilon]$ itself is an integral over t, $\frac{dS}{d\epsilon}$ remains invariant if the order between differentiation and integration is reversed. For that matter, the last two relations imply that

$$\frac{d}{d\epsilon}_{|\epsilon=0} \int_{t_1}^{t_2} L\big(q(t, \epsilon), \dot{q}(t, \epsilon), t\big) dt = 0 \iff \int_{t_1}^{t_2} \frac{d}{d\epsilon}_{|\epsilon=0} L\big(q(t, \epsilon), \dot{q}(t, \epsilon), t\big) dt = 0 \iff$$

$$\int_{t_1}^{t_2} \Big[\frac{\partial L}{\partial[q_0(t) + \epsilon\eta(t)]}\Big|_{\epsilon=0} \frac{d[q_0(t) + \epsilon\eta(t)]}{d\epsilon}\Big|_{\epsilon=0} + \frac{\partial L}{\partial[\dot{q}_0(t) + \epsilon\dot{\eta}(t)]}\Big|_{\epsilon=0} \frac{d[\dot{q}_0(t) + \epsilon\dot{\eta}(t)]}{d\epsilon}\Big|_{\epsilon=0}\Big] dt = 0$$

$$\iff \int_{t_1}^{t_2} \Big[\frac{\partial L}{\partial q_0(t)}\eta(t) + \frac{\partial L}{\partial \dot{q}_0(t)} \frac{d\eta(t)}{dt}\Big] dt = 0 \iff$$

$$\int_{t_1}^{t_2} \Big[\frac{\partial L}{\partial q_0(t)}\eta(t) + \frac{d}{dt}\big(\frac{\partial L}{\partial \dot{q}_0(t)}\eta(t)\big) - \big(\frac{d}{dt}\frac{\partial L}{\partial \dot{q}_0(t)}\big)\eta(t)\Big] dt = 0 \iff$$

$$\frac{\partial L}{\partial \dot{q}_0(t)}\eta(t)\Big|_{t_1}^{t_2} + \int_{t_1}^{t_2} \Big[\frac{\partial L}{\partial q_0(t)} - \frac{d}{dt}\frac{\partial L}{\partial \dot{q}_0(t)}\Big]\eta(t) dt = 0 \iff$$

$$\frac{\partial L}{\partial \dot{q}_0(t)}\eta(t_2) - \frac{\partial L}{\partial \dot{q}_0(t)}\eta(t_1) + \int_{t_1}^{t_2} \Big[\frac{\partial L}{\partial q_0(t)} - \frac{d}{dt}\frac{\partial L}{\partial \dot{q}_0(t)}\Big]\eta(t) dt = 0 \implies$$

$$\frac{\partial L}{\partial q_0(t)} - \frac{d}{dt}\frac{\partial L}{\partial \dot{q}_0(t)} = 0$$

The result (in the last line) of this brief calculation is the Euler-Lagrange equation for the generalized classical trajectory $q_0(t)$. In deriving it we have proceeded as follows:

(i) Since $\dot{q}(t)$ acts as an independent variable in $L(q(t), \dot{q}(t), t)$, the total derivative

$$\frac{d}{d\epsilon} L(q(t, \epsilon), \dot{q}(t, \epsilon), t)$$

concerns both, $q(t, \epsilon)$ and $\dot{q}(t, \epsilon)$. Considering that $df = \frac{df}{dx}dx$ generalizes to

$$df = \frac{\partial f}{\partial x_1} dx_1 + \frac{\partial f}{\partial x_2} dx_2$$

for a two-variable function $f(x_1, x_2)$ - with the *partial derivative* $\frac{\partial f}{\partial x_1}$ signifying differentiation of $f(x_1, x_2)$ with respect to x_1 keeping x_2 constant and with the partial derivative $\frac{\partial f}{\partial x_2}$ signifying differentiation of $f(x_1, x_2)$ with respect to x_2 keeping x_1 constant - the total derivative $\frac{d}{d\epsilon} L(q(t,\epsilon), \dot{q}(t,\epsilon), t)$ consists in two terms. They are the two terms in the second line.

(ii) Considering that for a *composite function* $f(x(u))$ (its independent variable x is itself a function of u), it is

$$df = \frac{df}{dx} \frac{dx}{du} du$$

(a direct consequence of $df = \frac{df}{dx} dx$) as a result of which the *chain rule of differentiation*

$$\frac{df}{du} = \frac{df}{dx} \frac{dx}{du}$$

follows, the total derivative of L with respect to ϵ is

$$\frac{d}{d\epsilon} L(q(t,\epsilon), \dot{q}(t,\epsilon), t) = \frac{\partial L}{\partial q} \frac{dq}{d\epsilon} + \frac{\partial L}{\partial \dot{q}} \frac{d\dot{q}}{d\epsilon}$$

This is manifest in the second line.

(iii) In view of the elementary rule for differentiating the product of two functions:

$$\frac{d}{dx}(f(x)g(x)) = (\frac{df}{dx})g(x) + (\frac{dg}{dx})f(x) \iff (\frac{df}{dx})g(x) = \frac{d}{dx}(f(x)g(x)) - (\frac{dg}{dx})f(x)$$

it follows that

$$\frac{\partial L}{\partial \dot{q}_0(t)} \frac{d\eta(t)}{dt}$$

in the third line, is equal to

$$\frac{d}{dt}(\frac{\partial L}{\partial \dot{q}_0(t)} \eta(t)) - \frac{d}{dt}(\frac{\partial L}{\partial \dot{q}_0(t)})\eta(t)$$

in the fourth line.

(iv) From the *Fundamental Theorem of Calculus*:

$$\int_a^b (\frac{df}{dx}) dx = f(b) - f(a)$$

(heuristically: $\int dx$ "offsets" $\frac{df}{dx}$), it follows that

$$\int_{t_1}^{t_2} dt \frac{d}{dt}(\frac{\partial L}{\partial \dot{q}_0(t)} \eta(t)) = \frac{\partial L}{\partial \dot{q}_0(t)} \eta(t_2) - \frac{\partial L}{\partial \dot{q}_0(t)} \eta(t_1)$$

resulting in the first term of the fifth line and, equivalently, in the difference of the first two terms of the sixth line.

(v) Since $\eta(t_2) = \eta(t_1) = 0$, each of the two terms on the right side of the equality in (iv) is equal to zero.

(vi) Since $\eta(t)$ satisfies the condition $\eta(t_1) = \eta(t_2) = 0$ but is in all other respects arbitrary, the sixth line implies the result in the seventh, the Euler-Lagrange equation.

The one-parameter family of trajectories $q(t, \epsilon) = q_0(t) + \epsilon\eta(t)$ defines the *variation* $\delta q(t)$ *of* $q(t)$ *about the trajectory* $q_0(t)$ which extremizes the action functional $S[q]$. This

$$\delta q(t) = q(t, \epsilon) - q_0(t) \equiv \epsilon\eta(t)$$

determines, in turn, the *functional derivative*:

$$\frac{\delta S[q]}{\delta q}\Big|_{q=q_0} \equiv \frac{dS}{d\epsilon}\Big|_{\epsilon=0}$$

This specifies the concrete mathematical content of our earlier qualitative statement to the effect that *a stationary point of the action functional is a point at which the functional derivative vanishes* and rigorously proves that the stationary point of the action functional in Fig.9 is indeed (q_0, S_0) with the minimum action S_0 being the action which corresponds to the classical trajectory $q_0(t)$, the *classical action*.

It is now a straightforward matter to generalize the Euler-Lagrange equation which we derived for a one-dimensional dynamical system to a dynamical system of an arbitrary number of generalized coordinates. Following the same procedure, we consider the one-parameter families of possible trajectories

$$q_i(t, \epsilon) = q_{0_i}(t) + \epsilon\eta_i(t) \quad ; \quad i = 1, 2, ...n$$

where, again, the functions $q_{0_i}(t)$ correspond to the stationary value of the action functional in (1.III) and the functions $\eta_i(t)$ are arbitrary and vanishing at t_1 and t_2. As a consequence, it is

$$\frac{dS}{d\epsilon}\Big|_{\epsilon=0} = 0 \Longleftrightarrow \int_{t_1}^{t_2} \Big[\frac{\partial L}{\partial q_{0_i}(t)} - \frac{d}{dt}\frac{\partial L}{\partial \dot{q}_{0_i}(t)}\Big]\eta_i(t)dt = 0$$

from which the Euler-Lagrange equations

$$\frac{\partial L}{\partial q_{0_i}(t)} - \frac{d}{dt}\frac{\partial L}{\partial \dot{q}_{0_i}(t)} = 0 \quad ; \quad i = 1, 2, ...n \tag{$E-L$}$$

for a dynamical system of an arbitrary number of degrees of freedom directly follow.

We have, thus, established the classical Euler-Lagrange equations of motion as a direct consequence of Hamilton's principle. It is important to note that the latter results in the equations of motion (E - L) only, not in the solution which specifies the classical trajectory. The general solution to (E - L) signifies a set of possible classical trajectories in configuration space. Which of them is realized depends on the initial conditions imposed on these differential equations. Inherent in Hamilton's principle to the effect that the trajectory which is realized is the trajectory which extremizes the action functional is, in fact, the existence of specified <u>external</u> (i.e. irrelevant to the action functional) conditions for the positions and the momenta at a certain initial moment. We shall return to this point.

An aspect of the action functional whose importance cannot be overstated is its relation to the symmetries of the physical system. The latter are <u>necessarily</u> the symmetries of the action functional itself. Such symmetries imply the existence of quantities which remain invariant in the course of the physical system's motion. It is, for that matter, necessary that we examine the crucial correspondence between the symmetries of S and the conserved quantities which characterize the dynamical behavior of the physical system, those attributes of the latter which remain invariant in the course of time. We have, in the first chapter, stated and stressed the three most fundamental conservation principles, namely that (i) time-translational invariance implies energy conservation, (ii) space-translational invariance (the homogeneity of space) implies momentum conservation and (iii) rotational invariance (the isotropy of space) implies angular-momentum conservation. In order to illustrate *Nöther's theorem*[40] to the effect that to *each continuous symmetry of the action functional corresponds a conservation law*, we consider the action functional for a free classical particle:

$$S[x] = \int_{t_1}^{t_2} dt \frac{1}{2} m \dot{x}^2(t)$$

where, as always, $\dot{x}(t) \equiv \frac{dx}{dt}$. Hamilton's principle then implies:

$$\frac{\delta S[x]}{\delta x}\bigg|_{x=x_0} \equiv \frac{dS}{d\epsilon}\bigg|_{\epsilon=0} = 0 \iff \int_{t_1}^{t_2} dt \frac{1}{2} m \frac{d}{d\epsilon}\bigg|_{\epsilon=0} \left(\dot{x}(t) + \epsilon \dot{\eta}(t)\right)^2 = 0 \iff$$

$$\int_{t_1}^{t_2} dt\, m\dot{x}(t) \frac{d}{dt}\eta(t) = 0 \iff \int_{t_1}^{t_2} dt \frac{d}{dt}\left(m\dot{x}(t)\eta(t)\right) - \int_{t_1}^{t_2} dt \frac{d}{dt}\left(m\dot{x}(t)\right)\eta(t) = 0$$

where, as always, it is $\eta(t_1) = \eta(t_2) = 0$. In advancing from the first to the second line, we have applied the elementary rule of differentiation to the effect that

$$\frac{dx^a}{dx} = ax^{a-1}$$

(a is a real number) followed by the chain rule and subsequently by the rule for the differentiation of the product of two functions. In turn, it is:

$$m\dot{x}(t)\eta(t)\big|_{t_1}^{t_2} - \int_{t_1}^{t_2} dt\, m\ddot{x}(t)\eta(t) = 0$$

where $\ddot{x}(t) \equiv \frac{d^2x}{dt^2}$.

Since now the first term on the left side of this equality trivially vanishes, it is

$$\int_{t_1}^{t_2} dt\, m\ddot{x}(t)\eta(t) = 0 \iff m\ddot{x}(t) = 0$$

As expected, this is merely the equation of motion for a free particle. We readily recognize it as the immediate consequence of Newton's second law to the effect that in the absence

---

[40]Amalie Emmy Nöther (1882 - 1935): *The most important woman in the history of Mathematics* according to A. Einstein, H. Weyl, P. Alexandrov and other prominent physicists and mathematicians of her time, was a German mathematician and physicist with significant contributions in both fields.

of any force, the acceleration $\ddot{x}(t)$ of the particle is zero: The homogeneity of empty space $(V(x) = 0)$ implies momentum conservation!

By way of further illustrating Hamilton's principle, we concretize these considerations in the more involved context of the simple classical harmonic oscillator which we specify here as two (rather than only one) particles of equal mass m respectively attached to the two ends of a freely oscillating spring. The displacement x of either particle from the point of equilibrium immediately determines the displacement of the other as $-x$. For that matter, there is only one degree of freedom described by the Cartesian coordinate x. The oscillating particle whose trajectory is $x(t) \equiv q(t)$, is subjected to acceleration $\vec{a}$ under the influence of force $\vec{F}$ according to Newton's second law $\vec{F} = m\vec{a}$. The oscillating motion is a consequence of the fact that the restoring force $\vec{F}$ is opposite in direction to the direction of the particle's displacement $\vec{x}$ from the point of equilibrium:

$$\vec{F} = -k\vec{x}$$

with *spring constant* k characterizing the spring. Since $\vec{F} = F(x)\vec{u}_1$ and $\vec{x} = x\vec{u}_1$, where $\vec{u}_1$ is the unit vector of the x-axis, the preceding relation may simply be stated as

$$F = -kx$$

with the minus sign always signifying the opposition between the direction of the restoring force and that of the displacement. For that matter:

$$\vec{F} = m\vec{a} \Rightarrow -kx = m\frac{d^2x}{dt^2}$$

that is

$$m\ddot{x} + kx = 0 \tag{1.V}$$

This is the classical equation of motion for the simple harmonic oscillator (equally valid for only one oscillating particle) derived through Newton's second law. We shall now use it in order to construct the action functional for this dynamical system. In turn, that action functional will independently result in (1.V).

We know that the potential for simple harmonic motion is the symmetric "parabolic" potential in Fig.4. A parabolic curve is the curve of a quadratic function, of a function the maximium power of whose independent variable is two. The parabolic potential for simple harmonic motion is indeed such a function:

$$V(x) = \frac{1}{2}kx^2 \tag{1.VIa}$$

At once, the kinetic energy of the simple harmonic oscillator is trivially equal to the non-relativistic $\frac{1}{2}mv^2$:

$$T(\dot{x}) = \frac{1}{2}m\dot{x}^2 \tag{1.VIb}$$

In turn, pursuant to (1.IV), the Lagrange function is

$$L = \frac{1}{2}m\dot{x}^2 - \frac{1}{2}kx^2 \qquad (1.VII)$$

with the direct consequence that the action functional for the simple harmonic oscillator is

$$S[x] = \int_{t_1}^{t_2} \frac{1}{2}\left(m\dot{x}^2 - kx^2\right)dt \qquad (1.VIII)$$

The graph of functional $S[x]$ in (1.VIII) is a parabolic curve. Note that initial conditions for the position and the momentum of either particle determine the position and the momentum of the other. Given such <u>external</u> initial conditions for that matter, the classical trajectory of the simple harmonic oscillator is that $x(t) = x_0(t)$ for which

$$\left. \frac{\delta S[x]}{\delta x} \right|_{x=x_0} = 0$$

The action which corresponds to that $x_0(t)$ is clearly the minimum value of $S[x]$. That, is the classical action $S_0[x_0]$. The variation δx of possible trajectories $x(t)$ about $x_0(t)$ yields a vanishing derivative of the action functional, a vanishing rate of change of the latter as the possible trajectories fluctuate about the classical $x_0(t)$. The Euler-Lagrange equation directly follows from (E - L):

$$\left. \frac{\delta S[x]}{\delta x} \right|_{x=x_0} = 0 \iff \frac{d(-V(x))}{dx} - \frac{d}{dt}\frac{dT(\dot{x})}{d\dot{x}} = 0 \qquad (1.IX)$$

As we shall see very soon, the physical essence of the stated δx is the quantum fluctuations of the trajectory about $x_0(t)$. In fact, the latter emerges as the classical trajectory at the limit of vanishing quantum fluctuations which, as we already know, formally corresponds to $\hbar \to 0$. This physical content was, of course, utterly unknown to William R. Hamilton when he introduced his famous principle in 1827 and to the generations of physicists over the subsequent hundred years. We reiterate that the Euler-Lagrange equations which we derived in detail are the direct consequence of Hamilton's principle

$$\left. \frac{\delta S}{\delta q} \right|_{q=q_0} = 0$$

in both, mechanical systems and fields. They are the classical equations of motion. By way of demonstrating this in the simple case of the harmonic oscillator, we advance the calculation which (1.IX) signifies.

From (1.VI) and the above-stated elementary rule of differentiation

$$\frac{dx^a}{dx} = ax^{a-1}$$

we obtain:

$$\frac{d(-V(x))}{dx} = -\frac{1}{2}k\frac{dx^2}{dx} = -kx \qquad (1.X)$$

Likewise, from (1.VIb) we obtain:

$$\frac{dT(\dot{x})}{d\dot{x}} = \frac{1}{2}m\frac{d\dot{x}^2}{d\dot{x}} = m\dot{x}$$

so that

$$\frac{d}{dt}\frac{dT(\dot{x})}{d\dot{x}} = m\ddot{x} \qquad (1.XI)$$

Replacing (1.X) and (1.XI) into (1.IX) we obtain (1.V), as expected.

Before we turn to the central significance which the action functional has in quantum theory, we stress that we have not derived the action functional from the classical equations of motion, only in order to derive the classical equations of motion from the action functional! The action functional precedes the classical equations of motion and Newton's laws despite the fact that the already known equations of motion can be used as a guide for the construction of the relevant action functional. Specifically, we could have constructed the action functional in (1.VIII) on purely dimensional arguments - that is, merely by exploiting the fact that the action functional has neither dimensions of length, nor dimensions of mass - with absolutely no reference to (1.V), then use Hamilton's principle in order to derive (1.V) from (1.VIII). We shall, in fact, demonstrate this procedure of *dimensional analysis* in the fifth chapter. In this respect, we also stress that the significance of the action functional by far transcends that of the mathematical device which yields the classical equations of motion. It is literally the central physical entity which conveys the dynamical behavior of a quantum system. As such, the action functional also captures the semi-classical limit and associated classical equations of motion. It is, in fact, impossible to consistently formulate quantum field theory with its inherent divergences exclusively as the result of the canonical quantization of the classical field equations of motion. A consistent formulation of quantum field theory is impossible without the action functional. In modern theories of quantum gravity, especially in string theory, it is impossible to even contemplate the classical equations of motion as the incipient point since the latter are unknown at the outset. The relevant action functional is a consequence of inherent physical symmetries (for example, proportional to the length of the cosmic line, or to the area of the world-sheet). It is on account of this physical significance that not only is the action functional of central importance in quantum theory, it lies at the very foundation of every physical theory.

2. The Transition Amplitude and the Path Integral

As we have already demonstrated, the proper specification of initial positions and initial momenta in a system of classical particles determines a unique trajectory for each of them. This is a direct consequence of the fact that initial conditions concerning the position and the momentum imposed on the equations of motion specify a unique position and a unique momentum for each particle at each subsequent moment. It follows, for that matter, that the transition from classical to quantum mechanics - which in canonical quantization is accomplished by the replacement at each moment of each unique value of position and momentum by probability distributions for these two observable quantities respectively - can be equivalently accomplished by replacing the unique classical trajectory by infinitely many possible *paths* connecting two arbitrarily specified points at two moments (Fig.10). Such an equivalent transition from classical to quantum mechanics - based on paths, as opposed to positions - signifies the *path-integral quantization*. The mathematical and physical content of this procedure and of the term *path* on which it is based will be the object of our analysis in this chapter. In this respect we stress from the outset that, unlike the observable probabilities for position in canonical quantization, neither is any of the infinitely many possible paths in path-integral quantization observable, nor for that matter is it associated with observable probabilities. The observable quantity is the probability that the quantum particle transit between the specified two points in space within a specified time interval.

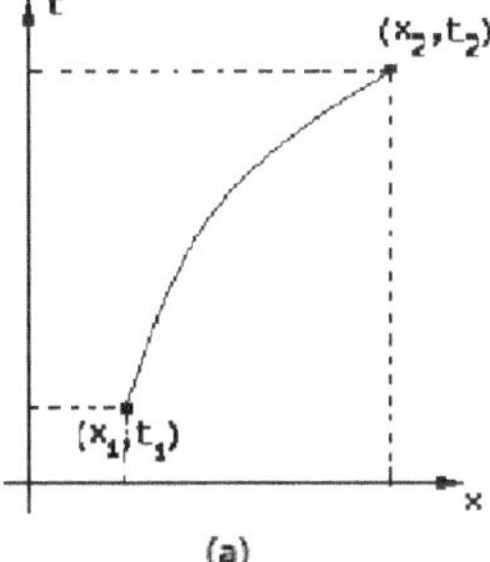
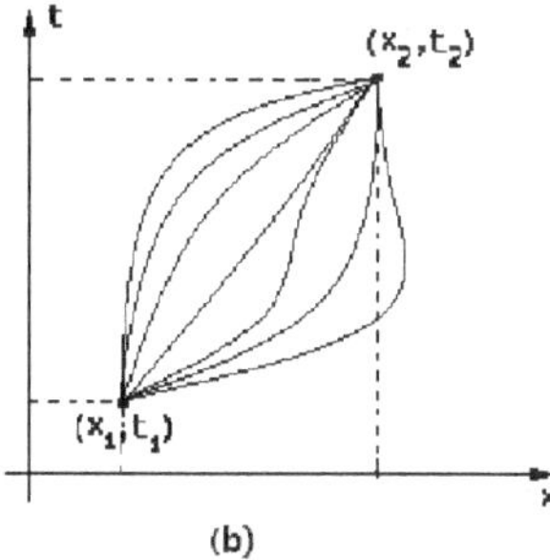

FIGURE 10. In the classical theory a particle follows a unique observable trajectory $x(t)$ from point x_1 at moment t_1 to point x_2 at moment t_2 (a). In the quantum theory the possible paths allowed are the infinitely many <u>unobservable</u> paths (including the path of the classical theory) which intercept x_1 and x_2 at moments t_1 and t_2 respectively. Each of those paths individually contributes to the probability that the quantum particle transit from x_1 at t_1 to x_2 at t_2 (b).

Such preliminary remarks as we have just made must not be interpreted as a statement to the effect that path-integral quantization is relevant only in quantum mechanics. The procedure of path-integral quantization is as central to quantum field theory as it is to quantum mechanics. In fact, it is in relativistic quantum field theory where manifest Lorentz invariance is an enormous advantage that the path-integral formulation transcends the level of mere equivalence to the - prohibitively tedious - canonical procedure. So much so that for all intents and purposes modern theories of quantum gravity such as superstring theory can only receive a path-integral formulation. In our study of quantum field theory we shall see that the path-integral formulation concerns the transition of a field from a specified field configuration on a three-dimensional surface of equal time to a specified field configuration on a three-dimensional surface of equal time. Here, we only remark that applied to physical systems of infinitely many degrees of freedom as fields are, the path-integral formulation is inevitably infinitely more involved than it is in quantum mechanics.

The *path integral* in quantum mechanics is the <u>continuous sum</u> over all possible paths which a quantum particle may follow between two given points specified at two moments. It is a highly non-trivial mathematical object of infinitely higher complexity than that of multiple integration in advanced calculus and equally infinite complexity associated with its rigorous definition. Nevertheless, it is not particularly difficult to understand as a concept. The ensuing approach combines mathematical rigor with a qualitative analysis of the physical concepts involved. As such, it will reveal the physical essence of the path integral without compromising its mathematical complexity. Interested readers can find a comprehensive and accessible treatment of this subject matter in [28], [31] and [32]. For the sake of simplicity, we continue here with our convention of a one-dimensional physical context on the understanding that the results we shall obtain can be directly generalized to the physical three-dimensional space.

The incipient point of our analysis is the qualitative observation that the description of a quantum particle as an evolving probability distribution for position is equivalent to the description of the quantum particle as a physical system with a predictable probability for transition from point x to point x'. For that matter, following Richard Feynman, we advance the following postulate: *The probability that a quantum particle transit from point x at moment t to point x' at moment t' is the square $|\langle x', t'|x, t\rangle|^2$ of the magnitude of function $\langle x', t'|x, t\rangle$.* Inherent in the concept of motion is the concept of trajectory. Since the concept of a coherent observable trajectory is utterly irrelevant to quantum particles, the implication in the above postulate is the set of infinitely many unobservable paths intercepting point x at moment t and point x' at moment t'. Before we establish the *transition amplitude* $\langle x', t'|x, t\rangle$ as a path integral however, we must elucidate what exactly we mean in stating that a particle transits between these two points. We recall in this respect that, pursuant to (3.VI) in the first chapter, at initial moment t wavefunction $\psi(t, x)$ is a superposition of position eigenstates each of which evolves <u>independently</u> of all the others. The statement concerning the transition of the quantum particle from (x, t) to (x', t') refers to an arbitrary but fixed once chosen position eigenstate $|x, t\rangle$, realized in the Hilbert space of the position operator $\hat{x}$ as $\delta_t(a - x)$, to the effect that *if the position of the quantum particle is point x at initial moment t then the probability that its position be point x' at future moment t' is determined <u>exclusively</u> by the evolution of $|x, t\rangle$*[41]. Recall in this respect that if the Hamiltonian operator $\hat{H}$ has no time dependence, the evolution operator $\hat{U}(t' - t) = e^{-\frac{i}{\hbar}\hat{H}(t'-t)}$ advances a state from moment t to moment t':

---

[41]Again, this does not imply that the particle itself is at (x, t). Instead, each x at t is a possible position, with the corresponding $|x, t\rangle$ determining the probability for position at (x', t') independently of all other $|x, t\rangle$. Note also that in $\delta_t(a - x)$, the position of the quantum particle is at $a = x$.

$$\psi(t', x) = e^{-\frac{i}{\hbar}\hat{H}(t'-t)}\psi(t, x) \tag{2.I}$$

Applying this to the initial position eigenstate $|x, t\rangle$, has the following direct implication for the transition amplitude:

$$\langle x', t'|x, t\rangle = \langle x'|e^{-\frac{i}{\hbar}\hat{H}(t'-t)}|x\rangle \tag{2.II}$$

that is: (i) the evolution operator advances any arbitrary, but fixed once chosen, position eigenstate $|x, t\rangle$ at initial moment t to a quantum state at final moment t' and (ii) the transition amplitude $\langle x', t'|x, t\rangle$ is consequent upon the evolution of $|x, t\rangle$ to a quantum state at future moment t'. Which quantum state? We give a specific example immediately below. For our purposes now, it suffices to know (as we already do) that, pursuant to (2.I), it is a smooth probability distribution described by some wavefunction $\tilde{\psi}(t', x)$ specific to the particular $\hat{H}$ of the quantum system. Considering that - pursuant to the eleventh section of the first chapter - $\langle x', t'|$ is realized as

$$\int_{-\infty}^{+\infty} dx \delta_{t'}(x - x')(\ )$$

in the Hilbert space of $\hat{x}$, it is:

$$\langle x', t'|x, t\rangle = \int_{-\infty}^{+\infty} dx \delta_{t'}(x - x')\tilde{\psi}(t', x) = \tilde{\psi}(t', x') \equiv \langle x', t'|\tilde{\psi}\rangle \tag{2.III}$$

which clearly reveals that the bra-state $\langle x', t'|$ in $\langle x', t'|x, t\rangle$ "picks" the value $\tilde{\psi}(t', x = x')$, so that $\langle x', t'|x, t\rangle$, *always for a given position x at initial moment t, is the probability amplitude $\tilde{\psi}(t', x')$ at a given point x' and final moment t'.* The transition probability from (x, t) to (x', t'), for that matter, is the probability that at future moment t' the position of the quantum particle be point x': $|\langle x', t'|x, t\rangle|^2 = |\tilde{\psi}(t', x')|^2$. Note that (2.III) is consistent with the definition of the Dirac bra-ket which we gave in the eleventh section of the first chapter. This justifies the notation $\langle x', t'|x, t\rangle$ for the transition amplitude from (x, t) to (x', t').

In order to illustrate the essence of (2.III) in a specific physical context, we consider the transition of a free quantum particle from point x at initial moment t to point x' at final moment t'. The evaluation of the transition amplitude $\langle x', t'|x, t\rangle$, necessitates the introduction of a complete set of momentum eigenstates $|p\rangle$ (plane waves of probability) which in view of

$$\int_{-\infty}^{+\infty} da\left(\frac{1}{\sqrt{2\pi\hbar}}e^{-\frac{i}{\hbar}pa}\right)\delta(x - a) = \frac{1}{\sqrt{2\pi\hbar}}e^{-\frac{i}{\hbar}px}$$

satisfy

$$\langle p|x\rangle = \frac{1}{\sqrt{2\pi\hbar}}e^{-\frac{i}{\hbar}px} \tag{2.IV\,a}$$

and likewise

$$\langle x|p\rangle = \frac{1}{\sqrt{2\pi\hbar}}e^{\frac{i}{\hbar}px} \tag{2.IVb}$$

In view of $\hat{H} = \frac{\hat{p}^2}{2m}$ for a free quantum particle and the theory which we introduced in the first chapter, we note that

$$\langle x',t'|x,t\rangle = \langle x'|e^{-\frac{i}{\hbar}\hat{H}(t'-t)}|x\rangle = \langle x'|e^{-\frac{i}{\hbar}\frac{\hat{p}^2}{2m}(t'-t)}|x\rangle = \int_{-\infty}^{+\infty} dp\langle x'|e^{-\frac{i}{\hbar}\frac{\hat{p}^2}{2m}(t'-t)}|p\rangle\langle p|x\rangle =$$

$$\int_{-\infty}^{+\infty} dpe^{-\frac{i}{\hbar}\frac{p^2}{2m}(t'-t)}\langle x'|p\rangle\langle p|x\rangle = \frac{1}{2\pi\hbar}\int_{-\infty}^{+\infty} dpe^{-\frac{i}{\hbar}\frac{p^2}{2m}(t'-t)}e^{\frac{i}{\hbar}px'}e^{-\frac{i}{\hbar}px} =$$

$$\frac{1}{2\pi\hbar}\int_{-\infty}^{+\infty} dpe^{-\frac{i}{\hbar}\frac{p^2}{2m}(t'-t)}e^{\frac{i}{\hbar}p(x'-x)} = \sqrt{\frac{m}{2\pi i\hbar(t'-t)}}e^{\frac{i}{\hbar}\frac{m}{2(t'-t)}(x'-x)^2} \tag{2.V}$$

where in order to arrive at the last equality we applied the technique of "completing the square" in the exponent followed by that of *Gaussian integration*. Both techniques are quite simple really, easily conveyed over the space of a few more lines. Nevertheless, in view of the rather demanding analysis in the present chapter and the fact that it is the final result in (2.V) which illustrates the essence of (2.III), it is probably best not to be unnecessarily ambitious at this stage. For this reason, we only digress, for the sake of completeness, to a brief illustration of these techniques without particular commitment to the elucidation of the mathematical operations involved. Readers who find it hard to follow may disregard this digression. Do note, however, that such techniques are indispensable in quantum field theory and will, for that matter, inevitably constitute an integral aspect of our analysis at an advanced stage.

Using the elementary identity $(a-b)^2 = a^2 - 2ab + b^2$, we recast the exponent on the left side of the last equality:

$$-\frac{i}{\hbar}\Big[\frac{t'-t}{2m}p^2 - p(x'-x)\Big] = -\frac{i}{\hbar}\Big[\sqrt{\frac{t'-t}{2m}}p - \frac{1}{2}\sqrt{\frac{2m}{t'-t}}(x'-x)\Big]^2 + \frac{i}{\hbar}\frac{m}{2(t'-t)}(x'-x)^2$$

Since the last term on the right side of this relation is independent of p, this relation recasts the integral on the left side of the last equality in (2.V) as

$$e^{\frac{i}{\hbar}\frac{m}{2(t'-t)}(x'-x)^2}\int_{-\infty}^{+\infty} dpe^{-\frac{i}{\hbar}\Big[\sqrt{\frac{t'-t}{2m}}p - \frac{1}{2}\sqrt{\frac{2m}{t'-t}}(x'-x)\Big]^2}$$

Clearly, this integral (continuous sum) remains invariant if we multiply and divide p in dp by the constant factor $\sqrt{\frac{t'-t}{2m}}$ and further subtract the constant $\frac{1}{2}\sqrt{\frac{2m}{t'-t}}(x'-x)$:

$$\int_{-\infty}^{+\infty} dpe^{-\frac{i}{\hbar}\Big[\sqrt{\frac{t'-t}{2m}}p - \frac{1}{2}\sqrt{\frac{2m}{t'-t}}(x'-x)\Big]^2} =$$

$$\sqrt{\frac{2m}{t'-t}}\int_{-\infty}^{+\infty} d\Big[\sqrt{\frac{t'-t}{2m}}p - \frac{1}{2}\sqrt{\frac{2m}{t'-t}}(x'-x)\Big]e^{-\frac{i}{\hbar}\Big[\sqrt{\frac{t'-t}{2m}}p - \frac{1}{2}\sqrt{\frac{2m}{t'-t}}(x'-x)\Big]^2}$$

This integral appears to be awfully complicated! That this is not really the case can be seen by setting

$$\left[\sqrt{\frac{t'-t}{2m}}\,p - \frac{1}{2}\sqrt{\frac{2m}{t'-t}}(x'-x)\right] = s$$

which renders the left side of the last equality in (2.V) as:

$$\frac{1}{2\pi\hbar}\int_{-\infty}^{+\infty} dp\, e^{-\frac{i}{\hbar}\frac{p^2}{2m}(t'-t)}e^{\frac{i}{\hbar}p(x'-x)} = \frac{1}{2\pi\hbar}\sqrt{\frac{2m}{t'-t}}\,e^{\frac{i}{\hbar}\frac{m}{2(t'-t)}(x'-x)^2}\int_{-\infty}^{+\infty} ds\, e^{-\frac{i}{\hbar}s^2}$$

Such integration as that which the integral over s signifies is known as *Gaussian*. Through an ingenious stratagem[42], the latter yields:

$$\int_{-\infty}^{+\infty} ds\, e^{-as^2} = \sqrt{\frac{\pi}{a}}$$

Thus, with $a = \frac{i}{\hbar}$, the left side of the last equality in (2.V) yields the expression on the right side of the latter. This concludes our brief digression into this essential aspect of quantum theory's mathematical foundation.

What (2.V) "tells us" is that the position eigenstate $|x,t\rangle$ at initial moment t developed into the smooth probability distribution described by the wavefunction

$$\tilde{\psi}(t',x') = \sqrt{\frac{m}{2\pi i\hbar(t'-t)}}\,e^{\frac{i}{\hbar}\frac{m}{2(t'-t)}(x'-x)^2} \tag{2.VI}$$

and that, for that matter, the amplitude that the quantum particle transit from point x at initial moment t, to point x' at final moment t' is given by the above probability amplitude for position $\tilde{\psi}(t',x')$ (with x' and t' respectively fixed at the chosen values). Note that there is nothing in the above brief calculation which even hints at the initial position eigenstate $|x,t\rangle$ as the quantum state of the position observed at point x and moment t. We stress again that unless point x has been specified as the observed position at moment t, the position eigenstate $|x,t\rangle$ is merely one of the infinitely many position eigenstates whose superposition at initial moment t yields a wavefunction of the free quantum particle, a fact which implies that the transition amplitude from an initial to a final point is invariably associated with an initial and a final wavefunction. We shall elucidate this statement in the next section, where it will become clear that the transition amplitude propagates a given wavefunction of the specified quantum system into the future.

Having thus explained a quantum particle's transition from (x,t) to (x',t') as the direct consequence of position eigenstate's $|x,t\rangle$ evolution, we may now establish the character of the transition amplitude $\langle x',t'|x,t\rangle$ as the integral over the infinitely many unobservable paths which intercept point x at moment t and point x' at moment t'. For that matter, we consider an intermediate moment t_0 ; $t < t_0 < t'$ and note that

---

[42]For the mathematically-literate reader: For $a > 0$, it is $\left(\int_{-\infty}^{+\infty} ds\, e^{-as^2}\right)^2 = \int_{-\infty}^{+\infty} dx\, e^{-ax^2}\int_{-\infty}^{+\infty} dy\, e^{-ay^2} = \int_{-\infty}^{+\infty}\int_{-\infty}^{+\infty} dxdy\, e^{-a(x^2+y^2)} = \int_0^{2\pi}\int_0^{+\infty} d\theta d\rho\rho e^{-a\rho^2} = 2\pi\int_0^{+\infty} d\rho\rho e^{-a\rho^2} = -\frac{\pi}{a}\int_0^{-\infty} d(-a\rho^2)e^{-a\rho^2} = \frac{\pi}{a}\int_{-\infty}^{0} du\, e^u = \frac{\pi}{a} \Longrightarrow \int_{-\infty}^{+\infty} ds\, e^{-as^2} = \sqrt{\frac{\pi}{a}}$. This result is also valid for $a = i$.

146

$\langle x', t'|x, t\rangle$ is the sum-total of all contributions stemming from all points in space at t_0. The expression which readily conforms with intuition (we formally arrive at the rigorous result immediately below) is:

$$\langle x', t'|x, t\rangle = \langle x', t'|x_0, t_0\rangle\langle x_0, t_0|x, t\rangle + \langle x', t'|x_0', t_0\rangle\langle x_0', t_0|x, t\rangle + \langle x', t'|x_0'', t_0\rangle\langle x_0'', t_0|x, t\rangle + ...$$

That is, $\langle x', t'|x, t\rangle$ is the <u>continuous sum</u> over the amplitude $\langle x', t'|x_0, t_0\rangle\langle x_0, t_0|x, t\rangle$ that the quantum particle transit from point x at moment t to point x' at moment t' through point x_0 at moment t_0, the amplitude $\langle x', t'|x_0', t_0\rangle\langle x_0', t_0|x, t\rangle$ that the quantum particle transit from point x at moment t to point x' at moment t' through point x_0' at moment t_0, the amplitude $\langle x', t'|x_0'', t_0\rangle\langle x_0'', t_0|x, t\rangle$ that the quantum particle transit from point x at moment t to point x' at moment t' through point x_0'' at moment t_0, ... (Fig.11(a)):

$$\langle x', t'|x, t\rangle = \int_{-\infty}^{\infty} dx_0 \langle x', t'|x_0, t_0\rangle\langle x_0, t_0|x, t\rangle \qquad (2.VII)$$

which, in turn, directly implies the relation

$$\int_{-\infty}^{\infty} dx_0 |x_0, t_0\rangle\langle x_0, t_0| \equiv \int_{-\infty}^{\infty} dx_0 |x_0\rangle\langle x_0| = 1 \qquad (2.VIII)$$

valid at any given moment.

We have derived (2.VIII) from the qualitative standpoint of the contribution to $\langle x', t'|x, t\rangle$ which each point in space signifies at a given intermediate moment. (2.VIII) is, of course, the completeness relation (11.XIV) for the position eigenstates. What we have actually done is to *insert a complete set* of position eigenstates $|x_0\rangle$ at intermediate moment t_0 , precisely as stated in the eleventh section of the first chapter. Since we appreciate physical and mathematical rigor, we resign the above-stated correct but qualitative standpoint in favor of the formal statement to the effect that *we express $\langle x', t'|x, t\rangle$ in terms of a complete set of position eigenstates.*

We consider now the relation between $\langle x', t'|x, t\rangle$ and all intermediate moments. Upon dividing the temporal interval $t' - t$ into N minute temporal intervals ($N >> 1$), (2.VII) directly implies that

$$\langle x', t'|x, t\rangle =$$

$$\int_{-\infty}^{\infty} dx_{N-1}... \int_{-\infty}^{\infty} dx_1 \langle x', t'|x_{N-1}, t_{N-1}\rangle\langle x_{N-1}, t_{N-1}|x_{N-2}, t_{N-2}\rangle...\langle x_2, t_2|x_1, t_1\rangle\langle x_1, t_1|x, t\rangle$$

$$(2.IX)$$

As it stands, (2.IX) is the expression of $\langle x', t'|x, t\rangle$ as a multiple integral. The evaluation of this integral depends on what $\langle x', t'|x_{N-1}, t_{N-1}\rangle...\langle x_2, t_2|x_1, t_1\rangle\langle x_1, t_1|x, t\rangle$ is. In order to determine this expression, we note first that as long as the N temporal intervals are of a certain length, however minute, we have not really made any progress toward the evaluation of $\langle x', t'|x, t\rangle$ since infinitely many paths always connect <u>any point in space</u> x_i

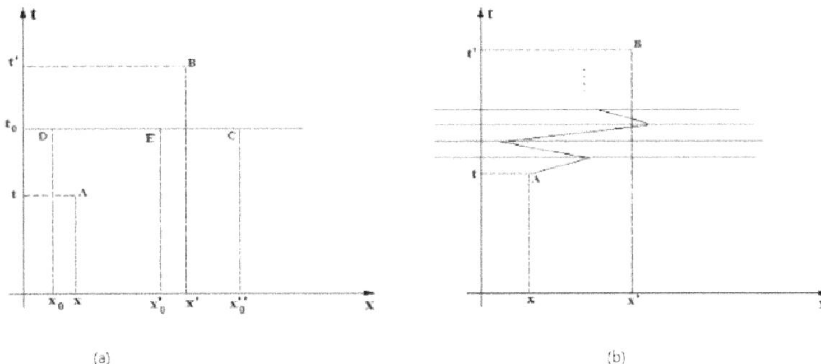

FIGURE 11. The transition amplitude between points A and B is the transition amplitude between A and B through intermediate point x_0 integrated throughout space at moment t_0 according to (2.VII) (a). The transition amplitude between A and B is the transition amplitude through $N \to \infty$ intermediate points integrated throughout space at successive moments in infinitesimal separation from each other. Each path in Fig.10(b) emerges as a continuous concatenation of the (necessarily straight) segments between successive moments (b).

at moment t_i with any point in space x_{i+1} at moment t_{i+1}, as can be seen in Fig.11(a). If, however, the temporal intervals are infinitesimally small so that $N \to \infty$ then any point x_i at t_i and any point x_{i+1} at t_{i+1} are necessarily connected by a single linear segment (Fig.11(b)). This is already an enormous advantage! For not only does it reveal that the transition amplitude is a continuous sum over all relevant paths, it reveals the precise sense in which this continuous sum is realized: *The transition amplitude* $\langle x', t' | x, t \rangle$ *naturally emerges as the path integral over all paths which intercept points x and x' respectively at moments t and t' at the limit at which all temporal intervals become infinitesimally small and $N \to \infty$.*

Before we proceed, we note that the most striking feature of the path integral is the infinite complexity of its *measure*. No longer are we dealing with a multiple integral. At the continuous limit $N \to \infty$, the multiple integral over x_1, x_2, ... x_{N-1} has given way to a continuous set of independent variables continuously integrated. Although, as we discussed in our analysis leading up to (2.IX), its emergence is easy to comprehend on grounds of physical consistency, the rigorous mathematical definition of the path integral is a formidable challenge. In order to hide our lack of knowledge as to its measure [28] we contrive the notation:

$$lim_{N\to\infty} \int_{-\infty}^{\infty} dx_{N-1}... \int_{-\infty}^{\infty} dx_2 \int_{-\infty}^{\infty} dx_1 \equiv \int \mathcal{D}[x] \qquad (2.X)$$

The expression to which the product of transition amplitudes in (2.IX) reduces at the continuous limit $N \to \infty$ is an expression established in retrospect by the mathematical and physical consistency of the result, rather than derived in advance from existing physical principles. We follow, for that matter, the course which Dirac and Feynman took and state that at the continuous limit $N \to \infty$, each transition amplitude between any point x_i at moment t_i and any point x_{i+1} at moment $t_i + \delta t$ differing from t_i by an infinitesimal lapse $\delta t \to 0$ is:

$$\langle x_{i+1}, t_i + \delta t | x_i, t_i \rangle = Ae^{-\frac{i}{\hbar}\delta t L(x_i, x_{i+1})} \qquad (2.XI)$$

with the Lagrange function L being a function of $x_i(t_i)$ and $x_{i+1}(t_{i+1}) \equiv x_{i+1}(t_i + \delta t)$ and with A being a constant. If we accept the validity of this statement for the moment then in view of $\delta x_i = \frac{\delta x_i}{\delta t}\delta t \equiv \dot{x}_i \delta t$ and $x_{i+1}(t_i + \delta t) = x_i + \dot{x}_i \delta t$, it follows that at $N \to \infty$ the product $\langle x', t' | x_{N-1}, t_{N-1} \rangle ... \langle x_2, t_2 | x_1, t_1 \rangle \langle x_1, t_1 | x, t \rangle$ in (2.IX) becomes[43]

$$e^{-\frac{i}{\hbar}\delta t L(x_{N-1}, x_{N-1} + \dot{x}_{N-1}\delta t)}...e^{-\frac{i}{\hbar}\delta t L(x_1, x_1 + \dot{x}_1\delta t)}e^{-\frac{i}{\hbar}\delta t L(x, x + \dot{x}\delta t)} =$$

$$e^{-\frac{i}{\hbar}\delta t(L(x_{N-1}, x_{N-1} + \dot{x}_{N-1}\delta t) + ... + L(x_1, x_1 + \dot{x}_1\delta t) + L(x, x + \dot{x}\delta t))} = e^{-\frac{i}{\hbar}\int_t^{t'} dt L(x, \dot{x})}$$

so that (2.IX), (2.X) and the above statement imply that

$$\langle x', t' | x, t \rangle = \int \mathcal{D}[x] e^{-\frac{i}{\hbar}S(t, t', [x])} \qquad (2.XII)$$

with

$$S(t, t', [x]) = \int_t^{t'} dt L(x, \dot{x}) \qquad (2.XIII)$$

being, of course, the action functional.

Before we comment on (2.XII), do note that had interpreted $e^{-\frac{i}{\hbar}S(t, t', [x])}$ as the transition amplitude $\langle x', t' | x, t \rangle$ - that is, over finite temporal separations in sharp contrast to (2.XI) - we would have contradicted the fundamental completeness relation (2.VIII) since, in view of

$$e^{-\frac{i}{\hbar}S(t, t', [x])} = e^{-\frac{i}{\hbar}(S(t_{N-1}, t', [x]) + S(t_{N-2}, t_{N-1}, [x]) + S(t_{N-3}, t_{N-2}, [x]) + ...)} =$$

$$e^{-\frac{i}{\hbar}S(t_{N-1}, t', [x])}e^{-\frac{i}{\hbar}S(t_{N-2}, t_{N-1}, [x])}e^{-\frac{i}{\hbar}S(t_{N-3}, t_{N-2}, [x])}...$$

such an interpretation would be tantamount to stating that

$$\langle x', t' | x, t \rangle = \langle x', t' | x_{N-1}, t_{N-1} \rangle .... \langle x_2, t_2 | x_1, t_1 \rangle \langle x_1, t_1 | x, t \rangle$$

---

[43]The - omitted, but ever present - constant A^N is eventually absorbed in $\int \mathcal{D}[x]$.

which differs from the correct relation (2.IX) by the absence of the intermediate integrations. In postulating (2.XI) as the transition amplitude over infinitesimal temporal separations, we have succeded in expressing a single path between any two points x_i and x_{i+1} in terms of $\delta t L(x_i, \dot{x}_i)$. In turn, any continuous concatination of such paths over a finite temporal separation $t' - t$ (Fig.11(b)) is expressed by the continuous sum $S(t, t', [x])$ over $\delta t L(x_i, \dot{x}_i)$ with the inherent implication that the corresponding path $x(t)$ which emerges is itself expressed by the action $S(t, t', [x])$. This result is, clearly, in perfect conformity with the physical content of the action functional (to each trajectory corresponds an action). For that matter, it is mathematically and physically consistent, thereby justifying our assumption as to the validity of (2.XI) and rendering (2.XII) the announced expression of the transition amplitude $\langle x', t'|x, t\rangle$ as the *Feynman path integral*, with $e^{-\frac{i}{\hbar}S(t,t',[x])}$ being the contribution of path $x(t)$ to $\langle x', t'|x, t\rangle$.

By way of a brief digression, we note at this point that if we wanted to be altogether consistent with our notation, we should keep the classical generalized coordinate $q(t)$ and accordingly write $S(t, t', [q])$, $L(q(t), \dot{q}(t), t))$ and $\langle q', t'|q, t\rangle$. The reason is that $q(t)$ emphasizes the position as a degree of freedom and thus as a dynamical variable. Nevertheless, for purposes of illustration we have opted for the somewhat "loose" notation of the Cartesian coordinate $x(t)$, as it is this notation which emphasizes the trajectory in physical space. Notation aside however, we state again that $e^{-\frac{i}{\hbar}S(t,t',[x])}$ is the contribution of path $x(t)$ to the transition amplitude $\langle x', t'|x, t\rangle$. We would interpret it, for that matter, as the probability amplitude that path $x(t)$ be realized, were it not for the fact that all infinitely many paths available to the quantum particle are clearly unobservable. Nevertheless, in order to render the ensuing analysis as accessible as possible, we shall refer to $e^{-\frac{i}{\hbar}S(t,t',[x])}$ as the "probability amplitude" associated with $x(t)$ where necessary.

An essential aspect of the path integral formalism which deserves attention in its own merit is the oscillatory character of $e^{-\frac{i}{\hbar}S(t,t',[x])}$. Whatever the precise expression of $S(t, t', [x])$ is, the oscillations of this exponential extend indefinetely, with infinitely many of its values being positive and infinitely many being negative, yet without any indication whatsoever as to how these values add up to a convergent result for $\langle x', t'|x, t\rangle$. In fact, the path integral in (2.XII) only has formal significance. By this we do not mean that the path integral is not defined, or that it is infinite. We make, instead, a statement to the effect that the path integral as it stands is ambiguous and that, for that matter, we must specify what exactly we mean by it - that is, in what sense the path integral yields a finite result and which result that is. We demonstrate this technique in a specific physical context in the next section.

The object of our attention now is the precise relation between the transition amplitude and the wavefunction. By way of initiating our advance in that direction, we stress that in establishing the transition amplitude as a direct consequence of a position eigenstate's evolution ((2.III), (2.V) and (2.VI)) and as a path integral, the preceding analysis has demonstrated the equivalence between the canonical formalism and the path-integral formalism.

3. The Transition Amplitude and the Wavefunction

Let us turn to relation (2.III) which essentially signifies the evolution of position eigenstate $|x,t\rangle$ to the effect that at moment t', the wavefunction is $\tilde{\psi}(t',x')$. The transition amplitude $\langle x',t'|x,t\rangle$ is merely the value of $\tilde{\psi}(t',x')$ at the specified x'. At initial moment t, the quantum system is in the superposition $\psi(t,x)$ of position eigenstates according to (3.VI) in the first chapter with each position eigenstate $|x,t\rangle$ yielding its own contribution $\tilde{\psi}(t',x')$ to the total probability amplitude for position at (x',t') which is, for that matter, the sum-total of all such contributions. This qualitative conclusion receives its rigorous expression in the simple statement:

$$\langle x',t'|\psi\rangle = \int_{-\infty}^{+\infty} dx \langle x',t'|x,t\rangle\langle x,t|\psi\rangle \tag{3.Ia}$$

which establishes that the total probability amplitude $\psi(t',x')$ at each x' at t' is the sum over all contributions $\tilde{\psi}(t',x')$ stemming from all position eigenstates $|x,t\rangle$. For that matter, (3.Ia) establishes that the transition amplitude acts as the *propagator*:

$$K(x',t';x,t) \equiv \langle x',t'|x,t\rangle$$

which advances quantum state $\psi(t,x)$ from moment t to moment t':

$$\psi(t',x') = \int_{-\infty}^{+\infty} dx\, K(x',t';x,t)\psi(t,x) \tag{3.Ib}$$

Clearly, at the limit $t' \to t$, the wavefunction $\tilde{\psi}(t',x')$ reduces to the initial position eigenstate $|x,t\rangle$ itself with the direct implication that the quantum-mechanical propagator $K(x',t';x,t)$ satisfies the *boundary condition*

$$lim_{t' \to t} K(x',t';x,t) = \delta_t(x'-x) \tag{3.II}$$

a statement to the effect that no instantaneous propagation is possible.

The evolution of $\psi(t,x)$ which (3.I) signifies is in perfect conformity with the evolution of a classical wave. Readers who have studied classical wave theory are familiar with *Huygens principle* which states that *each point on a wave front of a propagating wave acts itself as a source of spherical waves*. Not surprisingly, this principle also underlies the evolution of a probability distribution. The evolution of the initial $|x,t\rangle$ to the final probability distribution described by $\tilde{\psi}(t',x')$ is an expression of Huygens principle, $\tilde{\psi}(t',x')$ being the spherical wave of probability which emanates from source $|x,t\rangle$. The sum over all contributions from all spherical waves emanating from the propagating "wave of probability" at each point x at initial moment t, to the specific point x' at final moment t' - equivalently, the sum over all contributions from all $|x,t\rangle$ - yields the probability amplitude for position $\psi(t',x')$ at x' and, by extension, the entire wavefunction $\psi(t',x')$ at moment[44] t'. (3.I) is a statement to the effect that the evolution of a state is determined in analogy to Huygens principle - that is, in analogy to the evolution of a classical wave as the sum-total of all contributions respectively stemming from all

---

[44]In the presence of classical potentials in which space is in general not isotropic, each point on a wave front is a source of a spherical wave of infinitesimally small radius.

points along a wavefront. Not only does (3.I) make a statement to the effect that the wavefunction at future moment t' is the sum-total of all contributions stemming from all infinitely many position eigenstates throughout space at initial moment t, it reveals the precise significance of that statement: The contribution of each $|x,t\rangle$ realized as $\delta_t(a-x)$ is multiplied by the corresponding $c(x)$ (refer to (3.VI) in the first chapter) before it is summed over in (3.I). This ensures that each wavefunction $\psi(t,x)$ as it is at t has its own unique evolution pattern.

No physical context povides a better illustration of this classical analogy and its underlying consequence of the transition amplitude as a path integral than that of the *double-slit experiment* to which we briefly turn in what follows. Interested readers can find a comprehensive treatment of this subject matter in [1].

The set-up of the double-slit experiment involves a source which emits individual electrons (one at a time). The source is positioned before a diaphragm on which two tiny slits have been carved. The effect of the emission toward the diaphragm is observed by the spot which each electron registers on a screen behind it (Fig.12). Clearly, such a spot signifies the collapse of each individual electron wavefunction which occurs as a result of the instantaneous interaction between the probability distribution and the atoms of the screen throughout its extent (recall Schrödinger's statement quoted in the preceding chapter). Crucial to understanding the implication inherent in the pattern which eventually emerges on the screen is the observation that as the electrons are emitted individually, the result is identical to that obtained by directing only one electron respectively in each of infinitely many experimental contexts identical to the one described. This result is, of course, the probability distribution of a single electron at the moment it arrives on the screen. Clearly, for that matter, in this experimental context, the electron manifests the dialectical unity between the local aspect (its position which the spot on the screen signifies) and the global aspect (its probability distribution for position) stated in the second chapter.

If propagation through the slits is unhindered (no covers, detectors, or light sources placed on either slit), the gradual accumulation of spots on the screen results in an interference pattern identical to that obtained from the propagation of water waves through the two-slit diaphragm. Each slit acts as a source of spherical waves of probability in perfect conformity with Huygens principle. That interference pattern is identical to the probability distribution of a single electron according to the premise which we introduced in the first chapter: Absence of knowledge as to which slit the electron propagated through implies that the amplitude for electron propagation from point x at which the source is located and moment t at which the electron is emitted, to arbitrary point x' on the screen at moment t' is the particular superposition of the amplitude that the electron accomplish this transition through slit A at some intermediate moment t_1 and the amplitude that the electron accomplish this transition through slit B at the same moment:

$$\langle x',t'|x,t\rangle = \langle x',t'|A,t_1\rangle\langle A,t_1|x,t\rangle + \langle x',t'|B,t_1\rangle\langle B,t_1|x,t\rangle$$

which can, equivalently, be viewed as an immediate consequence of (2.VII) in the preceding section. In view of the fact that the diaphragm allows for only the two position

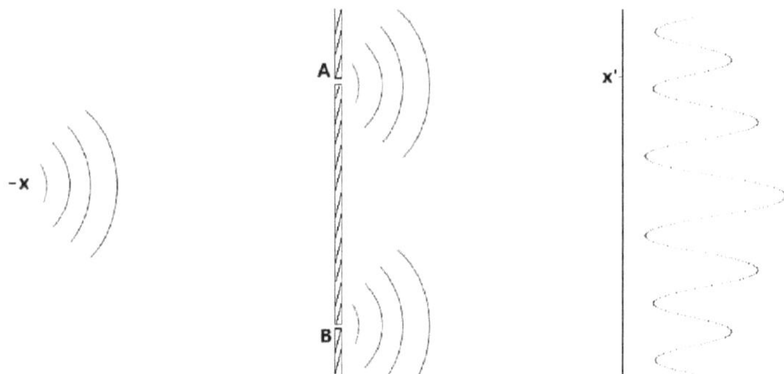

FIGURE 12. The double-slit experiment: Individual electrons emmited at point x propagate through the two slits to produce an interference pattern on the screen which imprints the probability distribution of a single emmited electron in line with Huygens principle.

eigenstates $|A, t_1\rangle$ and $|B, t_1\rangle$ which act as sources of spherical waves of probability toward the screen, this superposition is, at once, a clear demonstration of the premise to the effect that the probability amplitude for position $\psi(t', x')$ at point x' and final moment t' is the sum-total of all contributions $\tilde{\psi}(t', x')$ stemming from an earlier moment t_1. This, in turn, invites a direct extension of the double-slit experimental context to an arbitrary-slit experimental context.

If we open a third tiny slit on the diaphragm, the transition amplitude will be:

$$\langle x', t'|x, t\rangle = \langle x', t'|A, t_1\rangle\langle A, t_1|x, t\rangle + \langle x', t'|B, t_1\rangle\langle B, t_1|x, t\rangle + \langle x', t'|C, t_1\rangle\langle C, t_1|x, t\rangle$$

We can imagine, for that matter, an experimental set-up in which a diaphragm punctuated by an arbitrary number of tiny slits extends throughout infinite space, so that:

$$\langle x', t'|x, t\rangle = \langle x', t'|A, t_1\rangle\langle A, t_1|x, t\rangle + \langle x', t'|B, t_1\rangle\langle B, t_1|x, t\rangle + \langle x', t'|C, t_1\rangle\langle C, t_1|x, t\rangle + \ldots$$

then add a second infinitely-long diaphragm punctuated by an arbitrary number of tiny slits $A', B', \ldots$, so that:

$$\langle x', t'|x, t\rangle = \langle x', t'|A, t_1\rangle\langle A, t_1|A', t_2\rangle\langle A', t_2|x, t\rangle + \langle x', t'|B, t_1\rangle\langle B, t_1|A', t_2\rangle\langle A', t_2|x, t\rangle +$$

$$\langle x', t'|A, t_1\rangle\langle A, t_1|B', t_2\rangle\langle B', t_2|x, t\rangle + \langle x', t'|B, t_1\rangle\langle B, t_1|B', t_2\rangle\langle B', t_2|x, t\rangle + \ldots$$

If we continue to successively add infinitely long diaphragms, each punctuated with an increasing number of slits then at the limit at which the number of slits on each diaphragm becomes infinite with infinitesimal separation between successive slits and the distance between successive diaphragms becomes infinitesimally small, the infinitely many diaphragms vanish and each point in infinite space acts as a source of spherical waves of probability, thereby identically reproducing (2.IX) and (2.XII) in the preceding section. Moreover, the presence of the source which signifies an exclusive position eigenstate $|x, t\rangle$ at the moment of emission t, has itself become irrelevant at this infinite limit. Any point in infinite space - for example, A' - qualifies for the incipient point of propagation according to (3.I) and to the preceding considerations in this section. Thus, by indefinitely extending the set-up of the double-slit experiment both, in terms of slits and in terms of diaphragms, we have captured the essence of the transition amplitude as a path integral.

Thus, we have studied both aspects of the transition amplitude's... personality, that of a continuous sum over infinitely many paths intercepting two points and that of a propagator. As we have just seen, there is a deep relation between them. Either one of the two implies the other. We shall return to both aspects in quantum field theory in which the transition amplitude is of central importance. In anticipation of that we return to (2.II) in order to examine an inherent implication.

We assume that at past infinity a simple quantum harmonic oscillator is in its ground state $|\Omega_{-\infty}\rangle$ and inquire as to the probability that the oscillator emerge at future infinity in its ground state $|\Omega_{+\infty}\rangle$. Since the oscillator will not spontaneously transit to a higher energy state if left to itself, it is certain that it will make the transition from $|\Omega_{-\infty}\rangle$ to $|\Omega_{+\infty}\rangle$. That is

$$\langle\Omega_{+\infty}|\Omega_{-\infty}\rangle = 1 \qquad (3.III)$$

Pursuant to (3.I), this transition amplitude between the two quantum states has a very simple relation to the amplitude (always keeping in mind that we have yet to, rigorously, define the path integral over this purely oscillatory exponential):

$$\langle x'_{+\infty}|x_{-\infty}\rangle = \int \mathcal{D}[x] e^{-\frac{i}{\hbar}\frac{1}{2}\int_{-\infty}^{+\infty} dt(m\dot{x}^2 - kx^2)} \qquad (3.IV)$$

that the simple quantum harmonic oscillator transit from point $x_{-\infty}$ at past infinity to point $x'_{+\infty}$ at future infinity:

$$\langle\Omega_{+\infty}|\Omega_{-\infty}\rangle = \int_{-\infty}^{+\infty} dx'_{+\infty} \int_{-\infty}^{+\infty} dx_{-\infty}\langle\Omega_{+\infty}|x'_{+\infty}\rangle\langle x'_{+\infty}|x_{-\infty}\rangle\langle x_{-\infty}|\Omega_{-\infty}\rangle = 1 \quad (3.V)$$

This is, of course, a statement to the effect that $\langle x'_{+\infty}|x_{-\infty}\rangle$ propagates the oscillator ground state $|\Omega_{-\infty}\rangle$ at past infinity to the oscillator ground state $|\Omega_{+\infty}\rangle$ at future infinity. In fact, (3.V) allows for the extension of the path integral in (3.IV) to the ground state itself. We merely note, in this respect, that (3.IV) reduces to an infinite concatenation of Gaussian integrations resulting in

$$\langle \Omega_{+\infty} | \Omega_{-\infty} \rangle = N \int \mathcal{D}[x] e^{-\frac{i}{\hbar}\frac{1}{2}\int_{-\infty}^{+\infty} dt(m\dot{x}^2 - kx^2)} = 1 \qquad (3.VIa)$$

on the understanding that the "normalization" factor N ensures a finite result for $\langle \Omega_{+\infty} | \Omega_{-\infty} \rangle$ in general and can be adjusted so as to yield $\langle \Omega_{+\infty} | \Omega_{-\infty} \rangle = 1$ in particular.

The task now is to specify what exactly we mean by the ambiguous path integration over the oscillatory exponential in (3.IV), (3.V) and (3.VIa). A particular way of resolving this issue is to add a *convergence factor* $e^{-\frac{1}{2}\epsilon\int_{-\infty}^{+\infty} dt x^2(t)}$, ($\epsilon > 0$) in the action functional which will asymptotically damp the oscillations (Fig.13) yielding a finite result:

$$\langle \Omega_{+\infty} | \Omega_{-\infty} \rangle = N \int \mathcal{D}[x] e^{-\frac{i}{\hbar}\frac{1}{2}\int_{-\infty}^{+\infty} dt(m\dot{x}^2 - (k+i\epsilon)x^2)} = 1 \qquad (3.VIb)$$

on the understanding that once the finite result has been attained, that factor must be "switched off" by taking the limit $\epsilon \to 0$ (Fig.13)[45].

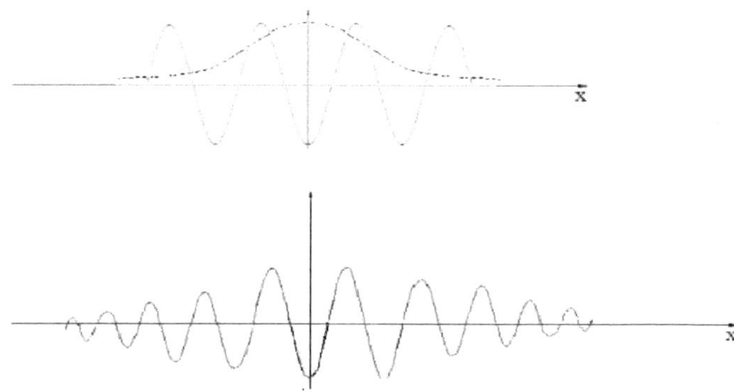

FIGURE 13. The figure at the top is a composition of two graphs. The continuous line represents $e^{-\frac{i}{\hbar}\frac{1}{2}\int_{-\infty}^{+\infty} dt(m\dot{x}^2 - kx^2)}$. The syncopated line represents $e^{-\frac{1}{2\hbar}\epsilon\int_{-\infty}^{+\infty} dt x^2(t)}$. The latter clearly acts as the amplitude of the former in $e^{-\frac{i}{\hbar}\frac{1}{2}\int_{-\infty}^{+\infty} dt(m\dot{x}^2 - (k+i\epsilon)x^2)} = e^{-\frac{1}{2\hbar}\epsilon\int_{-\infty}^{+\infty} dt x^2(t)} e^{-\frac{i}{\hbar}\frac{1}{2}\int_{-\infty}^{+\infty} dt(m\dot{x}^2 - kx^2)}$ represented by the graph at the bottom in which the oscillations asymptotically dwindle, yielding a path integral which converges to a number after the operation $\epsilon \to 0$.

---

[45]For reasons which will be explained in the next section, the frequency in the figure varies with x.

The state-to-state transition amplitude expressed as a path integral in (3.VI), foreshadows a crucial technique in quantum field theory developed by Julian Schwinger in 1969. That technique we shall examine at an advanced stage of our analysis on quantum field theory where the concept of propagator is of crucial importance. The basic concept which underlies that technique however, applies essentially also in quantum mechanics. For that matter, we convey it in what follows.

We consider the forced harmonic oscillator which we obtain by modifying the harmonic-oscillator potential through the addition of an external driving force $F(t)$. The transition amplitude from point x at moment t to point x' at moment t' is

$$\langle x', t' | x, t \rangle_F = \int \mathcal{D}[x] e^{-\frac{i}{\hbar} \int_t^{t'} dt (\frac{1}{2} m \dot{x}^2 - \frac{1}{2}(k+i\epsilon)x^2 + F(t)x(t))} \qquad (3.VII)$$

In turn, the amplitude that the quantum oscillator transit from a point x at past infinity to a point x' at future infinity is

$$\langle x'_{+\infty} | x_{-\infty} \rangle_F = \int \mathcal{D}[x] e^{-\frac{i}{\hbar} \int_{-\infty}^{+\infty} dt (\frac{1}{2} m \dot{x}^2 - \frac{1}{2}(k+i\epsilon)x^2 + F(t)x(t))} \qquad (3.VIII)$$

The dependence of this transition amplitude on the external force $F(t)$ can be explicitly worked out through a brief calculation. In the fifth chapter we shall advance a derivation of the result which we only cite here:

$$\langle x'_{+\infty} | x_{-\infty} \rangle_F = e^{-\frac{i}{\hbar}\frac{1}{2} \int_{-\infty}^{+\infty} dt' \int_{-\infty}^{+\infty} dt F(t')D(t'-t)F(t)} \int \mathcal{D}[x] e^{-\frac{i}{\hbar} \int_{-\infty}^{+\infty} dt (\frac{1}{2} m \dot{x}^2 - \frac{1}{2}(k+i\epsilon)x^2)} \qquad (3.IXa)$$

which is, of course:

$$\langle x'_{+\infty} | x_{-\infty} \rangle_F = \langle x'_{+\infty} | x_{-\infty} \rangle_{F=0} e^{-\frac{i}{\hbar}\frac{1}{2} \int_{-\infty}^{+\infty} dt' \int_{-\infty}^{+\infty} dt F(t')D(t'-t)F(t)} \qquad (3.IXb)$$

with $\langle x'_{+\infty} | x_{-\infty} \rangle_{F=0}$ being the transition amplitude in (3.IV) and with $D(t' - t)$ being a mathematical object which foreshadows the propagator in quantum field theory (not to be confused with the quantum-mechanical propagator $K(x', t'; x, t)$).

The concept which underlies (3.IX) relates directly to the external driving force. If we impose the, logically obvious (we turn $F(t)$ on and, respectively, off...), condition that the external $F(t)$ emerges at some unspecified moment in the far past and vanishes at some unspecified moment in the far future then at past and future infinity $t \to \mp\infty$ the vacuum states $|\Omega_{\mp\infty}\rangle$ do not depend on $F(t)$. We can, for that matter, express the amplitude $\langle \Omega_{+\infty} | \Omega_{-\infty} \rangle_F$ that the F-independent oscillator ground state at infinite past transit to the F-independent oscillator ground state at infinite future in the presence of $F(t)$ in terms of the amplitude $\langle x'_{+\infty} | x_{-\infty} \rangle_F$. It is a trivial matter to retrace the steps we took in order to arrive from (3.III) to (3.V), now in the presence of $F(t)$:

$$\langle \Omega_{+\infty} | \Omega_{-\infty} \rangle_F = \int_{-\infty}^{+\infty} dx'_{+\infty} \int_{-\infty}^{+\infty} dx_{-\infty} \langle \Omega_{+\infty} | x'_{+\infty} \rangle \langle x'_{+\infty} | x_{-\infty} \rangle_F \langle x_{-\infty} | \Omega_{-\infty} \rangle =$$

$$e^{-\frac{i}{\hbar}\frac{1}{2}\int_{-\infty}^{+\infty}dt'\int_{-\infty}^{+\infty}dt\,F(t')D(t'-t)F(t)}\int_{-\infty}^{+\infty}dx'_{+\infty}\int_{-\infty}^{+\infty}dx_{-\infty}\langle\Omega_{+\infty}|x'_{+\infty}\rangle\langle x'_{+\infty}|x_{-\infty}\rangle_{F=0}\langle x_{-\infty}|\Omega_{-\infty}\rangle$$

$$(3.Xa)$$

or, through (3.V):

$$\langle\Omega_{+\infty}|\Omega_{-\infty}\rangle_F = e^{-\frac{i}{\hbar}\frac{1}{2}\int_{-\infty}^{+\infty}dt'\int_{-\infty}^{+\infty}dt\,F(t')D(t'-t)F(t)} \qquad (3.Xb)$$

The statement which (3.X) makes is that in the presence of a driving force $F(t) \neq 0$, the state-to-state transition amplitude $\langle\Omega_{+\infty}|\Omega_{-\infty}\rangle_F$ is no longer equal to unity:

$$\langle\Omega_{+\infty}|\Omega_{-\infty}\rangle \neq 1$$

This is essentially because the application of a <u>time-dependent</u> external driving force $F(t)$ refutes the time-translational invariance which the quantum system had at past infinity. Thus, an energy eigenstate of the unperturbed Hamiltonian operator $\hat{H}$ (i.e. prior to the application of $F(t)$), no longer has the simple form (3.IIId) which ensures the time-independence of the corresponding probability distribution and of the quantum system in that energy eigenstate, as we saw in the first chapter. In the presence of a given $F(t)$, transitions between the energy eigenstates are inevitable and strictly probabilistic. In the presence of $F(t)$, the ground state $|\Omega_{-\infty}\rangle$ may transit to any of the infinitely many higher energy states of the oscillator, each such transition associated with a predictable probability. The probability amplitude for the particular transition from the ground state $|\Omega_{-\infty}\rangle$ to the ground state $|\Omega_{+\infty}\rangle$ in the presence of $F(t)$ is given by (3.Xb).

These considerations will be of particular relevance to the vacuum-to-vacuum transition amplitude $\langle 0|0\rangle$ in quantum field theory.

4. Semi-classical Limit and... Beyond Jupiter!

An essential aspect of path-integral quantization which testifies to the latter's physical consistency is the semi-classical behavior which emerges at the limit of vanishing quantum fluctuations. As we saw in the first chapter, the semi-classical limit is formally conveyed as $\hbar \to 0$. In order to understand what exactly we mean in the present context of path-integral quantization by this formal statement which treats the Plank constant $\hbar$ as a variable, we first note that the quantity $e^{-i\frac{1}{\hbar}S}$ in (2.XII) is, in fact, a (unobservable) wave which undulates with frequency $\frac{1}{\hbar}$ as the path $x(t)$, and thereby the action $S(t,t',[x])$, varies. A frivolous look at (2.XII) would give the impression that such a frequency is constant, very much like the fixed frequency $\frac{E_p}{\hbar}$ of a plane wave $\sim e^{\frac{i}{\hbar}px}e^{-i\frac{E_p}{\hbar}t}$. That this is not the case can be seen by considering that domain of infinitely many paths available to the quantum-mechanical particle in which the action functional $S(t,t',[x])$ receives very high values in comparison to the minimum classical action S_0 (Fig.9). For in that domain, $S \gg S_0$ is equivalent to treating $\hbar$ as a variable much smaller than 1 (so that, in turn, $\frac{1}{\hbar}$ is much bigger than some fixed value of S). The implication is that as we advance to progressively longer paths away from the classical trajectory $x_0(t)$, the amplitude in (2.XII) oscillates with progressively increasing frequency $\frac{1}{\hbar}$ which, in turn, gives rise to *destructive interference* at the limit $\hbar \to 0$ (Fig.14).

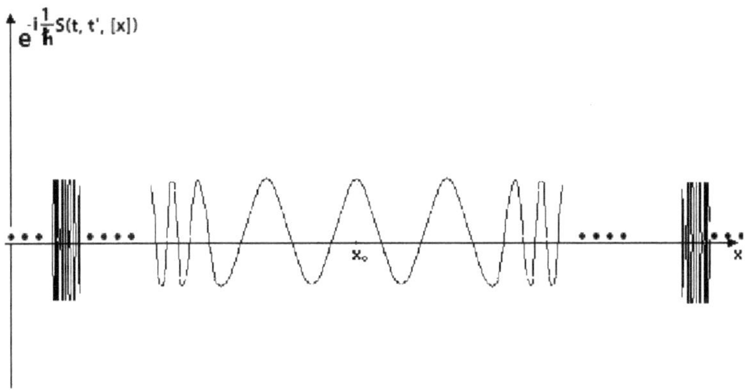

FIGURE 14. The horizontal axis represents the continuous set of unobservable paths which intercept x_1 at t_1 and x_2 at t_2. The vertical axis represents the contribution (weight) which each such path has to the path integral. The continuous set of contributions is a wave of constant amplitude but increasing frequency as the paths detract from the classical trajectory $[x_0]$. Arbitrarily high frequencies of oscillation and destructive interference occur for arbitrarily long paths (heavy lining). In the opposite direction, the paths which do not significantly detract from $[x_0]$ are associated with small frequencies and substantial contributions to the path integral. At the limit of vanishing quantum fluctuations ($\hbar \to 0$), $[x_0]$ emerges as the unique trajectory.

The meaning of destructive interference can be understood merely by considering that the integral, or "continous sum" as we have mostly referred to it, of a given function over a certain domain of its independent variable yields in that domain the total area bounded by the horizontal axis and the graph of that function itself, as can be confirmed by mere reflection. Area segments above the horizontal axis are positive. Those below the horizontal axis are negative. The total area which the integral of the function represents over a certain domain is the sum-total of the positive and the negative segments. If, over a certain domain, the number of positive-area segments of an oscillatory function is equal to the number of negative-area segments of that function and for each positive segment there is a negative segment of equal in magnitude area, the integral vanishes over that domain. Mere inspection of Fig.14 clearly reveals that - precisely in line with physical expectations - the path integral in (2.XII) does not vanish since the negative segments over any domain of $x(t)$ do not cancel against the positive segments in that domain on account of the progressive increase in the frequency of oscillation. Nevertheless, in the domain of excessively long paths away from $x_0(t)$ where the wavelength tends to zero and the frequency tends to infinity, the area bounded over the distance of half a wavelength accordingly tends to zero, asymptotically ($\hbar \to 0$) yielding an exact cancellation between the positive and the negative values of the "probability amplitude" $e^{-i\frac{1}{\hbar}S(t,t',[x])}$. This is the implication of destructive interference: Contributions of adjacent excessively long paths away from the classical trajectory $[x_0]$ "by and large" cancel against each other in the path integral.

As we move in the opposite direction, into the domain of progressively shorter paths intercepting the two given points, the effective frequency of oscillation becomes progressively smaller and the contributions of these paths to the path integral become progressively bigger (Fig.14). In the domain of paths about the classical trajectory $x_0(t)$, the action functional $S(t,t',[x])$ receives values *comparable in magnitude to that of the minimum classical action* S_0. A look at Fig.9 makes it obvious that in that domain, $S[x]$ is fairly constant (all action values $S(t,t',[x])$ are "roughly equal" to the classical action S_0). As a direct consequence, the amplitude $e^{-i\frac{1}{\hbar}S(t,t',[x])}$ is itself effectively constant in that domain, a fact which, in turn, implies that it is the paths in that domain which dominate the path integral in (2.XII). It is, for that matter, that domain the domain in which we expect the emergence of the semi-classical limit. On rigorous terms, the paths which dominate the path integral are the ones about the classical trajectory $x_0(t)$ for which $\frac{\delta S}{\hbar} \to 0$, δS being the infinitesimally small range of action values between some action value in the immediate vicinity of S_0 and S_0 itself. In order to demonstrate that $\frac{\delta S}{\hbar} \to 0$ is, indeed, the semi-classical limit, we multiply numerator and denominator in $\frac{\delta S}{\hbar}$ by the numerical value of S_0 (merely a dimensionless number). The denominator then becomes S_0 itself and the numerator is now δS multiplied by a constant. We note that at the limit $\delta S \to 0$: (i) that constant becomes utterly irrelevant, (ii) the ratio $\frac{\delta S}{S_0}$ is independent of $\hbar$ and (iii) $\frac{\delta S}{S_0} \to 0$ - that is, the quantum fluctuations about $x_0(t)$ vanish and a classical mechanical system following the unique trajectory $x_0(t)$ emerges. In line with the definition which we gave in the first chapter, this is precisely the semi-classical limit of the path integral.

We now arrive at the semi-classical limit from the formal perspective $\hbar \to 0$. Since the amplitude $e^{-i\frac{1}{\hbar}S(t,t',[x])}$ oscillates with infinite frequency at this limit, destructive interference in the path integral occurs throughout the path domain except for the infinitesimally small segment of paths in the immediate vicinity of $x_0(t)$ where the action functional $S(t,t',[x_0])$ is constant and no oscillations occur. As with the limit of vanishing quantum fluctutions $\frac{\delta S}{S_0} \to 0$, so with the formal limit $\hbar \to 0$ the classical trajectory $x_0(t)$ is singled out.

In certain texts the path integral is interpreted as a sum over quantum corrections to the classical trajectory on account of the dominant contributions which the paths about the classical trajectory yield. In my view, such an interpretation is not altogether consistent. For it can easily be perceived as implying that paths away from the semi-classical domain - all the more so, excessively long paths - have no contribution and are, thus, irrelevant. I take the view which Richard Feynman himself stated: "The electron does anything it likes. It just goes in any direction at any speed, forward or backward in time, however it likes and then you add up the amplitudes and it gives you the wavefunction"[46]! Let us, for that matter, consider the contribution which a path emanating from one corner of a laboratory and ending at the opposite corner of that room one second later has to the probability that the electron transit between these two points within one second if the path extends to maximum distance past planet Jupiter (idea and section-title borrowed from a milestone of a science-fiction film[47])! According to the preceding analysis, the exorbitant length of that unobservable path implies an accordingly exorbitant $\frac{1}{\hbar}$. Exorbitant enough, in fact, for that path to have its contribution in the path integral effectively cancelled by the contribution of another path less than one millimeter longer! Nevertheless, however minute the difference between the two contributions is, it is not strictly zero. Only paths of infinite length yield strictly zero contributions. Contributions to the path integral stemming from excessively long lengths are severely suppressed, but are not annihilated! In order to get a feeling as to the scale of such a suppression, we note that if (for illustration purposes only) after all cancellations in the path integral we associated the remaining contribution of the stated "laboratory-Jupiter-laboratory path" with an observable probability, we would have to wait many orders the present age of the Universe before we see that path realized!

An independent aspect of the preceding considerations deserves attention in its own merit. As the path which we have just considered "loops around" planet Jupiter in order to connect two points over a room's distance within one second, that path is necessarily associated with superluminal speeds. Speeds which exceed that of light call causality into question (can the effect precede the cause?). Causality is not an issue in non-relativistic quantum mechanics which tacitly assumes that the speed of light is just a speed like any other. However, the universal status which the speed of light actually has can never be ignored. Since we use non-relativistic quantum mechanics as a preamble to the physically rigorous relativistic quantum field theory, we must be concerned as to whether, to which extent and in what sense the results which we have obtained in the former, remain valid in the latter. Any inconsistency of the former with the latter, would suggest that quantum mechanics, although conceptually consistent, is physically irrelevant. In this respect, one could argue that since paths are unobservable, such paths as the one we have called into question are never an issue as long as the transition amplitude $\langle x', t' | x, t \rangle$ itself is not associated with superluminal speeds. Note, however, that, for given x and t, the transition amplitude in (2.V) is non-vanihising for all x' and t', a fact which implies that a quantum particle may propagate between any two points in an arbitrarily short time. In a relativistic theory, this would signal a violation of causality. By itself, this does not signify any physical inconsistency on behalf of quantum mechanics because we have obtained (2.V) through use of the non-relativistic Hamiltonian $\hat{H} = \frac{\hat{p}}{2m}$. If in the calculation which eventuated in (2.V) the replacement of the non-relativistic $\hat{H} = \frac{\hat{p}}{2m}$ by the relativistic $\hat{H} = \sqrt{\hat{p}^2 c^2 + m^2 c^4}$ (which we shall study in the seventh chapter) yielded a result which upheld causality in $\langle x', t' | x, t \rangle$, the

---

⁴⁶Quotation attributed to Freeman Dyson.
⁴⁷Stanley Kubric's "Space Odyssey 2001".

status of quantum mechanics as a physical theory would not be challenged. Yet, this is not the case. Such a replacement still yields a transition amplitude which extends out of the light cone (i.e. a transition amplitude for two space-time points communication beween which requires superluminal speeds) and causality is still violated. We shall see in the next chapter that only quantum field theory, the quantum theory of infinitely many degrees of freedom, solves this problem and shall study that solution.

We return now to the semi-classical limit in order to examine the conditions for its emergence, those which give rise to $\frac{\delta S}{S_0} \to 0$ formally expressed by $\hbar \to 0$. It would be instructive, in this respect, to briefly review the semi-classical limit in the context of canonical quantization first.

In the first chapter we examined the reason that quantum effects are more pronounced for light particles with the implication that the more massive a particle is, the closer to the semi-classical limit it lies. When we compare the electron, the prototype of quantum particle on account of its small rest mass, to a proton, to a nucleus, to a molecule then to a collection of molecules and finally to macroscopic objects such as a grain of sand, we see that as the mass increases, the wavelength of the plane wave of probability which describes the corresponding quantum system (treated as a quantum particle) in empty space decreases according to $p = \hbar k = \hbar \frac{2\pi}{\lambda}$. The distance of half wavelength $\frac{\lambda}{2}$ is, of course, a measure of the quantum fluctuations of position about the central point of maximum amplitude. As the mass becomes arbitrarily big (recall the example of the tennis ball), so does the momentum. Equivalently, the wavelength of the plane wave tends to zero with the same implication for the quantum fluctuations: $\frac{\lambda}{2} \to 0$. As we remarked in the first chapter, within each such half wavelength the quantum system is highly "localized" with the probability for position being maximum within the short range of each such half wavelength and negligible between any two successive half wavelengths. It follows, for that matter, that at the limit of vanishing quantum fluctuations the quantum system behaves like a classical object which moves at constant speed, precisely as expected.

Let us now examine the relation between mass and semi-classical limit in the context of path-integral quantization. Pursuant to the first section of the present chapter, the transition amplitude for a free quantum particle $(V(x) = 0)$ is

$$\langle x_2, t_2 | x_1, t_1 \rangle = \int \mathcal{D}[x] e^{-\frac{i}{\hbar} \frac{1}{2} m \int_{t_1}^{t_2} dt \dot{x}^2(t)} \tag{4.I}$$

The action functional $S(t_1, t_2, [x]) = \frac{1}{2} m \int_{t_1}^{t_2} dt \dot{x}^2(t)$ receives arbitrarily high values for particles of arbitrarily high mass m. Thus, the effect is exactly the same as that which we have already examined: Formally, at the limit $m \to \infty$ destructive interference occurs on account of the infinite frequency of oscillation which the mass signifies and the path integral vanishes unless $\int_{t_1}^{t_2} dt \dot{x}^2(t)$ is fairly constant, which is indeed the case in an infinitesimal range of values about the classical action $S_0(t_1, t_2, [x_0]) = \frac{1}{2} m \int_{t_1}^{t_2} dt \dot{x}_0^2(t)$. We conclude that $\hbar \to 0$ is a formal representation of the semi-classical limit which emerges for objects of very large mass.

This concludes our examination of the path-integral formulation of quantum mechanics. The results which we have obtained in this chapter, we shall extend to the infinitely many degrees of freedom of quantum fields. We reiterate that such an extension is non-trivial. Although the path-integral quantization of fields is infinitely more involved than the path-integral quantization of mechanical systems, it is, at once, of crucial importance in quantum field theory in which manifest Lorentz invariance is an enormous advantage.

PART II

FREE QUANTUM FIELDS - GENERAL CONSIDERATIONS

CHAPTER 4.

QUANTUM FIELD THEORY

1. The Classical Fields

The concept of continuity was a great leap forward for human thought. Before the sixteenth century, Nature had been perceived merely as a collection of objects in space. Continuity emerges in the sixteenth century as a concept inherent in the works of Nicolaus Copernicus, Galileo Galilei, Tycho Brahe and Johannes Kepler and subsequently receives its explicit and rigorous expression in the gravitational field of seventeenth century Newtonian physics.

We have already encountered the concept of field as a continuous set of degrees of freedom extending in space and time. The implications which this description has at classical level are quite straightforward. In contrast to a classical particle, which at each specific moment occupies a specific point in space, a classical field is a physical system which at each specific moment is characterized by a specific value of strength at each point in space. On rigorous terms, *a classical field is a physical system which at each space-time point is characterized by a specific value of a physical quantity (the field-strength).* The outstanding examples in this respect are, in historical order, the gravitational field of Newtonian physics which is produced by massive bodies and accounts for the interaction between them, the electromagnetic (*Maxwell*) field of *classical electrodynamics* which is produced by electric charges and accounts for the interaction between them and the gravitational field of general relativity which is produced by concentrations of matter (and energy) and accounts for the interaction between matter and space-time. In Newtonian theory, the acceleration due to gravity which determines the strength of the gravitational field receives, at each moment, a specific value at each point. In classical electrodynamics, the strength of the electromagnetic field is determined at each space-time point x by the strength of the electric field $\mathbf{E}(x)$ and that of the magnetic field $\mathbf{B}(x)$. In general relativity, the description of the gravitational field is substantially more involved, as the latter is not just another continuous set of degrees of freedom extending in space-time, but curved space-time itself. The field-theoretical concept, however, is the same. The strength of the gravitational field is expressed by the curvature of space-time at each space-time point.

All fields are quantum, of course! The underlying reason that the electromagnetic and the gravitational field are immediately recognized as classical is inherent in their long-range nature: The strength with which they interact vanishes only over infinite space-time separations. As we shall see, this property yields a semi-classical limit which effectively conceals the fundamental quantum nature of these two fields and conforms with the above-stated definition of a classical field. By contrast, the fields of the strong interactions (the *gluon fields*) and those of the weak interactions (*$W^{\pm}$-field, Z-field*) never appear as classical because - for independent reasons - they do not interact over arbitrarily large distances. Pursuant to the fourth premise in the Introduction, they respectively receive a classical formulation which constitutes the incipient point of quantization. Such fields as that formulation signifies are also classical, even though they do not correspond to actual semi-classical limits.

Thus, classical fields associated with interactions are not as intuitive as our experience of the electromagnetic and the gravitational field suggests. Yet, the really counterintuitive aspect of classical fields does not so much stem from interactions, as it does from particles! Pursuant to the second premise in the Introduction, all physical systems are essentially fields. Particles emerge only as an expression of underlying quantum fields. For example, the quarks and the electrons are respectively expressions of a quantum quark field and of a quantum electron field, as the photon and the graviton are respectively expressions of the quantum electromagnetic (photon) field and of the quantum gravitational field. Thus, we also define a classical electron field, a classical quark field and a classical field for each species of elementary particles observed in Nature. We can easily conceptualize the classical fields as smooth and extended entities in space-time. Yet, common sense revolts at the idea of doing the same for particles. How can we conceptualize the classical electron field or the classical quark field in a manner which conforms with the concept of "point-like"? We shall resolve this apparent contradiction when we discuss the crucial difference which exists between the statistical behavior of quantum fields which manifest themselves as "particles" and the statistical behavior of quantum fields which manifest themselves as interactions. The point of the present discussion is to emphasize the classical field as a category of thought which transcends the observed electromagnetic and gravitational fields. These two fields are only convenient examples of classical field because our intuition - essentially the product of centuries of thought based on the dichotomy between particles and fields - readily accepts them as smooth and extended physical entities. Aim instead at a more abstract level of thought which admits the classical field as a smooth entity of infinitely many degrees of freedom extending in space-time, regardless of whether that entity is observable or not, physical or formal, associated with interactions between particles or with particles themselves.

The mechanical systems which we have examined so far are non-relativistic. They are respectively treated as Newtonian and quantum-mechanical by observers with respect to whom the classical and the average quantum-mechanical speeds involved are insignificant in comparison to the speed of light c. Such a low-energy approximation also applies to classical fields. The field of Newtonian gravity and the electrostatic Coulomb field are examples. This, however, is only an approximation valid exclusively in the particular circumstance of insignificant speeds and energies. All physical systems are relativistic: *Their dynamical behavior remains invariant under transformations between observers all of whom agree that the speed of light is equal to c.* This is all the more valid for fields. As we shall see, there are no relativistic quantum particles! A relativistic quantum system is necessarily a relativistic quantum field! Nor is there a quantum field of Newtonian gravity, or a quantum Coulomb field! Special relativity is, for that matter, an integral aspect of field theory. In what follows, we only discuss some of its basic concepts necessary for the analysis on classical fields. However, the deeper the level at which the reader understands special relativity is, the deeper his or her insight on field theory will be. The essence of special relativity is entirely inherent in its space-time diagrams. The study of the Appendix, for that matter, is as essential to understanding field theory as it is essential to understanding the theory of relativity itself.

Each point in the Euclidean three-dimensional space is specified by three coordinates (x, y, z) with respect to an arbitrary coordinate system. In order to bring this in line with the notational conventions of relativity we shall use the superscripted notation and replace (x, y, z) by (x^1, x^2, x^3). We can now make a more rigorous statement to the effect

166

that each point in space is in one-to-one correspondence with vector $\vec{x} = (x^1, x^2, x^3)$. The magnitude $|\vec{x}|$ of that vector yields the distance of the corresponding point from the origin $(0, 0, 0)$ of the coordinate system and the orientation of that vector yields the direction in which the corresponding point is. In the most simple case of a Cartesian coordinate system, it follows from Pythagoras theorem that $|\vec{x}|^2 = (x^1)^2 + (x^2)^2 + (x^3)^2$. Thus both, the magnitude $|\vec{x}| = +\sqrt{(x^1)^2 + (x^2)^2 + (x^3)^2}$ and the orientation of each vector $\vec{x}$, are inherent in the three coordinates x^1, x^2 and x^3 and specify a point in three-dimensional space. Indeed, each point in three-dimensional space is essentially in identity with its corresponding vector $\vec{x} = (x^1, x^2, x^3)$ and the three-dimensional space itself is the collection of all points $\vec{x}$. The characteristic property of this space and, indeed, of every *metric space* (a space in which distance has been defined) is that the distance between any two points is independent of the choice of coordinates and observers. As the distance is an inherent property of space itself, it cannot possibly depend on who observes it, or with which coordinates he chooses to label the corresponding two points.

In the framework of special relativity, the three-dimensional space is unified with time in a four-dimensional space-time known as the *Minkowski space-time*, after Hermann Minkowski who introduced this concept in 1907. Minkowski space-time describes space-time in the absence of gravity and is, thus, flat. In the formalism of relativity the temporal coordinate is defined as $x^0 = ct$, with c being the speed of light in empty space and with t being the temporal lapse. Each point in Minkowski space-time is specified by three spatial coordinates (x^1, x^2, x^3) and one temporal coordinate x^0 with respect to an arbitrary coordinate system whose origin is $(0, 0, 0, 0)$. Each point in Minkowski space-time is in one-to-one correspondence with vector $x^\mu = (x^0, x^1, x^2, x^3) \equiv (x^0, \vec{x})$, with the index μ receiving the values $0, 1, 2, 3$ (so that $\mu = 0$ specifies the temporal coordinate $x^0 = ct$, $\mu = 1$ specifies the spatial coordinate x^1, $\mu = 2$ specifies the spatial coordinate x^2, and $\mu = 3$ specifies the spatial coordinate x^3). Clearly, each point in Minkowski space-time specifies an event[48] and the distance between any two events is independent of coordinates and observers. Thus, the statement that space and time are unified has the significance that different curves which connect the same pair of events may have different elapsed times, just as different paths between two points in ordinary space can have different lengths[49].

In Minkowski space-time observers may move with constant velocity and thus be in uniform motion, or they may be in non-uniform accelerating motion. The *inertial observers* of the first category register the absence of gravity. By contrast, the *non-inertial observers* of the second category register the "fictitious inertial forces" of the old Newtonian perspective. Of course, from the rigorous perspective of the principles which underlie the theory of relativity - namely, the *Equivalence Principle* and the associated principle of *General Covariance* - there is nothing fictitious in any observation, nor is

---

[48]The event "A bomb exploded in Paris at 7.20pm yesterday" is identical to that point x^μ whose $\vec{x} = (x^1, x^2, x^3)$ specifies point "Paris" and x^0 specifies moment "7.20pm yesterday". The news announcement could have equivalently stated: "Space-time point x^μ"!

[49]Do note that such curves are necessarily *time-like*, in the sense that the speed of the observer who delineates a curve is always below the maximum speed c. The above-stated effect "different elapsed times for different curves" resolves the famous *twin paradox*.

any observation more valid than any other. All observers are physically equivalent! Depending on how he moves, a non-inertial observer registers a very real gravitational field locally, in his immediate vicinity! At once, he himself also registers the zero space-time curvature of Minkowski space-time. This does not signify any contradiction. In his frame, the local gravitational field is the result of his curved world line in flat space-time, not the result of curved space-time (refer to the Appendix for further elucidation). Flat Minkowski space-time signifies the absence of gravity in the sense that there is always a coordinate transformation which reduces the curvilinear coordinate system of an accelerated observer to the "curvature-free" Cartesian coordinate system of an inertial observer. The latter is, clearly, a global coordinate system. It naturally covers the Minkowski space-time in its entirety. On the contrary, curved space-times are accordingly equivalent to gravitational fields in the sense that there are no coordinate systems which dispense with space-time curvature while at once covering the entire space-time. Thus, only the inertial observers register the absence of gravity which characterizes the Minkowski space-time. For that matter, although all observers are physically equivalent, the set of inertial observers constitutes a preferred set of observers in Minkowski space-time and will constitute the exclusive perspective of our approach to classical and quantum fields.

Transformations which relate different inertial frames are known as *Lorentz transformations*. Since inertial observers constitute a preferred set in Minkowski space-time, any two such observers must manifestly register the same dynamical behavior for a given physical system in their respective frames. On more rigorous terms, the equations which express the dynamical behavior of a physical system are *Lorentz covariant* - that is, they remain invariant under Lorentz transformations. A central consequence of the demand for Lorentz covariance is the *Lorentz-invariant magnitude* of every four-dimensional vector in Minkowski space-time. The two characteristic examples are the space-time vector $x^\mu = (x^0, \vec{x})$ and the *energy-momentum vector* p^μ. If the total energy of a particle as observed in an inertial frame is E and $\vec{p} = (p^1, p^2, p^3)$ is its corresponding momentum then $p^\mu = (p^0, p^1, p^2, p^3) = (p^0, \vec{p})$, with $p^0 = \frac{E}{c}$. The magnitudes of both four-dimensional vectors are Lorentz invariant. This invariance is most conveniently expressed in terms of the squares of these magnitudes $\eta_{\mu\nu}x^\mu x^\nu$ and $\eta_{\mu\nu}p^\mu p^\nu$ respectively (refer to the Appendix): The quantity $\eta_{\mu\nu}x^\mu x^\nu = (x^0)^2 - (x^1)^2 - (x^2)^2 - (x^3)^2 \equiv (x^0)^2 - |\vec{x}|^2$ and the quantity $\eta_{\mu\nu}p^\mu p^\nu = (p^0)^2 - (p^1)^2 - (p^2)^2 - (p^3)^2 \equiv (p^0)^2 - |\vec{p}|^2$ are respectively invariant under Lorentz transformations.

We can now put the intuitive concept of the classical field as a smooth entity extending in space and time in this physical context. In view of this objective, let us first briefly state that in classical field theory the observable field-strength is derived from a corresponding unobservable potential. The precise manner of this derivation need not concern us at this stage. We only note here that such a treatment is consistent with the analytical approach which replaces the Newtonian concept of force by that of the potential. A classical field, for that matter, is equivalently described by the potential which can be a scalar, a vector, or a tensor function of space-time (the two typical examples are the vector potential $A^\mu(x)$ of the Maxwell field and the metric tensor $g_{\mu\nu}(x)$ of space-time). Although these two concepts are very different in terms of physical and mathematical content, the distinction between them will, for all intents and purposes, not be essential to our analysis until much later. Until that advanced stage, since they both describe the same field, we shall take the *field-values* $\phi(x)$ to refer to either the field-strength, or to the corresponding potential and only state the distinction and precise relation between these two concepts where necessary. The notation $\phi(x)$ signifies a scalar field (on rigorous terms, the potential of a scalar field) - that is, a field which receives a numerical value ϕ at each space-time point x. On account of its comparative simplicity, this field will, in the present chapter, serve as the prototype for the analysis on the most fundamental aspects of all fields.

In classical mechanics, a particle occupies a specific point $\vec{x}$ in space at each moment x^0. The position $\vec{x}$ of the particle defines a certain number of degrees of freedom which evolve in the course of time. That is, in classical mechanics, $\vec{x}(x^0)$ is a dynamical variable. The concept is the same in classical fields. At each x^0 there are infinitely many field-values $\phi(x)$, one at each point $x^\mu \equiv x = (x^0, \vec{x})$ in space. Note, however, that the direct extension from mechanical systems to fields which this concept appears to signify is deceptive. For, the infinitely many degrees of freedom are now the field-values $\phi(x)$. Space and time have been demoted to mere labels! This is a central difference between particles and fields, a direct consequence of the central difference between a finite number of degrees of freedom and an infinite number of degrees of freedom: *Whereas, in mechanical systems the position $\vec{x}(x^0)$ is a dynamical variable, in fields each point $\vec{x}$ at each moment x^0 - on rigorous terms, each space-time point $x^\mu \equiv x = (x^0, \vec{x})$ - is only a label of the dynamical variable $\phi(x)$.*

In the Introduction, I stressed that from the rigorous perspective of quantum field theory the abstract classical relativistic wave fields are of primary importance. They constitute the incipient point in the formal procedure of field quantization. It is necessary, for that matter, to explore the implications which the preceding general considerations have for this specific class of classical fields in Minkowski space-time. With respect to an inertial observer and to his Cartesian coordinate system, a classical wave field $\phi(x)$ is, at each moment x^0, defined on that three-dimensional surface all of whose points $\vec{x}$ are characterized by x^0, in the sense that all synchronized clocks on that surface have registered the same lapse of time x^0. At each specific point $\vec{x}$ of that three-dimensional surface, $\phi(\vec{x})$ defines a finite number of degrees of freedom whose evolution to the future of that surface is described by $\phi(x^0, \vec{x})$. This, in turn, yields the concept of *field configuration*. At any given moment the field receives a unique value at each point on the corresponding three-dimensional surface. The collection of all such values defines a field configuration at that moment. A field configuration is naturally described by $\phi(x^0, x^1, x^2, x^3) \equiv \phi(x^0, \vec{x})$ on the understanding that, whereas the coordinates x^1, x^2 and x^3 span the entire three-dimensional surface, x^0 is fixed at the specific value of choice[50]. A configuration of a classical wave field can be best thought of as an infinite sheet which extends throughout three-dimensional space at a specific moment (Fig.15).

---

[50]Do note that this fixed value of x^0 does not have absolute significance. Events which are simultaneous in the frame of observer A, are not simultaneous in the frame of observer B in relative motion to A (see Appendix).

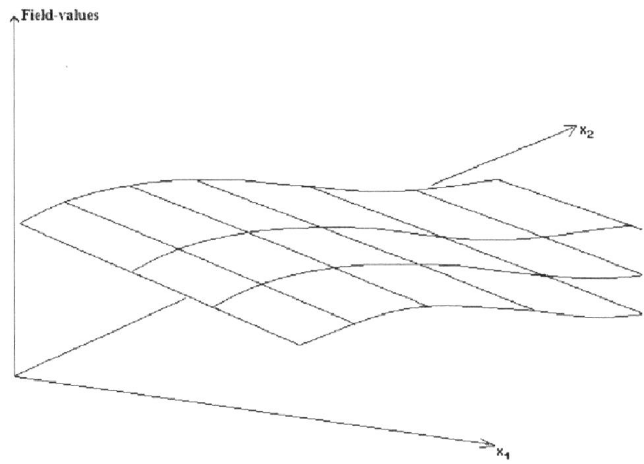

FIGURE 15. A configuration of a classical wave field at a specific moment. The (x^1, x^2)-plane represents the entire three-dimensional space with one dimension suppressed (each point in the (x^1, x^2)-plane corresponds to a straight line perpendicular to that plane).

It follows, for that matter, that the collection of all field configurations in one-to-one correspondence with each moment yields the entire dynamical behavior of a classical wave field. Accordingly, the dynamical behavior of a classical wave field can be best thought of as an infinite sheet undulating throughout three-dimensional space in the course of time (Fig.16). Such a dynamical behavior is, of course, identical to that of a freely propagating infinitely-extended three-dimensional wave. Given any two moments registered in an inertial frame, such an undulation defines in Minkowski space-time a four-dimensional geometric structure bounded by the two field configurations which correspond to these two moments and is the apparent field-theoretical analog to the trajectory of a classical relativistic particle within a specific time interval. As the trajectory of a classical particle is mathematically expressed by the solution to the corresponding equations of motion, so the collection of all field configurations in one-to-one correspondence with each moment, that *evolving field configuration*, is mathematically expressed by the solution to the corresponding field equations for free wave propagation.

Thus, knowledge of the solution to the equations of a classical wave field is equivalent to knowledge of that field's dynamical behavior. Is, however, the solution to the field equations always attainable? Clearly, this question transcends the category of the abstract free classical wave fields. In order to answer it, we must first address the precise relation between the concrete classical fields and the sources which, from the perspective of the classical theory, produce them. Electrostatic fields are produced by static electric charges, magnetic fields by electric currents, electromagnetic radiation by accelerating

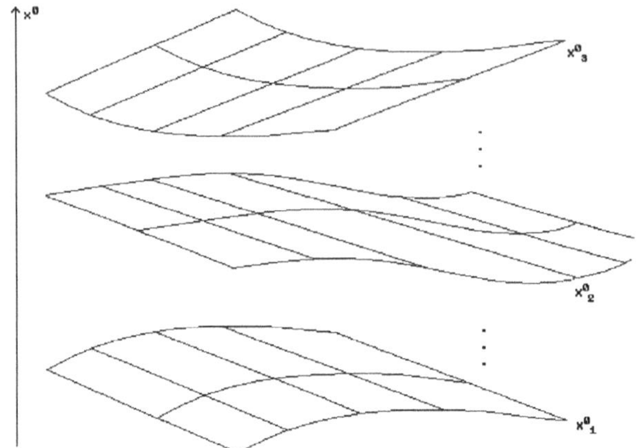

FIGURE 16. The evolution of a classical relativistic wave field in the course of time as observed in an inertial frame. The continuous set of field configurations bounded between the configuration specified at $x^0{}_1$ and that specified at $x^0{}_3$ is a four-dimensional geometric structure which describes the dynamical behavior of the classical wave field within the lapse $x^0{}_3 - x^0{}_1$.

electric charges and gravitational fields by concentrations of matter and energy. Yet, as I stressed in the Introduction, all fields - both, the "interactions between particles" and the "particles" themselves - are fundamentally quantum and their dynamical behavior is revealed to us only through the quantization of the corresponding abstract classical wave fields. From the rigorous perspective of quantum field theory, all fields are primary entities of independent existence. They interact with other fields, but they are never produced by them. For example, the photon field interacts with the electron field but it is not produced by the electron field any more than the electron field is produced by the photon field! At infinite space-time separations the two quantum fields decouple and the classical electromagnetic waves emerge as the semi-classical limit of the photon field. We shall, in due course, examine the specific circumstances which obscure the primary character of the electromagnetic field and generate the appearance of the latter as the product of sources. At this stage we only stress that it is the character of fields as fundamental and primary physical entities which motivates our approach to classical fields as wave fields dissociated from sources.

The analysis associated with Fig.16 constitutes, for that matter, a generic description of all classical fields as wave fields and, thus, as primary entities. We are referring to wave fields as free fields of course, infinitely-extended waveforms which freely propagate in space. Yet, we cannot in the least dispense with interactions. A theory of free fields alone has no physical content. The essence of the physical world is revealed to us only through the interactions between fields. It is precisely these interactions which crystallize in the appearance of material bodies and particles producing fields. In order to obtain precise knowledge of a system of interacting quantum fields, we need to quantize the corresponding system of interacting classical wave fields. For example, in order to understand the dynamical behavior of the system of the quantum electron field and the photon field in interaction with each other we need to dispense with the notion of the Maxwell field as the product of sources and quantize, instead, the system of the classical electron wave field in interaction with the Maxwell field considered as a primary wave field. If we could accomplish this task, we would have precise knowledge of the dynamical behavior of the interacting quantum system. We would then obtain classical electrodynamics, in whose context the Maxwell field is the product of sources, as a certain semi-classical limit of the interacting quantum system. Thus, upon recalling that the transition from classical to quantum physics signifies the replacement of the unique classical trajectory by infinitely many possible trajectories, it becomes clear that precise knowledge of the solution to the field equations which express the dynamical behavior of a system of interacting classical wave fields is required as a condition for quantization.

We may now answer our antecedent question. From both, the conventional perspective of interaction-mediating classical fields and from the rigorous perspective of primary classical wave fields the situation is as follows: In the absence of interactions, the exact solution to the field equations is invariably known. In the presence of interactions, the exact solution to the field equations is generally - but not invariably - unattainable. In general, the equations which describe interacting fields are highly non-trivial and their exact solution can only be attained in exceptional cases. Examples of such cases as allow for the possibility of an exact solution are that in which the fields remain free in vacuum (away from their sources) and that of high symmetry (N1).

The decisive impact which interactions have on the dynamical behavior of fields is reflected in the distinction between *free classical fields* and *interacting classical fields*, a distinction equally valid in the conventional and the rigorous approach. The solution to the equations of free classical fields is invariably attainable, with less than considerable effort at that! The reason is that such field equations are *linear differential equations*. Inherent in this technical term are the two characteristic properties which such equations have and which will suffice for the purposes of our analysis. The first, is that of an already familiar mathematical concept: The general solution is a linear combination of simple solutions which are easy to obtain. This property is of enormous physical importance as will become clear in what follows. The second property is essentially a consequence of the first and involves the boundary conditions which we shall discuss soon. The linear character of these equations determines that a small change in the boundary conditions imposed on the solution only have an accordingly small effect on the solution itself. In order to obtain a feeling of this situation consider a stretched spring set loose. It will naturally execute oscillations the amplitude of which is determined by the initial stretch. If, now, you repeat the procedure by stretching the spring a little more or a little less than before then the spring will execute oscillations the amplitude of which will, accordingly, be a little bigger or a little smaller than before.

In general, the solution to differential equations which describe fields in interaction with each other is associated with formidable difficulties. The reason is that such field equations are *non-linear differential equations*. The highly non-linear differential equations of general relativity are, indeed, a reflection of the self-interacting nature of the gravitational field. Non-linear differential equations do not have the property of superposition. It is, in fact, the absence of this property which underscores the non-trivial task of solving non-linear differential equations. Closely associated with this fact is the property which relates to boundary conditions. A small change in the boundary conditions imposed on the solution to a non-linear differential equation can have an enormous effect on the solution itself. Almost everybody is familiar with the avalanche effect which a shout can trigger on a snow-covered mountain slope, or with the anecdotal boisterous effect which the flutter of a butterfly can have on the weather miles away from the butterfly itself. Thus, it can be seen that, even though non-linear differential equations describe the vast majority of realistic situations (from the formation of a cloud, to a thought process in the human brain!), only rarely can the solution to such equations be obtained.

Let us now proceed to the implications of linearity. The character of the general solution to a linear differential equation as a linear combination of simple solutions, directly implies the physical property of superposition. This can be readily understood through the analogy with the undulating infinite sheet. The dynamical behavior which that paradigm signifies is in essence identical to an arbitrary waveform propagating along a perfectly elastic chord of infinite length, albeit in three-dimensional space. As an arbitrary one-dimensional waveform is a superposition of infinitely many harmonic (sinusoidal) waves of all wavelengths, so a classical wave field is a superposition of infinitely many plane waves of energy of all wavelengths. The mathematical expression of this superposition is the linear combination. A classical wave field is a linear combination of plane waves of energy. This heuristic result will, in fact, be so crucial to quantum field theory, that it is necessary to derive it from a rigorous mathematical perspective and explore its physical implications.

2. Classical Wave Fields and the Harmonic Oscillator

The *general solution* to a linear differential equation is always a linear combination of *partial solutions*. We are familiar with this fact. In the third section of the first chapter we saw that the eigenstates of an operator which expresses an observable physical attribute form a complete set. An arbitrary wavefunction $\psi(t, x)$ is a linear combination of that operator's eigenfunctions $\psi_1(t, x)$, $\psi_2(t, x)$, $\psi_3(t, x)$, ... This is not surprising. Quantum mechanics is a linear theory. The Schrödinger equation is an example of a linear differential equation. Both, $\psi(t, x)$ and the eigenfunctions $\psi_n(t, x)$ are solutions to it, with the former being the general solution and the latter being the partial solutions. Accordingly, the linear character of a free classical field $\phi(x^0, \vec{x})$ expresses the fact that the latter as the general solution to the corresponding field equation is a linear combination

$$\phi(x^0, \vec{x}) = c_1\phi_1(x^0, \vec{x}) + c_2\phi_2(x^0, \vec{x}) + ... + c_n\phi_n(x^0, \vec{x}) + ... \qquad (2.I)$$

of functions $\phi_1(x^0, \vec{x})$, $\phi_2(x^0, \vec{x})$, $\phi_3(x^0, \vec{x})$, ... each of which is a partial solution to that equation. In order to uniquely determine $\phi(x^0, \vec{x})$ we also need to specify the coefficients c_1, c_2, c_3, ... We have seen in the first section of the first chapter that the specification of boundary conditions specifies a unique solution to the field equations. Thus, these coefficients are determined by the specific boundary conditions which the free classical field $\phi(x^0, \vec{x})$ is expected to satisfy in the specific physical context. For example, certain fields are expected to vanish at arbitrarily large distances and certain others are required to vanish on boundary surfaces.

The simplest free classical fields are the classical wave fields. The partial solutions to a wave-field equation are the plane waves of energy. We know from the first chapter that the classical plane waves of energy are sinusoidal functions of space and time with temporal dependence of the form e^{-iwt}, w being the frequency of oscillation. Each plane wave is specified and labelled exclusively by the magnitude of its wave number. Unlike the one-dimensional physical contexts which we examined in the first chapter, we are now dealing with actual plane waves propagating in physical three-dimensional space. Since a plane wave may propagate in any direction in three-dimensional space, the quantity $\frac{2\pi}{\lambda}$, which in one-dimensional space expresses the wave number k in terms of the wavelength λ, now necessarily expresses the magnitude $|\vec{k}|$ of *wave vector* $\vec{k} = (k^1, k^2, k^3)$ oriented in the direction of propagation. This simply means that if $\vec{k}_1$, $\vec{k}_2$ and $\vec{k}_3$ are unit vectors (each, of magnitude equal to one) respectively aligned in the direction of coordinate x^1, of coordinate x^2 and of coordinate x^3, then $\vec{k} = k^1\vec{k}_1 + k^2\vec{k}_2 + k^3\vec{k}_3$ with $|\vec{k}| = +\sqrt{(k^1)^2 + (k^2)^2 + (k^3)^2} = \frac{2\pi}{\lambda}$. We may simplify the notation by setting $k \equiv |\vec{k}|$ and state that each wave number k defines a specific frequency w_k. As the plane waves of probability are of the form $e^{\frac{i}{\hbar}px}$ with $p = \hbar k$, it is clear that the spatial dependence of the classical plane waves of energy in physical three-dimensional space is of the form $e^{i\vec{k}.\vec{x}}$, with $\vec{k}.\vec{x} = k^1x^1 + k^2x^2 + k^3x^3$. In the next chapter we shall substantiate these qualitative arguments as to the mathematical expression of the classical plane waves of energy. Here we only add that solutions to wave-field equations which express classical plane waves of energy are real functions of space-time, as all functions which express observable quantities are. The use of the stated complex exponential expressions is merely a matter of convenience since the latter are more tractable than the cosines and sines. Eventually, however, it is the "real part" $cosx$ in $e^{ix} = cosx + isinx$ (or the "imaginary part" $sinx$) which alone constitutes the solution to a wave-field equation. By contrast, the plane waves of probability, being unobservable (as all wavefunctions are), are entirely complex functions of space and time as we have seen (refer to (3.IIc) in the first chapter).

It would appear, for that matter, that the partial solutions to a wave-field equation are the plane waves:

$$c_k e^{-i(w_k t - \vec{k}.\vec{x})} \;,\; c_{k'} e^{-i(w_{k'} t - \vec{k}'.\vec{x})} \;,\; c_{k''} e^{-i(w_{k''} t - \vec{k}''.\vec{x})}, \ldots$$

This, however, is not the complete set of partial solutions. The complex conjugate of each such partial solution is also a partial solution. The reason that wave-field equations have this fundamental property will also become clear in the next chapter. The essential information for our purposes now is the fact that the complete set of partial solutions to a wave-field equation also includes the plane waves:

$$c_{\vec{k}}^{*}e^{i(w_k t-\vec{k}.\vec{x})} \ , \ c_{\vec{k}'}^{*}e^{i(w_{k'} t-\vec{k}'.\vec{x})} \ , \ c_{\vec{k}''}^{*}e^{i(w_{k''} t-\vec{k}''.\vec{x})}, ...$$

Thus, in view of the fact that the set $k, k', k'', ...$ is continuous, the general solution $\phi(x^0, \vec{x})$ to a wave-field equation is the linear combination

$$\phi(x^0, \vec{x}) = \int d^3k [c(k)e^{-i(w_k t-\vec{k}.\vec{x})} + c^*(k)e^{i(w_k t-\vec{k}.\vec{x})}] \qquad (2.II)$$

where we state again and stress that this is the expression for the scalar (Klein-Gordon) field which we shall study in the next chapter, yet which we use here as generic to all free wave fields on account of its comparative simplicity. In general, alongside the continuous summation over k there is also a discrete summation over additional field degrees of freedom, as we shall see. What exclusively matters for our analysis here, is the fact that all wave fields receive the same decomposition in terms of plane waves as that in (2.II).

The character of the coefficients $c(k)$ and $c^*(k)$ in (2.II) remains to be determined. First, however, we elucidate the character of the integral in this expression. The significance of this three-dimensional integration is $\int d^3k = \int_{-\infty}^{+\infty} dk^1 \int_{-\infty}^{+\infty} dk^2 \int_{-\infty}^{+\infty} dk^3$. That is, continuously sum the integrand (i.e.: $c(k)e^{-i(w_k t-\vec{k}.\vec{x})} + c^*(k)e^{i(w_k t-\vec{k}.\vec{x})}$) over k^3 from $-\infty$ to $+\infty$ keeping k^2 and k^1 fixed, repeat for k^2 keeping k^1 fixed and, finally, repeat for k^1. Equivalently, $\int d^3k$ can be understood as the continuous summation over all magnitudes k in all possible directions of propagation in three-dimensional space. If this point appears to be somewhat obscure, think of a sphere of radius k and continuously sum the quantity $c(k)e^{-i(w_k t-\vec{k}.\vec{x})}$ over all points of that sphere. Having established this result, repeat the procedure on a sphere of infinitesimally bigger radius $k' = k + dk$ and continue in this manner indefinitely until all spheres of radii between zero and infinity have been considered. Next, continuously sum over these results. Repeat this procedure for $c^*(k)e^{i(w_k t-\vec{k}.\vec{x})}$. Finally, add up the two integrals.

At this point it is necessary that we briefly digress in order to elucidate an aspect of the relativistic notation which we shall introduce in what follows. We have reserved the symbol k for the magnitude $|\vec{k}|$ of the wave vector ($k \equiv |\vec{k}|$) and we shall accordingly use the symbol p for the magnitude $|\vec{p}|$ of the momentum ($p \equiv |\vec{p}|$). In relativistic Physics, it is standard practice to use the same notation k and p for the magnitudes of the 4-vectors k^μ, the frequency-wavelength vector which we introduce immediately below and p^μ, the energy-momentum vector. That is, $k = \sqrt{\eta_{\mu\nu}k^\mu k^\nu} = \sqrt{(k^0)^2 - |\vec{k}|^2}$ and $p = \sqrt{\eta_{\mu\nu}p^\mu p^\nu} = \sqrt{(p^0)^2 - |\vec{p}|^2}$. The context will always make it clear which of the two physical quantities k and p respectively refer to. An additional point of caution concerns expressions such as $k.x = \eta_{\mu\nu}k^\mu x^\nu = k^\mu x_\mu = k^0 x^0 - \vec{k}.\vec{x}$, $k^2 = k.k = \eta_{\mu\nu}k^\mu k^\nu = k^\mu k_\mu = (k^0)^2 - |\vec{k}|^2$ and the corresponding to them expressions obtained through the replacement of k^μ by p^μ. In all such expressions k, p and x respectively signify k^μ, p^μ and x^μ. This is always the case: When a symbol ($k, p, x, ...$) either precedes, or follows a dot, it signifies a corresponding 4-vector.

As we remarked, all inertial observers in Minkowski space-time must agree that $\phi(x)$ in (2.II) is a general solution to the relevant wave-field equation as a condition that $\phi(x)$ express a field. By this we do not mean that all inertial observers necessarily agree on a specific value ϕ at any given point x. The field which we here denote by the generic $\phi(x)$ may be a vector field, a spinor field, or a tensor field, a field which comprises components such that if a certain inertial observer registers only one of them, another one will register a different component, or a linear combination of components. If $\phi(x)$ is a field, all observations relate to each other through Lorentz transformations in such a manner that the corresponding wave-field equation has the same form for all inertial observers. It follows that if the coefficients $c(k)$ and $c^*(k)$ are merely a matter of choice then (2.II) is not Lorentz covariant. This is clearly evident in the three-dimensional character of the integral which purports to express the field $\phi(x)$ in four-dimensional Minkowski space-time. In order that the solution (2.II) be physically acceptable, $c(k)$ and $c^*(k)$ must be such as to ensure the Lorentz covariance of (2.II) in addition to being determined by whichever choice of field configuration.

Which requirements then must (2.II) satisfy that it may be Lorentz covariant? (i) It must signify a four-dimensional integral $\int d^4k$, instead of the three-dimensional $\int d^3k$, with $k^\mu = (k^0 = \frac{w}{c}, k^1, k^2, k^3) = (k^0, \vec{k})$ being a 4-vector in Minkowski space-time as $p^\mu = \hbar k^\mu$, $p^\mu = (p^0 = \frac{E}{c}, p^1, p^2, p^3) = (p^0, \vec{p})$ also is. (ii) It must be consistent with the relativistic relation

$$w^2 = |\vec{k}|^2 c^2 + m^2 c^4 \implies (k^0)^2 - |\vec{k}|^2 = m^2 c^2 \implies k^2 = m^2 c^2$$

between frequency and wave number the validity of which follows directly from the relativistic relation

$$E^2 = p^2 c^2 + m^2 c^4 \implies (p^0)^2 - |\vec{p}|^2 = m^2 c^2 \implies p^2 = m^2 c^2$$

between energy and momentum (refer to the Appendix). (iii) Since $w^2 = |\vec{k}|^2 c^2 + m^2 c^4 \implies w = ck^0 = \pm\sqrt{|\vec{k}|^2 c^2 + m^2 c^4}$, it must rule out the negative value of w. The reason is that a negative value of w implies a negative value of $E = \hbar w$. We shall see in the fifth chapter that systems which allow for negative energies while evolving from past infinity to future infinity are physically unacceptable! Instead of (2.II), we consider, for that matter

$$\phi(x^0, \vec{x}) = \frac{1}{\sqrt{(2\pi)^3}} \int d^4k \, \delta(k^2 - m^2 c^2) \theta(k^0) [C(k)e^{-i(w_k t - \vec{k}.\vec{x})} + C^*(k)e^{i(w_k t - \vec{k}.\vec{x})}] \quad (2.IIIa)$$

Allowing for the numerical factor which multiplies the integral, the δ-function $\delta(k^2 - m^2 c^2)$ enforces the second requirement since now all values of k^0 for which $(k^0)^2 - |\vec{k}|^2 \neq m^2 c^2$ yield zero contributions to the integral (qualitatively, this should already be obvious from $\delta(x - x_0)$ in Fig.2(a)). The *Heaviside function* $\theta(k^0)$ which we shall formally introduce in the next chapter receives the value 1 if $k^0 \geq 0$ and the value 0 if k^0 is negative, thus enforcing the third requirement.

Thus, $\phi(x^0, \vec{x})$ in (2.IIIa) is a Lorentz covariant expression. We would appreciate the significance of this statement all the more if we used the relation

$$w_k t - \vec{k}.\vec{x} = k^0 x^0 - \vec{k}.\vec{x} = \eta_{\mu\nu} k^\mu x^\nu \equiv k.x$$

in order to recast (2.IIIa) in the form

$$\phi(x) = \frac{1}{\sqrt{(2\pi)^3}} \int d^4 k \, \delta(k^2 - m^2 c^2) \theta(k^0) [C(k) e^{-ik.x} + C^*(k) e^{ik.x}] \qquad (2.IIIb)$$

which does not separate space from time, rendering Lorentz covariance explicit in view of the fact that invariant exponentials ($k.x$ is explicitly Lorentz invariant) are integrated over an invariant volume element.

The task now is to determine the conditions which $c(k)$ and $c^*(k)$ must satisfy in order that (2.III) be equivalent to (2.II). To this end, we shall need the identity

$$\delta(f(x)) = \frac{\delta(x - x_0)}{|f'(x_0)|} \; ; \; f(x_0) = 0 \qquad (2.IV)$$

on the understanding that if $f(x)$ vanishes at multiple values of x ($f(x_0) = f(x_0') = ... = 0$), we accordingly sum the contributions stemming from these values. The denominator on the right side is, of course, the absolute value of the first derivative of $f(x)$ at point $x = x_0$: $|f'(x_0)| = |\frac{df(x)}{dx}|_{x=x_0}|$.

We note that since $k^2 - m^2 c^2 = (k^0)^2 - |\vec{k}|^2 - m^2 c^2 = (k^0)^2 - (|\vec{k}|^2 + m^2 c^2)$, it follows that

$$\delta(k^2 - m^2 c^2) = \delta\big((k^0)^2 - (|\vec{k}|^2 + m^2 c^2)\big) = \delta\big((k^0 - \sqrt{|\vec{k}|^2 + m^2 c^2})(k^0 + \sqrt{|\vec{k}|^2 + m^2 c^2})\big)$$

where in order to advance the second equality, we used the elementary identity $a^2 - b^2 = (a+b)(a-b)$. Of the four independent variables k^μ which $k^2 - m^2 c^2$ comprises, the one of interest in (2.IV) is the "energy" k^0, since the values of k^0 for which $k^2 - m^2 c^2 = 0$ are the values which yield the classical relativistic energy-momentum relation $k^0 = \pm\sqrt{|\vec{k}|^2 + m^2 c^2}$. For that matter, (2.IV) implies

$$\delta(k^2 - m^2 c^2) = \frac{\delta(k^0 - \sqrt{|\vec{k}|^2 + m^2 c^2})}{2k^0} + \frac{\delta(k^0 + \sqrt{|\vec{k}|^2 + m^2 c^2})}{2k^0} \; ;$$

$$k^0 = +\sqrt{|\vec{k}|^2 + m^2 c^2} \qquad (2.V)$$

where the operation $\frac{d(k^2 - m^2 c^2)}{dk^0} = \frac{d((k^0)^2 - (|\vec{k}|^2 + m^2 c^2))}{dk^0} = 2k^0$ which yields the denominator in the two terms is an expression of the elementary rule of differentiation $(x^a)' = ax^{a-1} \Longrightarrow (x^2)' = 2x$.

The replacement of (2.V) in the four-dimensional integral in (2.III) now, yields:

$$\int d^4 k \, \delta(k^2 - m^2 c^2)\theta(k^0) = \int d^3 k \int_0^{+\infty} \frac{dk^0}{2k^0}\delta\left(k^0 - \sqrt{|\vec{k}|^2 + m^2 c^2}\right) =$$

$$\int \frac{d^3 k}{2k^0} \quad ; \quad k^0 = \sqrt{|\vec{k}|^2 + m^2} \qquad\qquad (2.VI)$$

on the understanding that, in enforcing the value $k^0 = -\sqrt{|\vec{k}|^2 + m^2 c^2}$, the second term in (2.V) yields zero contribution on account of $\theta(k^0)$. Replacing (2.VI) in (2.III) yields the <u>Lorentz covariant</u> expression

$$\phi(x^0, \vec{x}) = \int \frac{d^3 k}{\sqrt{(2\pi)^3 2k^0}}[a(k)e^{-i(w_k t - \vec{k}.\vec{x})} + a^*(k)e^{i(w_k t - \vec{k}.\vec{x})}] \qquad\qquad (2.VIIa)$$

or, in a more elegant form

$$\phi(x) = \int \frac{d^3 k}{\sqrt{(2\pi)^3 2k^0}}[a(k)e^{-ik.x} + a^*(k)e^{ik.x}] \qquad\qquad (2.VIIb)$$

where $a(k) = \frac{C(k)}{\sqrt{2k^0}}$, $a^*(k) = \frac{C^*(k)}{\sqrt{2k^0}}$. Thus, the general solution to a wave-field equation is of the form (2.VII). Comparing (2.VII) to (2.II) reveals that the latter is a Lorentz covariant expression only if $c(k) = \frac{a(k)}{\sqrt{(2\pi)^3 2k^0}}$, $c^*(k) = \frac{a^*(k)}{\sqrt{(2\pi)^3 2k^0}}$.

The general solution in (2.VII) expresses the evolution of a field configuration (Fig.16). Which field configuration is a matter of choice. Two distinct field configurations respectively specified on the same three-dimensional surface of equal time yield two distinct evolution patterns for $\phi(x^0, \vec{x})$ (two different undulation patterns in the pictorial analogue in Fig.16). The general linear combination in (2.VII) expresses, for that matter, all infinitely many possible evolution patterns. Which one the classical wave field will follow depends exclusively on the values of the coefficients $a(k)$ and their complex conjugates. These coefficients receive a very simple physical interpretation. Since each plane wave in the continuous summation in (2.VII) is specified exclusively by the magnitude of its wave number k, the corresponding coefficients $a(k)$ and $a^*(k)$ are respectively the maximum amplitudes of that wave and of the wave which the complex conjugate term expresses[51]. A choice of maximum amplitude $a(k)$ made for each k at a given moment, specifies a field configuration on the corresponding three-dimensional surface of equal time. Since as a consequence of the wave field's linear nature there is no interaction between the plane waves, once such a choice has been made all amplitudes remain fixed in their chosen values and uniquely specify an evolution pattern. We shall see that this has an important implication for the free quantum fields.

Inherent in (2.VII) is a crucial physical implication. We can elicit it by analogy to the waveform propagating along a perfectly elastic chord of infinite length. Any such

---

[51]For a plane wave of energy, the amplitude $A(k)$ is a real function. Measuring lengths and durations from an arbitrary space-time point results in $A(k)e^{-i(k.x+\phi)} = A(k)e^{-ik.x}e^{-i\phi} = (A(k)e^{-i\phi})e^{-ik.x} \equiv a(k)e^{-ik.x}$. Again, only the real part expresses the plane wave of energy.

waveform is a superposition of infinitely many harmonic waves of all wave numbers. Each harmonic wave in this superposition is propagating energy stemming from the oscillatory motion of the molecules throughout the elastic chord. Each molecule interacts with the molecules in its immediate vicinity imparting its oscillatory energy to them and causing their oscillation. The interaction which exists between the molecules of the chord results in coupled oscillations. This invites a direct association with the infinitely many degrees of freedom of a classical wave field. In direct analogy with each molecule of a vibrating chord, each degree of freedom $\phi(x^0, \vec{x})$ specified at a given point $\vec{x}$ behaves essentially as an oscillating particle. In turn, each plane wave in the resolution of a classical wave field $\phi(x^0, \vec{x})$ is the result of coupled oscillations throughout the continuous set of degrees of freedom. It is as though the classical wave field $\phi(x^0, \vec{x})$ consists of infinitely many oscillators coupled amongst themselves through an infinite succession of joint springs throughout space (Fig.17).

FIGURE 17. Each of the infinitely many field degrees of freedom $\phi(x^0, \vec{x})$ (dark blobs) of a classical wave field is an oscillator. The actual classical wave field emerges at the continuous limit of the infinite concatenation of joint oscillators.

Thus, by formally generalizing the dynamical behavior of oscillating particles at the continuous limit which the infinitely many degrees of freedom signify, we have arrived at a central aspect of field theory: A wave field is a set of infinitely many coupled oscillators! We have already studied the simple harmonic oscillator in the first chapter. It is one of the simplest, yet most important physical systems. The simplicity of the harmonic oscillator is manifest in the potential well against which it evolves. We have seen that a classical simple harmonic oscillator is a particle which executes oscillations within the parabolic potential well of Fig.18(a). Such dynamical behavior is expressed by a simple solution to an elementary linear differential equation. Systems of coupled oscillators, on the contrary, are not as simple. Although their equations of motion remain tractable for a small number of oscillators, the complexity of such equations increases as that number increases, yielding a thoroughly intractable problem for the continuous set of the infinitely many coupled oscillators in Fig.17. Yet, the solution to the linear differential equations of classical wave fields is, almost effortlessly, attainable! We expect, for that matter, this apparent contradiction to be resolved by an underlying reduction of the continuous set of coupled oscillators in Fig.17 to a continuous set of simple harmonic oscillators. Before we proceed in this direction, note that there is actually nothing exceptional in such couplings as are those which characterize the oscillators in Fig.17. Oscillating physical systems are, by nature, very social! They love to couple and interact amongst themselves! From the molecules of chords and the atoms in crystals to the atoms which constitute the molecules in chemistry and biology the dynamical behavior is that of coupled oscillators. It is, in fact, the simple harmonic oscillator which is the exception in Nature, rather than the rule. The reason that, despite its rather exceptional character, the simple harmonic oscillator is one of the most important physical systems stems from the fact that the dynamical behavior of all other oscillating systems is invariably reducible to that of the simple harmonic oscillator. We shall demonstrate this directly in the case of the classical wave fields.

We start from the physical content of (2.VII). A free classical wave field $\phi(x^0, \vec{x})$ is a superposition of infinitely many classical plane waves of energy of all wavelengths. Each plane wave as it is throughout space at a given moment has the same sinusoidal form which the behavior of a simple classical harmonic oscillator has in the course of time (Fig.18(b)). This is, of course, a direct consequence of the evolution of each field degree of freedom. If the evolving wave field $\phi(x^0, \vec{x})$ is a plane wave of energy (all $a(k)$ except for one are equal to zero), then at any specific point $\vec{x}$ the associated oscillator $\phi(x_0, \vec{x})$ is in the simple harmonic motion of Fig.18(b). The maximum amplitude of that harmonic motion clearly coincides with the unique maximum amplitude $a(k)$ of the plane wave and, as we saw in the first chapter, the square $|a(k)|^2$ determines the constant energy which that plane wave carries through a unit of area in the unit of time (recall in this respect that the energy of the harmonic oscillator is also exclusively determined by the square of its maximum displacement).

Thus, classical plane waves are sinusoidal waveforms throughout space at a given moment and throughout time at a given point. As a consequence of the first property, a classical plane wave is characterized by fixed wavelength λ and, thereby, by fixed wave number $k = |\vec{k}|$. As a consequence of the second property, a classical plane wave is characterized by constant (time-independent) frequency of oscillation w at each point in space. We stress again that the fixed wave number k directly determines the constant frequency of oscillation w common to all degrees of freedom $\phi(x^0, \vec{x})$ through the relativistic relation $w = \sqrt{(kc)^2 + m^2 c^4}$. Pursuant to the Appendix, the rest-mass m of the classical wave field vanishes for fields which propagate at the speed of light c, as is the case, for example, with the electromagnetic wave field. At each point $\vec{x}$ in space, $\mathbf{E}(x^0, \vec{x})$ and $\mathbf{B}(x^0, \vec{x})$ of an electromagnetic plane wave respectively oscillate with constant frequency $w = kc$ (Fig.18(b)).

From these considerations it follows that each plane wave in the resolution of a classical wave field $\phi(x^0, \vec{x})$ is the result of harmonic oscillations of constant frequency and amplitude common to all coupled oscillators in Fig.17. In turn, the resolution of a free classical wave field in infinitely many plane waves of different wave numbers directly implies the announced reduction: Each mode of oscillation, however complicated, which each coupled oscillator in Fig.17 has, is a superposition of infinitely many elementary modes of simple harmonic oscillation of different frequencies and amplitudes, as the dynamical behavior, however complicated, of $\phi(x^0, \vec{x})$ is a superposition of infinitely many plane waves of different wave numbers and amplitudes! Each coupled oscillator resolves itself in infinitely many simple harmonic oscillators of different frequencies and amplitudes!

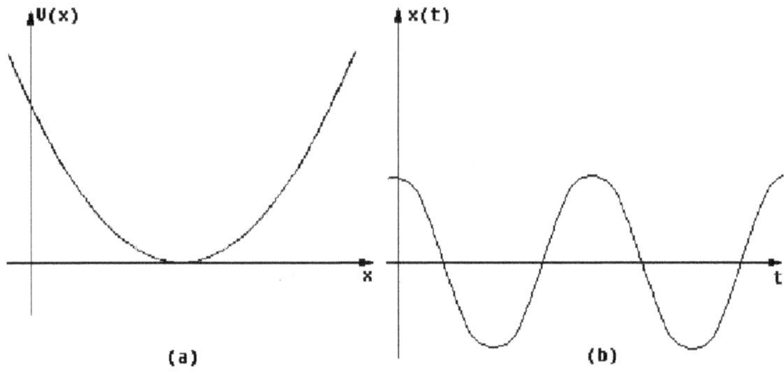

FIGURE 18. The parabolic potential well for simple harmonic motion (a) and simple harmonic motion (constant amplitude and constant frequency) in the course of time (b). The "steeper" the potential in (a) is, the higher the frequency in (b) is. (In both, (a) and (b), the location of the origin $(0,0)$ is arbitrary).

The conclusions to which we have arrived reveal that each plane-wave configuration in the resolution of a classical wave field behaves as though it were in its entirety a simple classical harmonic oscillator described by the graph in Fig.18(b). In fact, that simple classical harmonic oscillator is the abstract representation of the corresponding concrete plane wave and constitutes a generic representation of the infinitely many simple harmonic oscillators which produce that plane wave. Each plane-wave configuration corresponds to a specific wave number k and, thus, receives an abstract representation as a simple harmonic oscillator the frequency of which is determined exclusively by k and the amplitude of which (a matter of choice) is specific to k. The higher the value of k is, the higher the frequency w of the abstract simple harmonic oscillator is and, for that matter, the steeper the corresponding parabolic potential is (Fig.19), a fact which can be directly deduced from Fig.18. This result is of crucial importance! Now each such abstract simple harmonic oscillator is utterly independent from all others and evolves in complete dissociation from all others for the same reason that in the superposition of the infinitely many plane-wave configurations each configuration evolves independently from all others - that is, due to the linear nature of the free field. The direct implication is that a classical wave field resolves itself to infinitely many de-coupled abstract simple harmonic oscillators each of which uniquely corresponds to wave number k and generically represents infinitely many concrete simple harmonic oscillators - one at each point in physical space - of common frequency w determined exclusively by k and of common amplitude specified at the outset as a matter of choice. This global result mirrors the local result to which we arrived just above: A classical wave field is a superposition of infinitely many de-coupled abstract simple harmonic oscillators, as each coupled oscillator in Fig.17 is a superposition of infinitely many concrete simple harmonic oscillators!

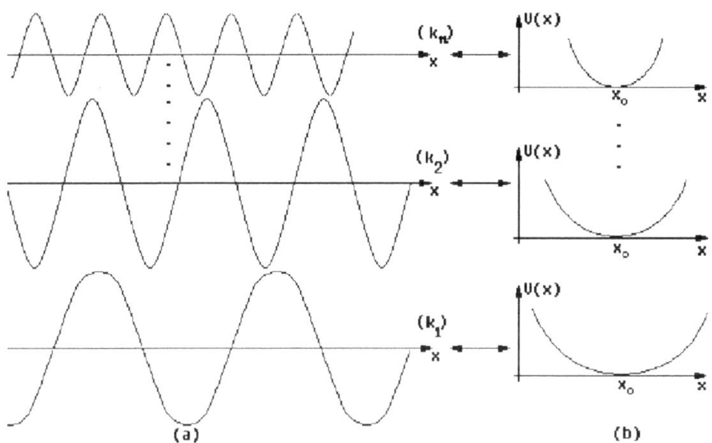

FIGURE 19. One-dimensional representation of the infinitely many plane-wave configurations which compose a classical wave field $\phi(x^0, \vec{x})$ at a certain moment (two dimensions of space have been suppressed). Each plane-wave configuration corresponds to a specific wave number $k_1 < k_2 < ... < k_n < ...$ (a). Abstract representation of each plane-wave configuration as a simple classical harmonic oscillator. The higher k is, the steeper the parabolic potential is. The abstract representation coincides with the concrete simple harmonic oscillator at each point $\vec{x} = \vec{x}_0$ (b).

Thus, by skillfully exploiting the property of superposition which characterizes the free fields we have managed to express the concrete set of coupled oscillators in physical three-dimensional space as an abstract set of de-coupled simple harmonic oscillators! These two sets constitute two distinct and equivalent representations of a classical wave field. In the continuous collection of concrete coupled oscillators throughout space (Fig.17), each oscillator is naturally labelled by its equilibrium point $\vec{x}$ about which it oscillates. In the continuous collection of abstract de-coupled harmonic oscillators, each oscillator is naturally labelled by the wave number $k = |\vec{k}|$ which characterizes the corresponding plane wave of energy (Fig.19(a)) (accordingly, each concrete simple harmonic oscillator in the resolution of a coupled oscillator is labelled by its frequency $w(k)$ and, consequently, also by k). A classical wave field amounts to the continuous set $\phi(x^0, \vec{x})$ of concrete coupled oscillators evolving in Minkowski space-time (Fig.17). Equivalently, it amounts to the continuous set $\phi(x^0, \vec{k})$ of abstract de-coupled simple harmonic oscillators depicted in Fig.19(b).

The collection of all wave vectors $\vec{k}$ is strongly reminiscent of the collection of all points $\vec{x}$ which formally defines the three-dimensional space. The only difference is that the latter collection is physical, whereas the former collection is abstract. Thus, if we define an abstract three-dimensional space which consists of all $\vec{k}$ then in that abstract space each magnitude $k = |\vec{k}|$ labels the corresponding abstract simple harmonic oscillator and each point $\vec{k}$ signifies, in addition, the orientation of that oscillator. This abstract three-dimensional space must necessarily be an aspect of a higher-dimensional entity. We are dealing with relativistic plane waves of energy propagating in four-dimensional Minkowski space-time. Since each $\vec{k}$ relates to three-dimensional space in terms of distance (wavelength) and orientation, it must be the three-dimensional component of a four-dimensional vector. We already know that the fourth component of that vector is essentially the frequency w, a quantity which signifies the number of oscillations over a given duration and thus clearly relates to time. Since, for that matter, $k^0 = \frac{w}{c}$, the collection of all points $k^\mu = (k^0, k^1, k^2, k^3)$ defines an abstract four-dimensional space. Each point in this space whose zero component k^0 relates to $\vec{k}$ through $k^0 = \sqrt{k^2 + m^2 c^2} \Leftrightarrow w = \sqrt{k^2 c^2 + m^2 c^4}$ corresponds to a plane wave of energy of wave four-vector k^μ. The essence of this abstract four-dimensional space will become transparent soon, when we introduce the concept of momentum space. Now, let us reiterate the central result to which we have arrived:

A classical wave field is a continuous collection of concrete coupled oscillators in Minkowski space-time (Fig.17). These oscillators de-couple only in the abstract four-dimensional space which we have just defined. In that abstract space, a classical wave field is a continuous collection of simple harmonic oscillators (Fig.19).

This reduction of the incipient set of coupled oscillators to a set of simple harmonic oscillators resolves the apparent contradiction between the inextricable difficulties implied by the former and the simple solution implied by the linear nature of the classical wave fields and will be of crucial importance to quantization.

We may now proceed to quantum field theory. As I remarked in the first chapter, the quantization of classical fields has been accorded the term "second quantization" because historically it followed the first quantization of quantum mechanics. There was, nevertheless, a peculiar twist in that course of events. The work of Max Planck on the radiation emitted by a black body and the subsequent work of Albert Einstein revealed that, in sharp contrast to the smooth electromagnetic waves of Maxwell theory, light consists in discrete photons. At the turn of the twentieth century, the profound implication which this fact has went unnoticed. Well before the development of quantum mechanics, the quantum character of the electromagnetic radiation strongly implied that the electromagnetic field is, in its entirety, a quantum field. In turn, this would eventually render the time-honored dichotomy between particles and fields physically irrelevant: All particles are expressions of underlying quantum fields!

3. Field Quantization - Fields as Particles

Let us first make an intuitive approach to quantum fields. Every high-school student is familiar with the electrostatic field, the field produced by stationary electrically-charged objects. Of course, all objects consist of particles, and particles are fundamentally quantum entities. Consider, for that matter, the Coulomb electrostatic field produced by an idealized classical point-like charge. That field is classical. Its strength receives a well defined value at each point throughout space which, in addition, remains constant in the course of time. If we now quantize that physical system, the idealized classical point-like particle will be replaced by an actual quantum particle - say, an electron. The electron is described by a probability distribution for position or, equivalently, by a probability distribution for momentum. This, in turn, directly implies that the field which the electron produces must itself be subjected to quantum fluctuations. The probabilistic character of the electron's position and momentum directly implies that, at each point in space and at each specific moment, the strength of the field which the electron produces has a probabilistic character itself. At each space-time point it consists in a continuous set of possible values each of which is realized with a specific probability. The electromagnetic field is indeed a quantum field, the photon field!

Having thus established the quantum character of the electromagnetic field, we can now dissociate it altogether from the electron which served as its source and study it as a free quantum field. Clearly, however, for all its intuitive value such an approach is not rigorous. It emanates from the classical notion of fields as products of sources and, thus, fails to account for the sources themselves as expressions of underlying fields. I have, in the Introduction, stressed the rationale concerning the association of each known particle with a classical relativistic wave field as the incipient point of field quantization. From the rigorous perspective of quantum field theory all fields are primary and fundamental physical entities, not products! For that matter, we start directly with an abstract classical relativistic wave field. Its quantization will yield the actual free quantum field. As was the case with the quantization of mechanical systems, there are two independent and equivalent procedures of quantization - namely, the canonical quantization and the path-integral quantization. We begin with the former.

The essence of canonical quantization for fields can be properly understood by considering the specific value which the classical relativistic wave field receives at a specific point in space at a given moment. In the first chapter we saw how the procedure of canonical quantization replaces, at each moment, the unique classical position by a continuous set of possible positions each of which is realized with a predictable probability $|\psi(t,\vec{x})|^2 dx^1 dx^2 dx^3$. The same procedure will, for that matter, also quantize a free classical wave field $\phi(x^0,\vec{x})$ if consistently applied to its infinitely many degrees of freedom. Clearly, the canonical procedure must now be independently applied on each degree of freedom at each point $\vec{x}$ in space and all infinitely many degrees of freedom $\phi(x^0,\vec{x})$ must be simultaneously quantized in an inertial frame. At each specific moment x^0 the canonical procedure of field quantization independently replaces the specific field-value $\phi(x^0,\vec{x})$ at each $\vec{x}$ by a continuous set of infinitely many possible field-values. Now, each point $\vec{x}$ on the three-dimensional surface of equal time x^0 merely labels the corresponding continuous set of possible field-values which describes the quantized dynamical variable[52]. The sharp difference with quantum mechanics is already manifest: It is because $\vec{x}$ merely labels the corresponding infinitely many possible values of what is now the dynamical variable, the quantized field, that $\vec{x}$ is not associated with probabilites. No probabilities for position exist! The probabilities are exclusively those of the quantized field $\phi(x^0,\vec{x})$. For that matter, each $\vec{x}$ labels at once the corresponding probability distribution for field-values. This sharp difference signifies the radical conceptual break with quantum mechanics which we discuss immediately below.

In the first chapter we saw that every quantized dynamical variable is represented by an operator. In quantum mechanics the position is a dynamical variable and the operator which represents it is the position operator $\hat{x}$. In quantum field theory the corresponding dynamical variable is the quantum field which, for that matter, is accordingly represented by an operator. Since at this stage we have agreed to denote a classical wave field by $\phi(x^0,\vec{x})$, the operator which represents the corresponding free quantum field is $\hat{\phi}(x^0,\vec{x})$. The physical content and mathematical significance of this operator will become clear as we advance the analysis in this and the next chapter.

Thus, it can be seen that the formal procedure of *equal-time canonical quantization* which has just been described results in the replacement of the entire classical configuration on a three-dimensional surface of equal time x^0 by infinitely many possible configurations on that surface. In sharp contrast to the classical situation in Fig.15, at each moment there are now infinitely many possible field configurations with each point $\vec{x}$ labeling infinitely many possible field-values (Fig.20). At each moment x^0 the free quantum field $\hat{\phi}(x^0,\vec{x})$ is a continuous set of possible field configurations each of which evolves in infinitely many distinct independent patterns to the future of x^0.

Up to this point we have applied the procedure of canonical quantization to a classical wave field in direct analogy to its application to a classical particle. As in "first quantization" the canonical procedure replaces the classically unique position of a particle by infinitely many possible positions at each moment, so in "second quantization" the canonical procedure replaces the classically unique value of a wave field at each point

---

[52]In direct analogy to the canonical procedure in quantum mechanics, the canonical procedure of field quantization equivalently accomplishes a replacement of the unique classical *field-momentum* $\pi(x^0,\vec{x})$ at each point $\vec{x}$ of the three-dimensional surface of equal time x^0. Although for the purposes of the ensuing analysis the exclusive reference to $\phi(x^0,\vec{x})$ suffices, do note that the field-momentum $\pi(x^0,\vec{x})$ is as physically important as the field $\phi(x^0,\vec{x})$ itself.

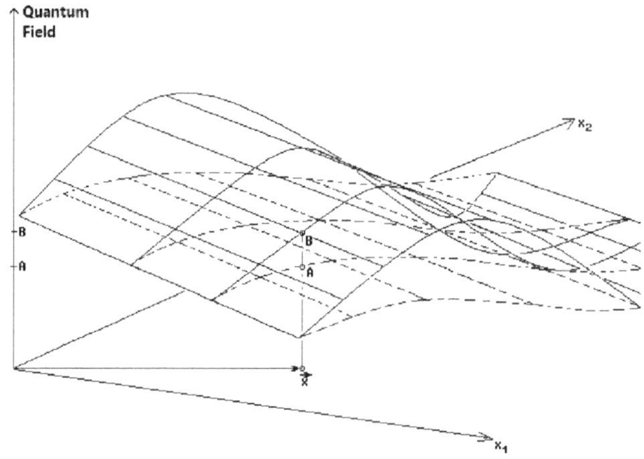

FIGURE 20. The sheets, respectively rendered by the continuous and syncopated lines, represent two of the infinitely many possible configurations of the free quantum field $\hat{\phi}(x^0, \vec{x})$ at a specific moment x^0. Each point $\vec{x}$ in space labels infinitely many possible field-values $\hat{\phi}(x^0, \vec{x})$, two of which (A and B) are depicted here.

of a three-dimensional surface of equal time by infinitely many possible field-values at that point. This, however, is as far as any analogy with quantum mechanics takes us. Inherent in the replacement of the classical field-value $\phi(x)$ at each space-time point $x \equiv (x^0, \vec{x})$ by infinitely many field-values of $\hat{\phi}(x)$ is the radical conceptual break which "second quantization" takes from "first quantization". Whereas quantum-mechanical particles are as immutable as their classical counterparts, in quantum field theory particles are manifestations of underlying quantum fields! A free quantum field expresses itself in terms of quantum particles!

We can grasp the essence of this fundamental consequence which the infinity of degrees of freedom has through the following conceptual argument. The infinitely many field configurations, themselves a direct consequence of the infinitely many quantized coupled degrees of freedom, must necessarily express the infinitely many quantized de-coupled degrees of freedom. In quantizing each coupled harmonic oscillator, the procedure of equal-time quantization equivalently quantizes each simple harmonic oscillator of the classical wave field in Fig.19 and, thereby, replaces each plane wave of energy by a plane wave of probability. This replacement will, in fact, constitute the foundation of our analysis on equal-time quantization. Yet, plane waves of probability are associated with momentum $\vec{p} \equiv (p^1, p^2, p^3)$ of fixed magnitude $|\vec{p}|$ and describe free quantum particles propagating with this momentum in three-dimensional space (Fig.2(b)). The relativistic momentum $|\vec{p}|$ determines a relativistic energy E. It follows that a free quantum field $\hat{\phi}(x)$ expresses itself in terms of *actual* relativistic free quantum particles of fixed momentum and energy!

This initial argument is inevitably crude. For each point $\vec{x}$ in space merely labels the dynamical variable, it is not a dynamical variable itself and cannot, for that matter, be associated with probability for position. The precise manner in which a quantum field manifests and expresses itself in terms of particles, as well as the precise significance of that manifestation and expression will be a central objective of the entire analysis in this and the next chapter. At this stage, let it suffice to state that particles are essentially an interpretation of the excitations of the underlying quantum field and that, for that matter, only those plane waves of probability which "tower" above the others in the spectrum of infinitely many labeled by $|\vec{p}|$ manifest the underlying quantum field as particles. It is in this sense that the semblance of a finite number of degrees of freedom with position $\vec{x}$ as a dynamical variable emerges when specific physical conditions are met. Not only does a quantum field express itself in terms of particles, the particles themselves express the underlying quantum field and its infinitely many degrees of freedom! In the words of Julian Schwinger[53]: *Experimentally, we are concerned with particles, yet the old equations describe fields... When you begin with field equations, you operate on a level where the particles are not there from the start. It is when you solve the field equations that you see the emergence of particles.*

We see, for that matter, that the physical essence of free quantum fields is independently described in two distinct languages, that of field configurations and that of quantum particles. These two languages are equivalent. A free quantum field is a collection of infinitely many evolving possible field configurations. Equivalently, a free quantum field is a collection of actual free quantum particles. Each distinct number of particles signifies a collection of infinitely many plane waves of probability the nature of which will be examined in the next section. It is that collection which, in fact, yields a set of infinitely many field-values at each point $\vec{x}$ at each moment x^0 - or, in the precise sense of relativistic covariance, at each space-time point $x \equiv (x^0, \vec{x})$. The set of infinitely many possible field configurations and the set of infinitely many plane waves of probability constitute the two equivalent and independent descriptions of a free quantum field.

This conclusion is abstract. It does not resolve the apparent contradiction between the infinitely many possible numbers of particles and the fact that only one number of particles is invariably realized. It does not specify the precise manner in which a free quantum field expresses itself in terms of quantum particles. It does neither specify the number of particles in terms of which a free quantum field expresses itself, nor the quantum states in which those particles are. It does not determine the energy spectrum of a free quantum field. By way of demonstrating the apparent contradictions which may arise from any misinterpretation of the above abstract conclusion, suffice it to pose the following question. Any field configuration specified on a three-dimensional surface of equal time x^0 will give rise to infinitely many distinct evolving field configurations. Yet, the plane waves of probability remain invariant in the course of time. They do not evolve into something else. How then can a plane wave of probability as it is at x^0 give rise to the infinitely many evolving field configurations of $\hat{\phi}(x^0, \vec{x})$? We defer the answer

---

[53]Nobel-prize winner Julian S. Schwinger (1918 - 1994). One of the pioneers of quantum field theory.

to this rather frivolous question and the resolution of the issues raised above to the fifth section of the present chapter, at which stage the formal procedure of equal-time canonical quantization will be clear. At this point note a central consequence of the theory.

An arbitrary collection of field configurations in one-to-one correspondence with each moment defines a possible evolution pattern for a field configuration specified at initial moment x^0 (for example, for either of the two configurations depicted in Fig.20). There are infinitely many such choices and, consequently, infinitely many possible evolution patterns for a given initial configuration. You can mentally visualize this situation by assigning infinitely many possible evolution patterns, as in Fig.16, to any of the infinitely many field configurations generically represented in Fig.20. The subset of infinitely many possible evolution patterns for a given initial field configuration (say, the configuration at $x^0{}_1$ in Fig.16) which is obtained through the demand that these evolution patterns also eventuate at a given field configuration at a specified moment (say, the configuration at $x^0{}_3$ in Fig.16) is conceptually identical to the set of infinitely many possible paths for a quantum particle which intercept two given points in space at two given moments (Fig.9). This replacement of the unique "trajectory" of the free classical wave field in Fig.16 by infinitely many "trajectories" specified by infinitely many evolving possible field configurations is the equivalent quantization procedure to the canonical field quantization, the *path-integral quantization* for fields.

It should be clear, for that matter, that the configuration aspect of a quantum field is not at all less important than its particle aspect, as may appear to be the case at first sight. In fact, in curved space-times where no preferred sets of observers exist and the concept of particle becomes really nebulous, it is the aspect of field configuration which receives prominent physical significance.

One of the essential differences between the canonical and the path-integral quantization in field theory concerns the symmetries of space-time. We have seen that the equations which express the dynamical behavior of classical fields are Lorentz covariant. Lorentz covariance also underlies both, the canonical and the path-integral quantization, of course. However, the canonical procedure with its inherent equal-time quantization on three-dimensional surfaces of simultaneity in inertial frame A breaks manifest Lorentz invariance since events which are simultaneous in inertial frame A are not simultaneous in another inertial frame B. This is a rather unattractive aspect of the procedure. Lorentz covariance expresses a fundamental symmetry of the physical system in Minkowski space-time. We would like, for that matter, to have Lorentz covariance manifest at all stages of a quantization procedure. By contrast, the path-integral quantization is manifestly Lorentz covariant since in the continuous sum over all possible evolution patterns which respectively emanate from and eventuate at two given field configurations each pattern is manifestly a function of space-time and no separation of time from space occurs.

Thus, the path-integral field quantization is mathematically more elegant than the canonical field quantization. However, the latter has a slight intuitive advantage over the former. The expression of free quantum fields in terms of particles naturally emerges also in the context of path-integral quantization. Yet, it is somewhat easier to conceptualize this expression in the context of the canonical procedure in which quantum particles emerge as excitations of simple quantum harmonic oscillators, than in the context of the path-integral procedure in which quantum particles emerge as transition amplitudes in space-time. For that matter, we shall primarily let the canonical quantization guide us to the essence of the free quantum fields. We shall examine the path-integral quantization of fields only at an advanced stage of our analysis. I mentioned above that the abstract conclusion concerning the expression of quantum fields in terms of quantum particles raises certain issues. The most significant of them concerns the energy spectrum of a free quantum field. The resolution to this issue resolves all others. In what follows, we shall pursue this issue in the context of the canonical procedure. Our approach will, as always, be conceptual.

4. From Abstract Energies to Concrete Particles

We have established two equivalent representations of a free classical wave field, that of the continuous set $\phi(x^0, \vec{x})$ of coupled oscillators in Fig.17 and that of the continuous set $\phi(x^0, \vec{k})$ of de-coupled simple harmonic oscillators in Fig.19. It is self-evident that the procedure of canonical quantization which was described in the previous section accordingly results in two independent and equivalent representations for the corresponding free quantum field $\hat{\phi}(x^0, \vec{x})$. Our objective now is the energy spectrum of this $\hat{\phi}(x^0, \vec{x})$. Not only does the energy spectrum of a free quantum field reveal the precise nature of the stated two representations, it reveals at once the full significance of the $\phi(x^0, \vec{k})$-representation. Indeed, it would be very difficult to determine the energy spectrum of a free quantum field from the quantization of the infinitely many coupled oscillators in Fig.17. However, the quantization of the set of infinitely many de-coupled simple harmonic oscillators in Fig.19 and, thereby, the determination of the free quantum field's energy spectrum are straightforward! The reason, as we have seen, is the linear nature of $\phi(x^0, \vec{x})$. Each degree of freedom $\phi(x^0, \vec{k})$ represents a de-coupled simple harmonic oscillator which evolves independently from all others, vibrating at its own frequency with its own amplitude. The most general solution to the field equation of $\phi(x^0, \vec{x})$ is a linear superposition of these simple harmonic oscillators. In order to quantize $\phi(x^0, \vec{x})$ we must simply quantize these infinitely many simple harmonic oscillators.

We start from the canonical replacement of each plane wave of propagating energy by a plane wave of probability. The former are observable and characterized by constant wave number $\vec{k} = (k^1, k^2, k^3)$ and constant frequency $w = \sqrt{k^2 c^2 + m^2 c^4}$. The latter are unobservable wavefunctions characterized by observable constant momentum $\vec{p} = (p^1, p^2, p^3)$ and observable constant energy E determined by momentum magnitude $p = |\vec{p}|$. In order to determine the relation between E and p we recall from the first chapter that the kinematic relation $E = \frac{p^2}{2m}$ between the energy E and the momentum-magnitude $p = |\vec{p}|$ of a non-relativistic free classical particle of mass m also characterizes the corresponding quantum-mechanical plane wave of probability and underlies the mathematical formalism of quantum mechanics. We expect, for that matter, that the relativistic kinematic relation $E^2 = p^2 c^2 + m^2 c^4$ will accordingly characterize the plane waves of probability of a free quantum field. This relation has been derived in the Appendix through a simple Lorentz transformation to the free particle's *rest frame* (the frame in which the relativistic particle is at rest). Nevertheless, as - through quantization - quantum fields correspond to classical fields, not to particles, we would like to avoid the reference to the latter. We note instead that the implication which the two basic relations $\vec{p} = \hbar\vec{k}$ and $E = \hbar w$ have for a free quantum field is that the infinitely many plane waves of probability in Fig.21 emerge from the quantization of the infinitely many classical plane waves of energy in Fig.19. In turn, in the context of field quantization, $E^2 = p^2c^2 + m^2c^4$ emerges as a direct consequence of the classical relation $w^2 = k^2c^2 + m^2c^4$. Each of the infinitely many plane waves of probability which compose a free quantum field of rest-mass m has energy $p^0 = \frac{E}{c}$ and momentum $\vec{p} = (p^1, p^2, p^3)$ which relate to each other through $E^2 = p^2c^2 + m^2c^4$. The fact that this relation also characterizes the relativistic particles merely reflects the fact that the latter are an expression of the former: A free quantum field expresses itself in terms of quantum particles.

Before we proceed, it is instructive to simplify these mathematical expressions. The fundamental constants which appear in relativistic quantum Physics are the speed of light c and the Planck constant $\hbar$. It is very convenient, for that matter, to introduce a system of units in which $c = \hbar = 1$. In this system of units the classical relativistic relation $E^2 = p^2c^2 + m^2c^4$ reduces to $(p^0)^2 = p^2 + m^2$ with $p^0 = E$. The Lorentz-invariant square of momentum-magnitude is always $\eta_{\mu\nu}p^\mu p^\nu = (p^0)^2 - |\vec{p}|^2 \equiv (p^0)^2 - p^2$. As we have already remarked, we shall, for notational convenience, reserve the symbol p also for that magnitude on the understanding that the physical context will invariably render the significance of p transparent. For that matter, it is $p^2 = \eta_{\mu\nu}p^\mu p^\nu = (p^0)^2 - |\vec{p}|^2$ and $p = +\sqrt{\eta_{\mu\nu}p^\mu p^\nu}$ (note that, pursuant to the analysis in the Appendix, for a space-like p^μ it is $\eta_{\mu\nu}p^\mu p^\nu < 0$, rendering p an imaginary number). Thus, in the system of units $c = \hbar = 1$, the relation $E^2 = p^2c^2 + m^2c^4$ reduces to the simple statement $p^2 = m^2$. Such simplicity as this system of units allows for will serve us at an advanced stage of our analysis. Until then, we shall adhere to the - less versatile, but more conceptual - explicit use of c and $\hbar$.

The canonical replacement of the classical plane waves of energy by plane waves of probability is equivalent to the canonical quantization of the formers' abstract representation as simple harmonic oscillators in Fig.19(b). The quantization of each simple classical harmonic oscillator implies the replacement of the continuous energy spectrum by the discrete energy spectrum $E_n = \hbar w(n + \frac{1}{2})$ of the simple quantum harmonic oscillator with its characteristic constant distance $\hbar w$ between successive energy levels (Fig.21(b)). Each simple quantum harmonic oscillator now is an abstract representation of a corresponding plane wave of probability and constitutes a generic representation of the infinitely many concrete simple quantum harmonic oscillators at each point $\vec{x} = \vec{x}_0$ (Fig.21).

Thus the quantization of each plane wave of energy of wave four-vector $k^\mu = (k^0, \vec{k})$, $k^0 = \frac{w(k)}{c}$ has yielded a plane wave of probability of momentum four-vector $p^\mu = (p^0, \vec{p})$, $p^0 = \frac{E(p)}{c}$. As the former corresponds to a point in an abstract four-dimensional space defined by the collection of all $k^\mu = (k^0, \vec{k})$, so the latter corresponds to a point in an abstract four-dimensional space defined by the collection of all $p^\mu = (p^0, \vec{p})$. The infinitely many de-coupled quantum harmonic oscillators to which we have arrived through the

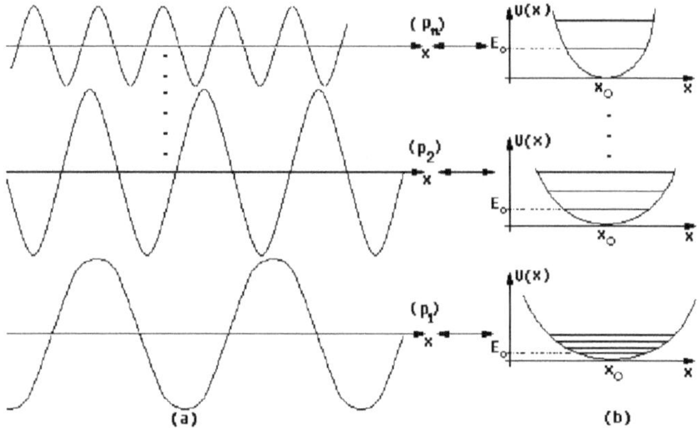

FIGURE 21. Canonical quantization of a classical wave field. Each of the infinitely many plane waves of probability which emerge from the quantization of the plane waves of energy in Fig.19 and compose the free quantum field $\hat{\phi}(x^0, \vec{x})$ corresponds to a specific value of momentum-magnitude $p_1 < p_2 < ... < p_n < ...$ (a). Abstract representation of each plane wave of probability as a simple quantum harmonic oscillator. The characteristic constant separation between successive energy levels in each oscillator is manifest. As p increases so does that separation and, pursuant to (6.I) in the first chapter, the ground-state energy E_0 (b).

canonical procedure (Fig.21(b)) are in one-to-one correspondence with the points of a four-dimensional *momentum space* defined by the collection of all four-momenta p^μ: *The infinitely many degrees of freedom of a free quantum field $\hat{\phi}(x^0, \vec{x})$ de-couple in momentum space. In that abstract space, a free quantum field is a continuous collection of simple quantum harmonic oscillators which evolve independently of each other.*

We proceed now to a simple observation. The linear nature of $\hat{\phi}(x^0, \vec{x})$ and the concomitant absence of mutual interactions amongst the infinitely many simple quantum harmonic oscillators imply that upon quantization the classical energy will necessarily become a continuous sum over the infinitely many energy eigenvalues of those infinitely many simple quantum harmonic oscillators. In order to clearly understand the essence of this statement randomly select two simple quantum harmonic oscillators respectively labelled as A and B in Fig.21. Now label the infinitely many energy levels of oscillator A by the integer $n = 0, 1, 2, 3, ...$ and, accordingly, the infinitely many energy levels of oscillator B by the integer $k = 0, 1, 2, 3, ...$ Oscillator A is necessarily in one of the infinitely many energy eigenstates in its parabolic potential. Label that eigenstate as $\psi_n^A(x^0, \vec{x})$ and its corresponding energy eigenvalue as E_n^A. Accordingly, oscillator B is in the energy eigenstate $\psi_k^B(x^0, \vec{x})$ with energy eigenvalue E_k^B. Then, the total energy of the system of the two quantum oscillators is $E_n^A + E_k^B$. Next, repeat the procedure for three oscillators and then for four and continue indefinitely until you exhaust all infinitely many of them. The energy of the free quantum field $\hat{\phi}(x^0, \vec{x})$ is merely the continuous sum-total $E_n^A + E_k^B + \ldots$ of all infinitely many individual energy eigenvalues.

The task now is to determine the energy eigenstates of the free quantum field. This is somewhat more involved. Pursuant to the analysis in the third section of the first chapter, it is

$$\hat{H}_A \psi_n^A(x^0, \vec{x}) = E_n^A \psi_n^A(x^0, \vec{x}) \qquad (4.I)$$

and

$$\hat{H}_B \psi_k^B(x^0, \vec{x}) = E_k^B \psi_k^B(x^0, \vec{x}) \qquad (4.II)$$

with $\hat{H}_A$ and $\hat{H}_B$ respectively being the energy operators for oscillator A and oscillator B. Note now that, since the two oscillators do not interact with each other, the energy operator $\hat{H}_A$ acts exclusively on its own eigenfunctions, on the energy eigenfunctions of oscillator A, with the implication that

$$\hat{H}_A \psi_n^A(x^0, \vec{x})\psi_k^B(x^0, \vec{x}) = \psi_k^B(x^0, \vec{x})\big(\hat{H}_A \psi_n^A(x^0, \vec{x})\big) = \psi_k^B(x^0, \vec{x})\big(E_n^A \psi_n^A(x^0, \vec{x})\big) =$$

$$E_n^A \psi_n^A(x^0, \vec{x})\psi_k^B(x^0, \vec{x}) \qquad (4.III)$$

Likewise,

$$\hat{H}_B \psi_n^A(x^0, \vec{x})\psi_k^B(x^0, \vec{x}) = E_k^B \psi_n^A(x^0, \vec{x})\psi_k^B(x^0, \vec{x}) \qquad (4.IV)$$

These two simple relations are telling us something very important! If $\psi_n^A(x^0, \vec{x})$ is an energy eigenstate of oscillator A corresponding to energy eigenvalue E_n^A and $\psi_k^B(x^0, \vec{x})$ is an energy eigenstate of oscillator B corresponding to energy eigenvalue E_k^B, then the product $\psi_n^A(x^0, \vec{x})\psi_k^B(x^0, \vec{x})$ is also an energy eigenstate of oscillator A corresponding to energy eigenvalue E_n^A. That is, the *eigenfunctions* $\psi_n^A(x^0, \vec{x})$ and $\psi_n^A(x^0, \vec{x})\psi_k^B(x^0, \vec{x})$ describe the same energy eigenstate of oscillator A. Likewise, the energy eigenfunctions $\psi_k^B(x^0, \vec{x})$ and $\psi_n^A(x^0, \vec{x})\psi_k^B(x^0, \vec{x})$ describe the same energy eigenstate of oscillator B corresponding to energy eigenvalue E_k^B. Did we not already expect this result? Since oscillators A and B do not interact with each other, their energy eigenvalues, in whichever eigenstates these oscillators may respectively be, must be simultaneously observable with absolute certainty. Recall from the analysis in the third section of the first chapter that physical attributes of a quantum system which are simultaneously observable with absolute certainty correspond to common eigenstates. This is what the above two simple relations reveal.

It is not difficult to see now that the energy eigenfunction of the system of the two oscillators is again the product $\psi_n^A(x^0, \vec{x})\psi_k^B(x^0, \vec{x})$. This follows from the formal expression

$$\hat{H}_{A,B}(\ ) = (E_n^A + E_k^B)(\ ) \tag{4.V}$$

of the statement "the system's total energy is $E_n^A + E_k^B$" and the fact that the two non-interacting and independent oscillators enforce the condition $\hat{H}_{A,B} = \hat{H}_A + \hat{H}_B$. Thus, the energy eigenfunction of the system of the two oscillators which must be replaced in ( ) is the stated product of the two energy eigenfunctions since

$$(\hat{H}_A + \hat{H}_B)\psi_n^A(x^0,\vec{x})\psi_k^B(x^0,\vec{x}) = \hat{H}_A\psi_n^A(x^0,\vec{x})\psi_k^B(x^0,\vec{x}) + \hat{H}_B\psi_n^A(x^0,\vec{x})\psi_k^B(x^0,\vec{x}) =$$

$$(E_n^A + E_k^B)\psi_n^A(x^0,\vec{x})\psi_k^B(x^0,\vec{x}) \tag{4.VI}$$

Having thus determined the eigenvalue equation for a system of two simple quantum harmonic oscillators, it is a straightforward matter to repeat this procedure for a system of three simple quantum harmonic oscillators and then for four and continue indefinitely. The energy eigenfunction of the free quantum field $\hat{\phi}(x^0,\vec{x})$ which corresponds to a given distribution $E_n^A, E_k^B, \ldots$ of oscillator energies is the product $\psi_n^A(x^0,\vec{x})\psi_k^B(x^0,\vec{x})\ldots$ of the corresponding infinitely many energy eigenfunctions and, itself, corresponds to energy eigenvalue $E_n^A + E_k^B + \ldots$. Of course, this is only one of the infinitely many energy eigenfunctions of the free quantum field. For, it is as physically legitimate that $n,k,\ldots$, respectively receive the value 10^9, as it is that they receive the value 1. There are infinitely many simple quantum harmonic oscillators, each with infinitely many available energy levels. In turn, there are infinitely many energy eigenstates for $\hat{\phi}(x^0,\vec{x})$ corresponding to all infinitely many combinations between the infinitely many $n,k,\ldots$. For example (refer to Fig.21(b)), if the oscillator labelled by momentum magnitude p_1 is in its first energy eigenstate, the oscillator labelled by momentum magnitude p_2 is in its 10^9-th energy eigenstate, the oscillator labelled by momentum magnitude p_3 is in its 100-th energy eigenstate, ... then the corresponding energy eigenfunction is $\psi_1^{(p_1)}(x^0,\vec{x})\psi_{10^9}^{(p_2)}(x^0,\vec{x})\psi_{100}^{(p_3)}(x^0,\vec{x})\ldots$ corresponding to energy eigenvalue $E_1^{(p_1)} + E_{10^9}^{(p_2)} + E_{100}^{(p_3)} + \ldots$. Another possibility is that oscillator (p_1) is in the 985-th energy eigenstate, oscillator (p_2) is in the 1000-th energy eigenstate, oscillator (p_3) is not excited at all (0-th energy eigenstate), ... Then the corresponding eigenfunction is $\psi_{985}^{(p_1)}(x^0,\vec{x})\psi_{1000}^{(p_2)}(x^0,\vec{x})\psi_0^{(p_3)}(x^0,\vec{x})\ldots$ corresponding to energy eigenvalue $E_{985}^{(p_1)} + E_{1000}^{(p_2)} + E_0^{(p_3)} + \ldots$.

We must now turn to the momentum spectrum which we choose to pursue without recourse to the eigenvalue equation for the momentum operator. Unlike the non-relativistic quantum-mechanical simple harmonic oscillator, the simple quantum harmonic oscillators in Fig.21 are as relativistic as the plane waves of probability which they represent. Since the energy E which characterizes a plane wave of probability in an inertial frame is a component of four-momentum $p^\mu = (\frac{E}{c}, \vec{p})$, each energy eigenvalue $E_1, E_2, \ldots, E_n, \ldots$ of the corresponding simple quantum oscillator is accordingly a component of four-momentum $p_n^\mu = (\frac{E_n}{c}, \vec{p}_n)$ such that the relativistic relation $E^2 = (pc)^2 + (mc^2)^2$ implies the relativistic relation $E_n^2 = (p_n c)^2 + (mc^2)^2$. Indeed, in the relation $E_n = \hbar w(n + \frac{1}{2})$ which expresses the energy spectrum of a single quantum harmonic oscillator, the classical frequency w is determined by the wave number k and the fixed energy $E = \hbar w$ of the plane wave of probability is determined by that wave's fixed momentum $\vec{p} = \hbar \vec{k}$. It follows that if $E_n = \hbar w(n + \frac{1}{2})$ is an energy eigenvalue of a simple quantum harmonic oscillator, then $\vec{p}_n = \hbar \vec{k}(n + \frac{1}{2})$ is a momentum eigenvalue of the same oscillator. In addition, since E_n and $\vec{p}_n$ are simultaneously observable with absolute certainty, they correspond to the common eigenfunction $\psi_n(x^0, \vec{x})$. Consequently, $p_n^\mu = (\frac{E_n}{c}, \vec{p}_n)$ is an energy-momentum eigenvalue of an oscillator in Fig.21(b) corresponding to energy-momentum eigenstate $\psi_n(x^0, \vec{x})$. Thus, $E_n^2 = (p_n c)^2 + (mc^2)^2$ trivially follows from $E^2 = (pc)^2 + (mc^2)^2$.

It is now a trivial matter to extend this result to the entire set of de-coupled degrees of freedom in Fig.21. The energy-momentum eigenfunction of the entire free quantum field which corresponds to a given energy-momentum distribution $p_n^{\mu(p_1)}, p_k^{\mu(p_2)}, \ldots$ with oscillator (p_1) excited in its n-th energy-momentum eigenstate, oscillator (p_2) excited in its k-th energy-momentum eigenstate, ... is the product $\psi_n^{(p_1)}(x^0, \vec{x})\psi_k^{(p_2)}(x^0, \vec{x})$... and itself corresponds to energy eigenvalue $p_n^{0(p_1)} + p_k^{0(p_2)} + ...$, to eigenvalue $p_n^{1(p_1)} + p_k^{1(p_2)} + ...$ for the $\mu = 1$-component of each p^μ, to eigenvalue $p_n^{2(p_1)} + p_k^{2(p_2)} + ...$ for the $\mu = 2$-component of each p^μ and to eigenvalue $p_n^{3(p_1)} + p_k^{3(p_2)} + ...$ for the $\mu = 3$-component of each p^μ. It is, of course, hardly necessary to state that not all oscillators can be excited. Each energy-momentum eigenstate of the free quantum field comprises infinitely many oscillators in their ground state.

Of special importance is the state of lowest energy, the ground state of $\hat{\phi}(x^0, \vec{x})$. We shall call it the *vacuum state*, or simply the *quantum vacuum*, of the free quantum field and, for reasons which we shall examine in the next section, denote it by $|0\rangle$. The vacuum state $|0\rangle$ is that energy-momentum eigenstate in which all oscillators in Fig.21(b) are in their ground state. It is, for that matter, the product $\psi_0^{(p_1)}(x^0, \vec{x})\psi_0^{(p_2)}(x^0, \vec{x})$.... The quantum vacuum is a central aspect of a quantum field. All higher quantum states (Fig.21) are excitations of it. All actual observable particles and virtual non-observable particles emerge from it in ways which we shall thoroughly investigate. However, the striking aspect of $|0\rangle$ which demands immediate attention is an apparent contradiction! Its energy eigenvalue $\frac{1}{2}\hbar w^{(p_1)} + \frac{1}{2}\hbar w^{(p_2)} + ...$ can immediately be seen to be infinite - or, in more formal language, *badly divergent*, being the sum of ground-state energies for the infinite number of oscillators in Fig.21(b).

Divergences are the hallmark of quantum field theory! They are all the exclusive consequence of an underlying infinity of degrees of freedom. The divergence in the vacuum energy of a free quantum field is the first divergence which we encounter and its origin in an infinity of degrees of freedom is manifestly obvious. Infinities and divergent quantities are physically meaningless, of course. However, they are not symptomatic of any pathology in the theory. Quantum field theory is conceptually and mathematically sound! Instead, divergences signify the limitations of quantum field theory at space-time scales which call into question the continuous character of space-time, as we shall see. The divergence in the vacuum energy is the simplest one and can be dispensed with in a more direct manner which does not tackle the issue of space-time scales. The ground-state energy of a simple quantum harmonic oscillator is an observable quantity exclusively in the sense that the difference between that energy and the minimum energy (conventionally set equal to zero) of the classical harmonic oscillator at rest (at the bottom point of the same parabolic potential) is observable. This is a result of general significance. Absolute energies have no physical content. Only relative energies (differences in energy) are observable. Since none of the infinitely many oscillators in Fig.21 is observable when the quantum field is in its vacuum state, no oscillator ground-state energy is an observable quantity. For that matter, neither is the sum-total of oscillator ground-state energies, a fact which implies that no difference between the energy of an excited energy-momentum eigenstate of the quantum field and the energy of the quantum field's vacuum state can be defined since the latter is unobservable. Thus, it is physically legitimate to dispense with that formally divergent vacuum energy by merely subtracting from it an infinite constant, thereby redefining it by an infinite amount. The result of such a *renormalization of the vacuum energy* is a finite value for that energy which we are at liberty to set equal to zero (N2).

We have, thus, quantized the infinitely many de-coupled degrees of freedom and determined the energy-momentum spectrum of a free quantum field. For the infinite redefinition of the vacuum energy has this implication: Now the energy eigenvalue of oscillator energy-momentum eigenstate $\psi_n^{(p)}(x^0, \vec{x})$ is merely $E_n = nE = n\hbar w$. Accordingly, the momentum eigenvalue of the same oscillator energy-momentum eigenstate is $\vec{p}_n = n\vec{p} = n\hbar\vec{k}$. In turn, the energy-momentum eigenvalue of oscillator energy-momentum eigenstate $\psi_n^{(p)}(x^0, \vec{x})$ is $p_n^\mu = (\frac{E_n}{c}, \vec{p}_n) = (n\frac{E}{c}, n\vec{p})$ - that is, $p_n^\mu = np^\mu$. Thus, each momentum state (a term which typically describes a plane wave of probability) in Fig.21 may be in any of infinitely many energy-momentum eigenstates, each of four-momentum np^μ describing n quanta of four-momentum $p^\mu = (\frac{E}{c}, \vec{p})$ and rest mass m each. As each such quantum reproduces the relativistic relation $E^2 = (pc)^2 + (mc^2)^2$, we interpret them as quantum particles. In turn, n is the *occupation number* of energy-momentum eigenstate $\psi_n^{(p)}(x^0, \vec{x})$ of momentum state p. Specification of the occupation number for each momentum state yields a complete description of the energy-momentum eigenstate $\psi_n^{(p_1)}(x^0, \vec{x})\psi_k^{(p_2)}(x^0, \vec{x})\psi_j^{(p_3)}(x^0, \vec{x})....$ of the free quantum field. The corresponding energy-momentum eigenvalue is now the sum of energy-momentum eigenvalues of excited oscillators

$$p_n^{\mu(p_1)} + p_k^{\mu(p_2)} + p_j^{\mu(p_3)} + ... = (n\frac{E^{(p_1)}}{c} + k\frac{E^{(p_2)}}{c} + j\frac{E^{(p_3)}}{c} + ... , \; n\vec{p}^{\,(p_1)} + k\vec{p}^{\,(p_2)} + j\vec{p}^{\,(p_3)} + ...)$$

which is clearly discrete since, as a direct consequence of the renormalization of the vacuum energy, only the excited momentum states contribute.

The interpretation of the quanta as particles may cause some perplexity. Is it not obvious that plane waves of probability describe free quantum particles? Why "interpret" the obvious? Such a question smacks of "quantum-mechanical" mentality. In quantum mechanics, free particles are, indeed, plane waves of probability for position precisely because the position $\vec{x}$ is a dynamical variable. As we have established, however, in quantum field theory the position $\vec{x}$ is merely a label. In energy-momentum eigenstate $\psi_n^{(p_1)}(x^0, \vec{x})\psi_k^{(p_2)}(x^0, \vec{x})....$ in which the infinitely many independent degrees of freedom are manifest in $\psi_n^{(p_1)}(x^0, \vec{x})$, $\psi_k^{(p_2)}(x^0, \vec{x})$, ... with $\vec{x}$ as a label, the emergence of particles is a matter of a non-trivial interpretation. In principle, there are no particles in $\psi_n^{(p_1)}(x^0, \vec{x})\psi_k^{(p_2)}(x^0, \vec{x})$..... There is only the concrete expression of the free quantum field $\hat{\phi}(x^0, \vec{x})$ and its infinitely many degrees of freedom. It is only after we single out the excited degrees of freedom in $\psi_n^{(p_1)}(x^0, \vec{x})\psi_k^{(p_2)}(x^0, \vec{x})$.... , it is only when in this energy-momentum eigenstate we acknowledge the excited momentum states as distinct from the infinitely many void momentum states, it is only when we declare that the former "stand out" in the entire set of momentum states, that a physically acceptable particle picture emerges. Such an interpretation of the excited momentum states in the entire set as particles is physically consistent. As it follows the renormalization of the vacuum energy, it is, in addition, supported by the fact that no void momentum state contributes to the energy and the momentum of the purported particle system. At ultimate, empirical level it is supported by observation and experiment. It is, for that matter, that the semblance of a finite number of degrees of freedom emerges in the domain of very low energies. Indeed, as we shall soon see, such a particle interpretation gives rise to the semblance of quantum mechanics at the non-relativistic limit of very low energies. It is imperative to understand, however, that a finite number of degrees of freedom is only a semblance. Not only does the interpretation of excitations as particles not imply a finite number of degrees of freedom, it reveals the particles in $\psi_n^{(p_1)}(x^0, \vec{x})\psi_k^{(p_2)}(x^0, \vec{x})$.... as an expression of the underlying free quantum field $\hat{\phi}(x^0, \vec{x})$. The infinitely many degrees of freedom which the void plane waves of probability represent are ever present and determine the dynamical behavior of $\hat{\phi}(x^0, \vec{x})$ in a variety of highly non-trivial ways, as we shall amply demonstrate. We shall constantly return to this premise and the physical perspective which it provides as our analysis progresses, most notably in the sixth and eighth sections of this chapter and in the seventh section of the next chapter in which we shall examine the precise relation of quantum field theory to quantum mechanics and its probabilities. Let it suffice to state at this point that adherence to this perspective is crucial in order to preclude the common misconception that quantum field theory is essentially a many-particle theory which uses the "quantum field" merely as an artifact of its procedure.

In order to better understand the physical content of the excited momentum states, those of occupation number other than zero, note that underlying the formal procedure of conferring higher significance upon them, of elevating them to particles is, of course, the linear nature of free quantum fields. Each of the infinitely many plane waves of probability exists independently of all others with $\vec{x}$ as a dynamical variable associated with a probability density for position. For that matter, an excited momentum state "stands out" in terms of its comparatively higher probability density with respect to the void momentum states which compose the vacuum state. Equivalently, it stands out in terms of its comparatively higher maximum amplitude. The abstract representation of each momentum state as a simple quantum harmonic oscillator implies the abstract representation of each momentum state's maximum amplitude as the energy-momentum level of the corresponding oscillator. That is, each momentum state specified by $E = \hbar w$ and $\vec{p} = \hbar \vec{k}$ is a plane wave of probability which admits infinitely many maximum amplitudes labelled by n ($n = 1, 2, 3, ...$). Each momentum state of fixed n has energy $E_n = nE$ and momentum $\vec{p}_n = n\vec{p}$. Each such oscillator energy-momentum eigenstate is occupied by n particles, each of energy E and momentum $\vec{p}$. It follows that the equal spacing of each oscillator's energy-momentum eigenvalues constitutes the abstract representation of the equal spacing of corresponding momentum state's occupation particle numbers (Fig.22). Nevertheless, we stress again that such a particle interpretation only has formal significance. The physical essence of a quantum field lies in the totality of the infinitely many momentum states, in that infinity of degrees of freedom which renders $\vec{x}$ merely a label irrelevant to probabilities.

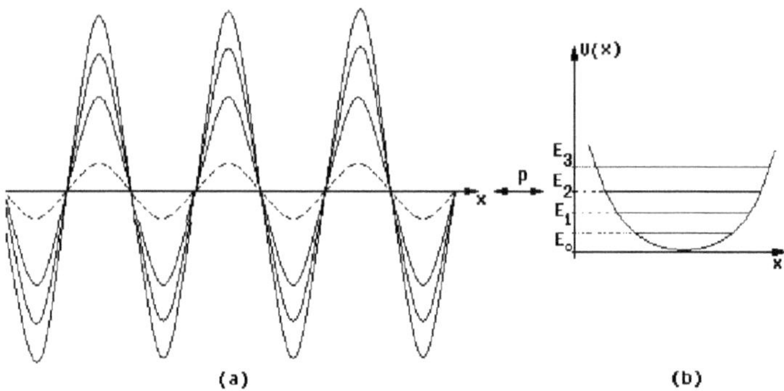

FIGURE 22. A plane wave of probability of momentum $\vec{p}$ and energy E admits a discrete spectrum of infinitely many possible maximum amplitudes. Depicted are those corresponding to occupation numbers $n = 0$, $n = 1$, $n = 2$, $n = 3$. The syncopated plane wave of lowest maximum amplitude corresponds to $n = 0$ and is void of particles. The first increment yields a description of one quantum particle of momentum $\vec{p}$ and energy E. The second increment yields a description of two quantum particles, each of momentum $\vec{p}$ and energy E. The third increment yields a description of three quantum particles, each of momentum $\vec{p}$ and energy E.... (a). Each such momentum state of fixed n is represented by an energy-momentum eigenstate of the simple quantum harmonic oscillator which itself constitutes the abstract representation of momentum state $(E, \vec{p}\,)$ (b).

These conclusions render precise physical meaning to the earlier abstract statement that quantum fields express themselves in terms of particles. We may now regard these conclusions as general. In principle, they apply to all free quantum fields. Why only in principle? Because our analysis has yet to take into consideration possible dynamical aspects which may restrict the occupation numbers. Such a possibility concerns the intrinsic angular momentum. Quantum fields express themselves in terms of quantum particles. Yet, not all quantum particles have zero intrinsic angular momentum. We shall see in the seventh section of this chapter that quantum fields fall into two major categories. These are, the category of fields whose integer spin does not affect the conclusions to which we have arrived and the category of fields whose half-integer spin places severe constraints on the occupation numbers.

From the abstract classical relativistic wave fields, to the abstract representation of quantum particles as eigenstates of simple quantum harmonic oscillators, the insight which we have hitherto gained into the fundamental reality of quantum fields is the result of our advance to ever higher levels of abstraction. Before we take the next step of representing free quantum fields as abstract mathematical spaces, we turn to the physical significance of our results.

5. Quantum Fields and Symmetries

Since all observable attributes are represented by operators, we begin by introducing an operator for the observable occupation number of each momentum state such that the latter is an eigenvalue of the former. The occupation number n_{p_i} of momentum state (p_i), $i = 1, 2, 3, ...$ is represented by the corresponding *number operator* $\hat{N}_{p_i}$. That is,

$$\hat{N}_{p_i} \psi^{(p_i)}_{n_{p_i}}(x^0, \vec{x}) = n_{p_i} \psi^{(p_i)}_{n_{p_i}}(x^0, \vec{x}) \qquad (5.I)$$

which is merely the formal expression of the statement "the momentum state (p_i) contains n_{p_i} particles". It is high time that we simplify the laden notation which our analysis thus far has inevitably introduced by putting the Dirac notation into effect. Pursuant to the eleventh section of the first chapter, the Dirac notation replaces $\psi^{(p_i)}_{n_{p_i}}(x^0, \vec{x})$ by $|n_{p_i}\rangle$ and recasts (5.I) in the form

$$\hat{N}_{p_i} |n_{p_i}\rangle = n_{p_i} |n_{p_i}\rangle \qquad (5.II)$$

The immediate implication of this statement is

$$\hat{N}_{p_i} \psi^{(p_1)}_{n_{p_1}}(x^0, \vec{x}) \psi^{(p_2)}_{n_{p_2}}(x^0, \vec{x}) ... \psi^{(p_i)}_{n_{p_i}}(x^0, \vec{x}) ... =$$

$$n_{p_i} \psi^{(p_1)}_{n_{p_1}}(x^0, \vec{x}) \psi^{(p_2)}_{n_{p_2}}(x^0, \vec{x}) ... \psi^{(p_i)}_{n_{p_i}}(x^0, \vec{x}) ... \qquad (5.III)$$

which is itself the formal expression of the statement "of the total number of particles which the energy-momentum eigenstate $\psi^{(p_1)}_{n_{p_1}}(x^0, \vec{x}) \psi^{(p_2)}_{n_{p_2}}(x^0, \vec{x}) ... \psi^{(p_i)}_{n_{p_i}}(x^0, \vec{x}) ...$ of the free quantum field contains, n_{p_i} particles are in momentum state (p_i) and, thus, each of them is of four-momentum $p^{\mu (p_i)} = (E^{(p_i)}, \vec{p}^{(p_i)})$". The Dirac notation recasts this statement in the form

$$\hat{N}_{p_i} |n_{p_1}\rangle |n_{p_2}\rangle ... |n_{p_i}\rangle ... = n_{p_i} |n_{p_1}\rangle |n_{p_2}\rangle ... |n_{p_i}\rangle ... \qquad (5.IV)$$

Note that (5.IV) is a statement as to the occupation number n_{p_i} of plane wave (p_i) exclusively. It should be clear, for that matter, that the total number of particles in each energy-momentum eigenstate of the free quantum field is represented by the number operator

$$\hat{N} = \hat{N}_{p_1} + \hat{N}_{p_2} + \hat{N}_{p_3} + ... \qquad (5.V)$$

As a simple exercise on the number operator, consider the energy-momentum eigenfunction $\psi_{985}^{(p_1)}(x^0, \vec{x})\psi_{1000}^{(p_2)}(x^0, \vec{x})\psi_0^{(p_3)}(x^0, \vec{x})...$ which the Dirac notation recasts in terms of subscripted occupation numbers as $|985_{p_1}\rangle|1000_{p_2}\rangle|0_{p_3}\rangle.....$ It should be an easy matter to write the eigenvalue statement respectively for occupation numbers 985, 1000 and 0 yourselves. It should be just as easy to write the eigenvalue statement for the number operator $\hat{N}$ in the event that the total occupation number of this energy-momentum eigenstate is $1985 = 985 + 1000$.

We are now in position to appreciate the infinite wealth of possibilities which quantum fields have. Each of the infinitely many energy eigenstates of a quantum-mechanical system corresponds to a single energy eigenfunction. On the contrary, each of the infinitely many energy-momentum eigenstates of a free quantum field is, itself, a product of infinitely many energy-momentum oscillator eigenfunctions with each such eigenfunction describing a certain number of quantum particles. The implications are obvious! In quantum mechanics the quantum particles are as fundamental physical entities as the classical particles in classical mechanics are. By contrast, in quantum field theory the quantum particles are expressions of underlying quantum fields. In quantum mechanics the number of quantum particles is as immutable as the number of classical particles is in classical mechanics. By contrast, in quantum field theory there are infinitely many oscillators and corresponding momentum states for a free quantum field and thus infinitely many possibilities as to the number of quantum particles in terms of which a free quantum field expresses itself. Which such possibility is realized, which occupation numbers in which momentum states, is a matter of choice and boundary conditions. It is imperative to note, however, that once such a choice has been made for a free quantum field, the distribution of particles in momentum states and the corresponding occupation number for each momentum state remain invariant in the course of time. This point is important enough to merit closer examination.

The absence of interactions has the direct implication of time-translational invariance, the consequences of which we have examined in the first chapter. Since the infinitely many degrees of freedom $\hat{\phi}(x^0, \vec{p})$ evolve independently from each other, the evolution of a free quantum field affects neither the classical parabolic potentials in Fig.21(b), nor the energy-momentum eigenstates of the quantum oscillators within them. In particular, the concomitant energy conservation implies that a simple quantum harmonic oscillator which has been specified in the n-th energy eigenstate $(n = 1, 2, ...)$ in an inertial frame will permanently remain in that eigenstate with the same energy eigenvalue. Thus, if the free quantum field $\hat{\phi}(x^0, \vec{x})$ has been specified in energy-momentum eigenstate $|n_{p_1}\rangle|n_{p_2}\rangle....$, it will permanently remain in that eigenstate with energy-momentum eigenvalue $p_{n_{p_1}}^{\mu(p_1)} + p_{n_{p_2}}^{\mu(p_2)} + ... = n_{p_1}p^{\mu(p_1)} + n_{p_2}p^{\mu(p_2)} + ...$ The direct implication is that, once specified, the occupation numbers $n_{p_1}, n_{p_2}, \ldots$ remain constant in the course of time as registered in an inertial frame. Moreover, a simple quantum harmonic oscillator which has been specified in the n-th energy eigenstate in an inertial frame will obviously remain in that eigenstate under a Lorentz transformation. Consequently, all inertial observers agree on occupation numbers $n_{p_1}, n_{p_2}, \ldots$ and on the total number of particles $n_{p_1} + n_{p_2} + \ldots$. Clearly, this central aspect of the dynamical behavior of free quantum fields is not sustained in the presence of interactions[54]. Note now that we could have arrived at this result directly from the classical theory. In the preceding second section we saw that once a choice of maximum amplitude c_k has been made for each wave number k on a three-dimensional surface of equal time, all amplitudes remain fixed in their chosen values and uniquely specify an evolution pattern for the classical wave field. This is a direct consequence of time translational invariance. Since no symmetry of time or space distinguishes between classical and quantum behavior, the corresponding free quantum field has the same property. Once a choice of maximum amplitude has been made for each plane wave of probability, all maximum amplitudes - and, thus, occupation numbers - remain fixed in their chosen values and uniquely specify an energy-momentum eigenstate of the free quantum field.

How does this result conform with the fact that a free quantum field consists in infinitely many evolving field configurations? The answer is inherent in the infinitely many degrees of freedom. The energy-momentum eigenstate $|n_{p_1}\rangle|n_{p_2}\rangle...|n_{p_i}\rangle...$ necessarily involves a finite number of particles but an infinite number of momentum states since infinitely many oscillators remain in their ground state and infinitely many plane waves of probability remain void of particles. We observe a finite number of particles but the underlying infinitely many degrees of freedom are always at work, translating this specific energy-momentum eigenstate to infinitely many field configurations evolving in Minkowski space-time. At a given moment x^0, each point $\vec{x}$ in space labels infinitely many amplitudes in the infinitely many plane waves of probability which compose $|n_{p_1}\rangle|n_{p_2}\rangle...|n_{p_i}\rangle...$ (pictorially, they are the points at which a line perpendicular to the x-axis intercepts the infinitely many plane waves in Fig.22(a)). The infinitely many evolving field configurations emerge through the interpretation of these amplitudes as possible field-values labeled by $\vec{x}$, as the same values which in the context of equal-time canonical quantization replace the classical value $\phi(x^0, \vec{x})$. The choice of an amplitude at each $\vec{x}$ at moment x^0 yields a field configuration throughout space. The totality of the infinitely many field configurations at x^0 is the quantum field $\hat{\phi}(x^0, \vec{x})$ at that moment. We have already stressed, of course, that each $\vec{x}$ labels at once a set of infinitely many field-values and the corresponding to them predictable probability distribution.

We see again, for that matter, that free quantum fields are independently described in the language of field configurations and in the language of energy-momentum eigenstates - the language of quantum particles, as it were. Each energy-momentum eigenstate is equivalent to infinitely many evolving field configurations. It, itself, is neither a field

---

[54]Although beyond the scope of this book, I deem it necessary to mention in passing that accelerating observers in Minkowski space-time will not agree on the total number of particles. For example, where all inertial observers will agree that a free quantum field contains no particles (vacuum state), accelerating observers will register particles according to their state of motion. Can you sense the reason?

configuration, nor a set of infinitely many field configurations. The resolution of an arbitrary classical field configuration in infinitely many plane waves of energy, the nature of the former as a superposition of the latter, may give rise to the misconception that an arbitrary configuration of a quantum field is a superposition of infinitely many energy-momentum eigenstates and that, for that matter, the latter are themselves field configurations. The fact is, however, that in the classical theory fields and their configurations stand in opposition to particles. Plane waves of energy are not associated with classical particles! Quantum field theory, on the contrary, refutes the opposition between fields and particles, expressing quantum fields either in terms of configurations, or in terms of particles. Following Julian Schwinger, we stress again that field configurations precede particles - that is, field configurations precede the energy-momentum eigenstates which receive the interpretation of quantum particles. The direct implication is that it is logically inconsistent to interpret a field configuration as a superposition of plane waves of probability, or of energy-momentum eigenstates of the quantum field. As we shall see in detail in the next section, the latter, being a superposition of particle states by dint of interpretation, necessarily yields a general particle state which is equivalent - but not at all identical - to a set of infinitely many field configurations. We provide a concrete example in what immediately follows.

A remarkable demonstration of the premise to the effect that an energy-momentum eigenstate is equivalent to a set of infinitely many evolving field configurations is the no-particle state, the quantum vacuum which we shall examine in detail in the next chapter. Although void of actual quantum particles, the vacuum state comprises infinitely many amplitudes at each space-time point which receive the equivalent interpretation of field-values. The vacuum state is equivalent to infinitely many field configurations and it is not, for that matter, the zero-field state which a naive analogy with the classical vacuum would suggest.

These considerations also provide the resolution to the apparent contradiction stated in the preceding third section. In describing a quantum field, caution should always be exercised that the two distinct equivalent and independent languages - that of field configurations and that of energy-momentum eigenstates - not be confused. The infinitely many field configurations precede field quantization and the emergence of quantum particles. In specifying a field configuration prior to equal-time canonical quantization we deal with a free quantum field as the primary entity which the free quantum field actually is, not as the result of quantization of the solution to classical field equations. At this level there are no plane waves of probability and no simple quantum harmonic oscillators. We stress again, for that matter, that we have no right to claim a resolution of a field configuration specified on a three-dimensional surface of equal time in a multiplicity of plane waves of probability. Such a field configuration admits infinitely many possibilities for evolution and gives rise to infinitely many evolving field configurations. For that matter, this is also the case for the result of equal-time canonical quantization if the latter is interpreted as the replacement of a classical field configuration by infinitely many possible field configurations. When, on the contrary, we arrive at the free quantum field through equal-time canonical quantization interpreted as the replacement of classical plane waves of energy by plane waves of probability, we arrive at the level of energy-momentum eigenstates and of quantum particles. At this level, the infinitely many field configurations are replaced by the equivalent energy-momentum eigenstates and their unique evolution. Each of the infinitely many energy-momentum eigenstates of which the stated plane wave of probability is an aspect is equivalent to infinitely many field configurations. Equivalent, but not identical! It is because each such energy-momentum eigenstate is, itself, neither a field configuration, nor a set of infintely many field configurations that, once specified, each such energy-momentum eigenstate remains invariant in the course of time.

In order to further illustrate the essence of the matter, consider the most simple classical field configuration. It is self-evident that the mere replacement of that plane wave of energy by a plane wave of probability would not yield the infinitely many energy-momentum eigenstates of the free quantum field. The incipient plane wave of energy is that classical configuration which emerges when all c_k except for one in (2.II) of the present chapter are equal to zero. Field quantization is predicated upon the replacement of all infinitely many plane waves of energy in (2.II) by plane waves of probability, regardless of whether the corresponding amplitudes vanish or not. The simple reason is that field quantization equivalently replaces the infinitely many abstract simple classical harmonic oscillators by infinitely many abstract simple quantum harmonic oscillators the ground state of which is not the zero-energy state (Fig.21). The quantization of the classical field yields infinitely many plane waves of probability even when the classical field is expressed by a single classical plane wave of energy. Since the amplitudes of these plane waves of probability are arbitrary, each choice of amplitudes uniquely specifies an energy-momentum eigenstate of the free quantum field and the infinitely many field configurations which that eigenstate represents.

Let us now list the infinitely many energy-momentum eigenstates of a free quantum field. The basic one is the vacuum state. We have seen that it is that energy-momentum eigenstate in which all simple quantum harmonic oscillators are in their ground state and that, for that matter, it is expressed by the product $\psi_0^{(p_1)}(x^0, \vec{x})\psi_0^{(p_2)}(x^0, \vec{x})\psi_0^{(p_3)}(x^0, \vec{x})....$ In the Dirac notation this is recast as $|0_{p_1}\rangle|0_{p_2}\rangle|0_{p_3}\rangle.....$ We have also seen what the precise physical significance of this product is. Since the ground state of an oscillator in Fig.21(b) is the abstract representation of a zero occupation number for the corresponding plane wave of probability (Fig.22), all plane waves of probability which constitute the de-coupled degrees of freedom of the free quantum field in Fig.21(a), all such momentum states, are void of particles in the event that the free quantum field is in its vacuum state. Thus, the vacuum state of a free quantum field is defined as

$$|0_{p_1}\rangle|0_{p_2}\rangle|0_{p_3}\rangle...|0_{p_n}\rangle... \equiv |0\rangle \qquad (5.VI)$$

and clearly satisfies the condition

$$\hat{N}|0\rangle = 0|0\rangle \qquad (5.VII)$$

which serves as an equivalent definition. Note that the physical content of (5.VI) is always the above-stated premise: Unlike the classical vacuum, the quantum vacuum

202

$|0\rangle$ is not the zero-field state. It is characterized by the absence of actual observable quantum particles, but not by the absence of probabilities for field values.

Next, we consider the one-particle energy-momentum eigenstate

$$|0_{p_1}\rangle|0_{p_2}\rangle...|1_{p_k}\rangle|0_{p_{k+1}}\rangle... \equiv |1_{p_k}\rangle \qquad (5.VIII)$$

of energy-momentum

$$p_1^{\mu(p_k)} = (1E^{(p_k)} \, , \, 1p^{1^{(p_k)}} \, , \, 1p^{2^{(p_k)}} \, , \, 1p^{3^{(p_k)}})$$

This is the energy-momentum eigenstate obtained when all infinitely many simple quantum harmonic oscillators are in their ground states except for the oscillator (p_k) which is in its first excited state. Equivalently, when all infinitely many plane waves of probability are void of particles except for that of momentum $\vec{p}_k$ which is occupied by one particle. In fact, the abbreviated notation $|1_{p_k}\rangle$ merely emphasizes the particle interpretation of the excited momentum state (p_k).

Accordingly, the two-particle energy-momentum eigenstate

$$|0_{p_1}\rangle|0_{p_2}\rangle...|1_{p_k}\rangle|0_{p_{k+1}}\rangle...|1_{p_{k+n}}\rangle|0_{p_{k+n+1}}\rangle... \equiv |1_{p_k},1_{p_{k+n}}\rangle \qquad (5.IX)$$

of energy-momentum

$$p_1^{\mu(p_k)} + p_1^{\mu(p_{k+n})} = (E^{(p_k)} + E^{(p_{k+n})} \, , \, p^{1^{(p_k)}} + p^{1^{(p_{k+n})}} \, , \, p^{2^{(p_k)}} + p^{2^{(p_{k+n})}} \, , \, p^{3^{(p_k)}} + p^{3^{(p_{k+n})}})$$

is the energy-momentum eigenstate obtained when all infinitely many plane waves are associated with zero occupation numbers and are, thus, void of particles except for that of momentum $\vec{p}_k$ and that of momentum $\vec{p}_{k+n}$ which are respectively occupied by one particle. Proceeding in this manner indefinitely, we construct the entire set of a free quantum field's energy-momentum eigenstates $|n_p, n'_{p'}, n''_{p''}, ...\rangle$, $|m_k, m'_{k'}, m''_{k''}, ...\rangle$, ... respectively signifying n particles in momentum state p, n' particles in momentum state p', n'' particles in momentum state p'' ... and so on indefinitely.

6. Fock Space

We may now explore the implications which the linear nature of $\hat{\phi}(x^0, \vec{x})$ has for these energy-momentum eigenstates. To this end we shall exploit the theory which we developed in the third section of the first chapter. Recall that absence of knowledge as to the specific energy eigenvalue in a static attractive potential implies that the quantum particle is in a superposition of energy eigenstates according to (3.V). There are infinitely many possible such superpositions signifying the infinitely many possible quantum states $\psi(t, x)$ which give rise to the entire Hilbert space of the energy operator. Pursuant to (5.X) in the first chapter, the square of each coefficient's magnitude in each such superposition yields the probability that the quantum particle have the corresponding energy eigenvalue.

Although quantum fields are physical systems of infinitely higher complexity, the fundamental premise which (3.V) expresses remains identically the same. If at a given moment a free quantum field is in an energy-momentum eigenstate, it permanently remains in that eigenstate with the corresponding energy-momentum eigenvalue. That is, it permanently remains in a state characterized by a specific number of particles specifically distributed in the infinitely many plane waves of probability, or more rigorously, in an energy-momentum eigenstate which, as we have seen, is interpreted as n particles in specified momentum states. It is as physically legitimate, however, to merely inquire as to the expression of the free quantum field $\hat{\phi}(x^0, \vec{x})$ in terms of the specific number of particles n in the event that the actual energy-momentum eigenvalue $n_{p_1} p^{\mu(p_1)} + n_{p_2} p^{\mu(p_2)} + ...\; ;\; n_{p_1} + n_{p_2} + ... = n$ is unknown. Pursuant to (3.V), absence of knowledge as to the actual energy-momentum eigenvalue which the specified n particles signify implies that the free quantum field $\hat{\phi}(x^0, \vec{x})$ is in a superposition of all infinitely many energy-momentum eigenstates characterized by that particle number. Since each energy-momentum eigenstate may contribute to whatever extent, there are infinitely many possible linear combinations specifying infinitely many *n-particle wavefunctions* $\psi_n(x^0, \vec{x})$ for the expression of $\hat{\phi}(x^0, \vec{x})$ in terms of the specified number of particles n. Should we, for that matter, conclude that in the event $\hat{\phi}(x^0, \vec{x})$ is in a specific n-particle state $\psi_n(x^0, \vec{x})$, each energy-momentum eigenvalue has a specific probability for being the actual energy-momentum eigenvalue of $\hat{\phi}(x^0, \vec{x})$ determined by the corresponding coefficient in that linear combination?

The n-particle wavefunctions $\psi_n(x^0, \vec{x})$ to which we have arrived are fundamentally different from the quantum-mechanical wavefunctions. Whereas each term in a quantum-mechanical linear combination is a wavefunction as we have seen in the first chapter, each term in a linear combination which yields a wavefunction $\psi_n(x^0, \vec{x})$ for the expression of quantum field $\hat{\phi}(x^0, \vec{x})$ in terms of a specified number of particles n is a product of the infinitely many momentum states which the quantum field comprises. A quantum-mechanical wavefunction comprises a finite number of degrees of freedom and the position $\vec{x}$ as a dynamical variable. On the contrary, each $\psi_n(x^0, \vec{x})$ expresses the infinitely many degrees of freedom of the underlying $\hat{\phi}(x^0, \vec{x})$ and the position $\vec{x}$ as a label. No longer does $|\psi_n(x^0, \vec{x})|^2$ receive the interpretation of probability density for position precisely because in the n-particle state described by $\psi_n(x^0, \vec{x})$, $\vec{x}$ is merely a label. We have already stressed that only upon singling out the excited momentum states, only upon acknowledging them as distinct from the infinitely many void momentum states, do we arrive at the interpretation of each energy-momentum eigenstate of $\hat{\phi}(x^0, \vec{x})$ as quantum particles in specific momentum states and, thereby, at the interpretation of $\psi_n(x^0, \vec{x})$ as a wavefunction for n quantum particles. We have, at once, stressed that such a particle picture which bears the semblance of a finite number of degrees of freedom expresses those void momentum states and the underlying quantum field itself. The task before us now, is to construct all possible particle states of the quantum field $\hat{\phi}(x^0, \vec{x})$. In the course of such a construction, we shall demonstrate the precise physical significance of these remarks. An essential aspect of this procedure concerns the representation of the particle states. It is precisely because $\vec{x}$ is now merely a label that the abstract and flexible Dirac notation is best suited for their description.

We start with the one-particle state $\psi_1(x^0, \vec{x})$ which in the Dirac notation is rendered as $|1\rangle$. The one-particle state $|1\rangle$ constitutes the expression of the free quantum field in terms of a single particle. If $\hat{\phi}(x^0, \vec{x})$ is in $|1\rangle$ then the excitation of $|0\rangle$ interpreted as a particle may have the momentum eigenvalue p_1 as it may have any other momentum eigenvalue. As the actual energy-momentum eigenvalue is unknown, $|1\rangle$ is a superposition of infinitely many energy-momentum eigenstates characterized by $n = 1$. There are, in fact, infinitely many such superpositions signifying infinitely many one-particle states $|1\rangle$. That is, there are infinitely many linear combinations

$$|1\rangle = c_1|1_{p_1}\rangle|0_{p_2}\rangle|0_{p_3}\rangle... + c_2|0_{p_1}\rangle|1_{p_2}\rangle|0_{p_3}\rangle... + c_3|0_{p_1}\rangle|0_{p_2}\rangle|1_{p_3}\rangle... + ... \qquad (6.I)$$

of the infinitely many possible energy-momentum arrangements for a single quantum particle with each choice of complex coefficients $c_1, c_2, c_3, ...$ determining a distinct one-particle state $|1\rangle$.

Not all such choices are of immediate physical interest and mathematical significance. For, the semblance of quantum mechanics is upheld only on the necessary condition of a finite magnitude. Pursuant to the third section of the first chapter, the collection of all $|1\rangle$ defines a Hilbert space on condition that their magnitude $\langle 1|1\rangle$ be finite. Of immediate physical interest, for that matter, are those sets of $c_1, c_2, c_3, ...$ which produce one-particle states in conformity with the demand

$$\langle 1|1\rangle = 1 \qquad (6.II)$$

or, concretely

$$\int d^3x |\psi_1(x^0, \vec{x})|^2 = \int d^3x |\psi_1(\vec{x})|^2 = 1$$

So strong is the apparent consistency of (6.II) with the quantum-mechanical demand for 100% total probability over all space that we would indeed interpret $|\psi_1(x^0, \vec{x})|^2$ as the probability density for position, were it not that $\vec{x}$ is not a dynamical variable. It is merely a label for infinitely many evolving field degrees of freedom. Nevertheless, (6.I) and (6.II) are entirely "analogous" to the quantum-mechanical superposition of plane waves in (2.II) and corresponding total probability (3.VIII) in the first chapter. This has the following concrete significance: At the non-relativistic limit of very low energies the excited momentum state in each term of (6.I) satisfies the non-relativistic Schrödinger equation with its inherent probabilistic interpretation of the wavefunction. As a result, $\vec{x}$ appears as a dynamical variable giving rise to the semblance of probabilities for position and, for that matter, to the <u>semblance</u> of quantum-mechanical behavior. The precise manner in which the semblance of quantum mechanics emerges from a quantum field we shall examine in the seventh section of the next chapter at which stage we shall have the fire power of the relevant mathematical formalism at our disposal. The fact of the matter here is that at the non-relativistic limit (6.I) and (6.II) appear as the stated (2.II) and (3.VIII). Of course, under no circumstances do probabilities for position emerge! Under no circumstances is $\vec{x}$ a dynamical variable and the quantum field "quantum-mechanical"! The semblance of probabilities for position and of quantum mechanics itself emerges at the non-relativistic limit because it is that limit which allows for the description of the excited momentum states as solutions to the quantum-mechanical Schrödinger equation. Away from the non-relativistic limit, the formal but physically consistent particle interpretation is incompatible even with the semblance of probabilities. The particle which $|1\rangle$ or any energy-momentum eigenstate signifies is not at all quantum-mechanical. It is a particle which identically expresses the underlying quantum field.

Note now that each $|1\rangle$ in (6.I) is an eigenfunction of the number operator $\hat{N}$ since from (5.IV) and (5.V) it follows that

$$\hat{N}|1\rangle = (N_{p_1} + N_{p_2} + ...)(c_1|1_{p_1}\rangle|0_{p_2}\rangle|0_{p_3}\rangle... + c_2|0_{p_1}\rangle|1_{p_2}\rangle|0_{p_3}\rangle... + ...) = 1|1\rangle$$

Thus, the collection of all $|1\rangle$ which satisfy (6.II) defines the Hilbert space for the number operator $\hat{N}$ in the event that $\hat{\phi}(x^0, \vec{x})$ express itself in terms of a single particle, or simply, the *Hilbert space of one-particle states*. In particular, the set of energy-momentum eigenfunctions $|1_{p_1}\rangle|0_{p_2}\rangle|0_{p_3}\rangle..., |0_{p_1}\rangle|1_{p_2}\rangle|0_{p_3}\rangle..., |0_{p_1}\rangle|0_{p_2}\rangle|1_{p_3}\rangle...,...$ gives rise to the Hilbert space of one-particle states since any vector in that Hilbert space emerges as a specific linear combination in (6.I) satisfying (6.II). Pursuant to the analysis in the first chapter (refer to (3.X)), each vector in this Hilbert space satisfies the relation $\langle 1|1\rangle = |c_1|^2 + |c_2|^2 + |c_3|^2 + ...$ so that

$$\langle 1|1\rangle = |c_1|^2 + |c_2|^2 + |c_3|^2 + ... = 1 \qquad (6.III)$$

We note now that as the left side of (6.III) does not signify total probability, the $|c_1|^2$, $|c_2|^2$, $|c_3|^3$, ... on the right side do not signify probabilities either. Accordingly, the semblance of probabilities for the corresponding energy-momentum eigenvalues of $\hat{\phi}(x^0, \vec{x})$ emerges only at the non-relativistic limit of very low energies. The statement in (6.III) is indeed the precise significance of the formal statement in (6.II). The discrete character of the right side in (6.III) is merely a reflection of our approach. The set of momentum states $(p_1), (p_2), (p_3),...$ in Fig.21 is, of course, continuous. It is only for the sake of conceptual clarity that we discretized that continuous set. Reverting to the continuous approach, for that matter, we recast the above relation in the form

$$\langle 1|1\rangle = \int d^3p |c(p)|^2 = 1 \qquad (6.IV)$$

where we have consistently taken into consideration the fact that each coefficient c is now a function of $\vec{p}$, not just a function of $|\vec{p}| \equiv p$.

We proceed now to the two-particle state $\psi_2(x^0, \vec{x})$ which in the Dirac notation is rendered as $|2\rangle$. The two-particle state $|2\rangle$ constitutes the expression of the free quantum field in terms of two particles. If $\hat{\phi}(x^0, \vec{x})$ is in $|2\rangle$ then the excitations of $|0\rangle$ interpreted as two particles may respectively have the momentum eigenvalues p_k and p_l as they may have any other momentum eigenvalues. As the actual energy-momentum eigenvalue is unknown, $|2\rangle$ is a superposition of infinitely many energy-momentum eigenstates characterized by $n = 2$. There are, in fact, infinitely many linear combinations

$$|2\rangle = c_{1,1}|2_{p_1}\rangle|0_{p_2}\rangle|0_{p_3}\rangle... + c_{1,2}|1_{p_1}\rangle|1_{p_2}\rangle|0_{p_3}\rangle... + c_{1,3}|1_{p_1}\rangle|0_{p_2}\rangle|1_{p_3}\rangle... + ...$$

$$+c_{2,3}|0_{p_1}\rangle|1_{p_2}\rangle|1_{p_3}\rangle... + ... \tag{6.V}$$

of the infinitely many possible energy-momentum arrangements for the two quantum particles with each choice of complex coefficients $c_{1,1}$, $c_{1,2}$, $c_{1,3}$, ..., $c_{2,3}$, ... determining a distinct two-particle state $|2\rangle$.

Of immediate physical interest are those sets of coefficients in (6.V) which produce two-particle states in conformity with the demand

$$\langle 2|2\rangle = 1 \tag{6.VI}$$

The collection of all $|2\rangle$ which satisfy (6.VI) defines the *Hilbert space of two-particle states*, as you can confirm yourselves. In particular, the set of energy-momentum eigenfunctions $|2_{p_1}\rangle|0_{p_2}\rangle|0_{p_3}\rangle...$, $|1_{p_1}\rangle|1_{p_2}\rangle|0_{p_3}\rangle...$, $|1_{p_1}\rangle|0_{p_2}\rangle|1_{p_3}\rangle...$, ... gives rise to the Hilbert space of two-particle states. Each vector in this Hilbert space of two-particle states satisfies the relation[55]

$$\langle 2|2\rangle = \int d^3p_1 \int d^3p_2 |c(p_1, p_2)|^2 = 1 \tag{6.VII}$$

We continue in this manner indefinitely.

Having thus constructed the infinitely many Hilbert spaces of the infinitely many particle states in terms of which the free quantum field expresses itself, we proceed to an even higher level of abstraction. The specification of the infinitely many particle states of $\hat{\phi}(x^0, \vec{x})$ which we have just accomplished invites the more abstract question as to the expression of $\hat{\phi}(x^0, \vec{x})$ in terms of an arbitrary number of particles, without reference to the actual particle state of the free quantum field. Absence of knowledge as to the actual eigenvalue of $\hat{N}$ implies that $\hat{\phi}(x^0, \vec{x})$ is in a superposition of all infinitely many particle states. There are infinitely many possible linear combinations specifying infinitely many wavefunctions $\Psi(x^0, \vec{x})$ for the expression of $\hat{\phi}(x^0, \vec{x})$ in terms of an arbitrary number of particles. Since in the Dirac notation we are at liberty to express $\Psi(\vec{x})$ merely as $|\Psi\rangle$ (refer to (11.X) in the first chapter), an arbitrary such $\Psi(x^0, \vec{x})$ is

$$|\Psi\rangle = c_0|0\rangle + \Big[c_1|1_{p_1}\rangle|0_{p_2}\rangle|0_{p_3}\rangle... + c_2|0_{p_1}\rangle|1_{p_2}\rangle|0_{p_3}\rangle... + c_3|0_{p_1}\rangle|0_{p_2}\rangle|1_{p_3}\rangle... + ...\Big]+$$

$$\Big[c_{1,1}|2_{p_1}\rangle|0_{p_2}\rangle|0_{p_3}\rangle... + c_{1,2}|1_{p_1}\rangle|1_{p_2}\rangle|0_{p_3}\rangle... + c_{1,3}|1_{p_1}\rangle|0_{p_2}\rangle|1_{p_3}\rangle... + ...\Big] + ... + ... \tag{6.VIII}$$

with each choice of complex coefficients c_0, c_1, c_2, ..., $c_{1,1}, c_{1,2}$, ... determining a distinct wavefunction $\Psi(x^0, \vec{x})$ for an arbitrary number of quantum particles.

---

[55]In order to evaluate the double continuous sum in (6.VII), we merely keep variable p_1 fixed at value p_1 and sum over variable p_2 from fixed value p_1 onward, then repeat with variable p_1 fixed at momentum value p_2, ... In the end we sum these infinitely many results. This continuous process is implicit in the discrete set $(c_{1,1}, c_{1,2}, c_{1,3}, ..., c_{2,3}, ...)$.

Of immediate physical interest are those sets of coefficients for which

$$\langle \Psi | \Psi \rangle = 1 \qquad (6.IX)$$

We always adhere to the preceding analysis. $|\Psi(x^0, \vec{x})|^2$ does not signify probabilities for position because $\vec{x}$ is merely a label. It is because we have singled out the excited momentum states on the right side of (6.VIII) that $|\Psi\rangle$ is consistently interpreted as a state of an arbitrary number of quantum particles on account of (6.VIII) and (6.IX) - not at all quantum-mechanical particles, but particles which express the underlying quantum field.

The collection of all $|\Psi\rangle$ which satisfy (6.IX) defines a special type of Hilbert space. It is the Hilbert space of an arbitrary number of particles. From (6.VIII) it is obvious that all infinitely many Hilbert spaces which we have respectively constructed for the infinitely many particle states of the free quantum field are subspaces of the Hilbert space to which we have just arrived. It is, in fact, this Hilbert space which expresses the essence of quantum fields. As particles themselves are not fundamental entities, as they are instead only expressions of underlying quantum fields which are themselves the fundamental physical reality, there are infinitely many possibilities as to the number of quantum particles in terms of which a free quantum field expresses itself. Any choice of particle number is as physically legitimate as any other. This fundamental premise is not compatible with any specific Hilbert space of n-particle states, since any such Hilbert space arises only once a choice as to the number of particles in terms of which a free quantum field expresses itself has been made, but is in perfect conformity with the ultimate Hilbert space which emerges from (6.VIII) and (6.IX). This ultimate Hilbert space is known as[56] *Fock space* and lies at the mathematical foundation of quantum field theory.

As is the case with all Hilbert spaces which we have constructed, each vector in Fock space satisfies the relation

$$\langle \Psi | \Psi \rangle = |c_0|^2 + \int d^3p |c(p)|^2 +$$

$$\int d^3p_1 \int d^3p_2 |c(p_1, p_2)|^2 + \int d^3p_1 \int d^3p_2 \int d^3p_3 |c(p_1, p_2, p_3)|^2 + ... = 1 \qquad (6.X)$$

At this stage we have given concrete physical meaning to the expression of free quantum fields in terms of particles, determined the energy spectrum of free quantum fields, settled the issues which we raised at the outset and arrived at an abstract mathematical space which encompasses all possible expressions of free quantum fields in terms of particles. That is, all except one!

7. The Democratic State and the Absolute Monarchy: Bosons and Fermions

---

[56]After Soviet theoretical physicist Vladimir A. Fock (1898 - 1974).

All quantum fields have inherent properties manifest in the particles which express them. The Lorentz-invariant *rest mass* (registered in a frame with respect to which the momentum of the excited quantum state is zero), the spin, the electric charge and the *colour* (a fundamental charge of the quark fields) are examples of such inherent properties. Of particular importance to our analysis in this section is the spin.

Classical particles are trivially distinguished by their different properties. Any difference in mass, size, colour, etc., directly distinguishes one particle from the other. This is also the case with quantum particles. If observation distinguishes one quantum particle from another then such a distinction necessarily relies on differences in the intrinsic physical properties of the two particles, such as rest mass, electric charge, spin, magnetic moment and whichever other charges. Nevertheless, the issue of distinguishing between quantum particles is fundamentally different from that of distinguishing between classical particles. Two classical particles which appear to be identical in all respects may still be distinguished by their respective positions and trajectories. With quantum particles this is clearly not the case. Neither do quantum particles occupy distinct positions, nor do they delineate observable trajectories. If at a certain moment observation has yielded position $\vec{x}$ and position $\vec{x}'$ for two electrons respectively then it is a matter of time before the two probability distributions spread out (as in Fig.11) and eventually overlap, rendering any subsequent measurement void of any capacity to determine which of the two electrons had, at the moment of observation, position $\vec{x}$ and which had position $\vec{x}'$. Quantum particles of the same intrinsic properties are *identical particles*. Since such particles express the same underlying quantum field we may restate this conclusion from the rigorous perspective of quantum field theory: *The necessary and sufficient condition that quantum particles be identical is that they express the same underlying quantum field.*

This property of quantum fields has deep physical implications. In order to uncover them we shall need the theory on the intrinsic angular momentum which we developed in the first chapter. Recall in this respect that only one of the three spin-components can be determined at a given moment and that, consequently, the quantum number s of the intrinsic angular momentum determines the magnitude $\hbar s$ of the latter. The statement to the effect that the spin of a quantum field is s signifies the fact that the magnitude of that field's spin is $\hbar s$. Our analysis on the free quantum fields has, up to this point, been predicated on the tacit assumption of zero spin. As a direct consequence, each momentum state is an independent degree of freedom. If, on the contrary, the spin of a quantum field is not zero, its possible z-projections signify additional degrees of freedom. Since, for that matter, different eigenvalues of $\hat{s}_z$ signify different quantum states, each quantum state is no less determined by s_z than it is determined by the fixed momentum of the corresponding plane wave. For example, a plane wave of momentum magnitude p and eigenvalue s_z for the projection of the spin on the z-axis is a distinct quantum state from the plane wave of momentum magnitude p and eigenvalue $-s_z$. Clearly, the introduction of the spin degree of freedom enlarges the Fock space, with each plane wave of probability in Fig.21 now signifying more degrees of freedom than one. The formalism which we have developed applies identically on the understanding that each momentum state must now be replaced by a number of quantum states determined by the same momentum as well as by the eigenvalues s_z.

Let us concretize these considerations in a specific physical context of $|s, s_z\rangle \equiv |\frac{1}{2}, \pm\frac{1}{2}\rangle$. Let us postulate two non-interacting protons in the following arrangement. One of them is part of a cosmic ray, the product of a supernova explosion at a distance of ten billion light years. The other is a proton which has just been produced in a particle accelerator. Both protons register somehow their existence simultaneously in a detector somewhere on Earth. As they are both observed simultaneously, there is no way of telling which proton was in the cosmic ray and which in the accelerator. The fundamental reason the two protons are identical, the reason that no errors have appeared in the independent production of the two protons despite the fact that one of them has been produced ten billion light years away from the other and ten billion years before the other is precisely the fact that there is a universal proton quantum field, a field which extends across space and time. Both protons are excitations of the same universal proton quantum vacuum and express the same universal proton quantum field. We can understand the implications of this identity if we disregard all other protons (and anti-protons) in the Universe and merely consider the corresponding two-particle state of the free proton quantum field. The underlying reason for this consideration is always the fundamental premise: Since we cannot tell which proton was in the cosmic ray and which in the accelerator, the actual energy-momentum eigenvalue of the proton field is unknown. As the energy-momentum eigenvalue of the proton field is unknown, the expression of the latter in terms of two protons is a superposition of infinitely many energy-momentum eigenstates characterized by $n = 2$. Using the Dirac notation $|0_{p_i}, s_z\rangle$ and $|1_{p_i}, s_z\rangle$ for each quantum state of momentum magnitude p_i ; $i = 1, 2, ...$, occupation number 0_{p_i} and 1_{p_i} respectively, and $s_z = \pm\frac{1}{2}$ (we stress again that these two values of s_z signify two distinct quantum states), it is:

$$|2\rangle = c_{1\uparrow,1\downarrow}|1_{p_1}, \tfrac{1}{2}\rangle|1_{p_1}, -\tfrac{1}{2}\rangle|0_{p_2}, \tfrac{1}{2}\rangle|0_{p_2}, -\tfrac{1}{2}\rangle...+$$

$$c_{1\uparrow,2\uparrow}|1_{p_1}, \tfrac{1}{2}\rangle|0_{p_1}, -\tfrac{1}{2}\rangle|1_{p_2}, \tfrac{1}{2}\rangle|0_{p_2}, -\tfrac{1}{2}\rangle... + c_{1\uparrow,2\downarrow}|1_{p_1}, \tfrac{1}{2}\rangle|0_{p_1}, -\tfrac{1}{2}\rangle|0_{p_2}, \tfrac{1}{2}\rangle|1_{p_2}, -\tfrac{1}{2}\rangle... + ...$$

$$+...c_{i\downarrow,j\uparrow}|0_{p_1}, \tfrac{1}{2}\rangle|0_{p_1}, -\tfrac{1}{2}\rangle|0_{p_2}, \tfrac{1}{2}\rangle|0_{p_2}, -\tfrac{1}{2}\rangle...|0_{p_i}, \tfrac{1}{2}\rangle|1_{p_i}, -\tfrac{1}{2}\rangle...|1_{p_j}, \tfrac{1}{2}\rangle|0_{p_j}, -\tfrac{1}{2}\rangle...+... \quad (7.I)$$

where the reason the occupation number 2_{p_i} is precluded is pending and will become obvious in what follows. The crucial observation for our analysis is that for each energy-momentum eigenstate in which either proton occupies a quantum state characterized by p_i and either value of s_z and the other proton occupies a quantum state characterized by p_j and either value of s_z there is an energy-momentum eigenstate in which the latter proton occupies the former quantum state and the former proton occupies the latter quantum state. Thus, if in (7.I) we agree to always reserve the first entry for the proton which does not advance to the next entry before the other proton has "run through" all infinitely many entries then for each energy-momentum eigenstate in which the proton of the first entry occupies a certain quantum state and the other proton a certain other quantum state there is an energy-momentum eigenstate in which this specific order is reversed. For example, the first term in (7.I) implies the term

$$c_{1\downarrow,1\uparrow}|1_{p_1},-\tfrac{1}{2}\rangle|1_{p_1},\tfrac{1}{2}\rangle|0_{p_2},\tfrac{1}{2}\rangle|0_{p_2},-\tfrac{1}{2}\rangle\cdots$$

This premise applies likewise to each term of (7.I). For that matter, the general term

$$c_{i\downarrow,j\uparrow}|0_{p_1},\tfrac{1}{2}\rangle|0_{p_1},-\tfrac{1}{2}\rangle|0_{p_2},\tfrac{1}{2}\rangle|0_{p_2},-\tfrac{1}{2}\rangle\cdots|0_{p_i},\tfrac{1}{2}\rangle|1_{p_i},-\tfrac{1}{2}\rangle\cdots|1_{p_j},\tfrac{1}{2}\rangle|0_{p_j},-\tfrac{1}{2}\rangle\cdots$$

implies the term

$$c_{j\uparrow,i\downarrow}|0_{p_1},\tfrac{1}{2}\rangle|0_{p_1},-\tfrac{1}{2}\rangle|0_{p_2},\tfrac{1}{2}\rangle|0_{p_2},-\tfrac{1}{2}\rangle\cdots|1_{p_j},\tfrac{1}{2}\rangle|0_{p_j},-\tfrac{1}{2}\rangle\cdots|0_{p_i},\tfrac{1}{2}\rangle|1_{p_i},-\tfrac{1}{2}\rangle\cdots$$

Since the two protons are identical it is clear that the contribution $|c(p_i,-\tfrac{1}{2};p_j,\tfrac{1}{2})|^2 \equiv |c_{i\downarrow,j\uparrow}|^2$ of energy-momentum eigenstate

$$|0_{p_1},\tfrac{1}{2}\rangle|0_{p_1},-\tfrac{1}{2}\rangle|0_{p_2},\tfrac{1}{2}\rangle|0_{p_2},-\tfrac{1}{2}\rangle\cdots|0_{p_i},\tfrac{1}{2}\rangle|1_{p_i},-\tfrac{1}{2}\rangle\cdots|1_{p_j},\tfrac{1}{2}\rangle|0_{p_j},-\tfrac{1}{2}\rangle\cdots$$

to $\langle 2|2\rangle$ (refer to (6.VII) in the preceding section) is the same as the contribution $|c(p_j,\tfrac{1}{2};p_i,-\tfrac{1}{2})|^2 \equiv |c_{j\uparrow,i\downarrow}|^2$ of energy-momentum eigenstate

$$|0_{p_1},\tfrac{1}{2}\rangle|0_{p_1},-\tfrac{1}{2}\rangle|0_{p_2},\tfrac{1}{2}\rangle|0_{p_2},-\tfrac{1}{2}\rangle\cdots|1_{p_j},\tfrac{1}{2}\rangle|0_{p_j},-\tfrac{1}{2}\rangle\cdots|0_{p_i},\tfrac{1}{2}\rangle|1_{p_i},-\tfrac{1}{2}\rangle\cdots$$

to $\langle 2|2\rangle$:

$$|c(p_i,-\tfrac{1}{2};p_j,\tfrac{1}{2})|^2 = |c(p_j,\tfrac{1}{2};p_i,-\tfrac{1}{2})|^2 \tag{7.II}$$

As we showed in the preceding section, (7.II) is not a statement on probabilities. If, however, we stress the formal analogy between (6.VII) and quantum mechanics in order to qualitatively - but not rigorously - interpret the former then (7.II) is a statement to the effect that the probability the proton in the cosmic ray occupy the former quantum state and the proton in the accelerator occupy the latter quantum state is identical to the probability the proton in the accelerator occupy the former quantum state and the proton in the cosmic ray occupy the latter quantum state. As the two protons are indistinguishable, this in turn has the inherent implication that the probability the emergence of either proton precede the emergence of the other by ten billion years is the same as the probability the emergence of the latter proton precede the emergence of the former by ten billion years!

There is nothing special about the two-particle state of the proton quantum field, of course. The result which we have just obtained is identically valid for an arbitrary particle state $|n\rangle$ of any quantum field. For each energy-momentum eigenstate in the superposition which yields $|n\rangle$ there is an energy-momentum eigenstate in which the order of any two particles (anti-particles, or particle - anti-particle) in two quantum states is reversed with the implication that if in an energy-momentum eigenstate we interchange the order in which we specify two particles in two quantum states, if we reverse our choice and first create a particle in quantum state $|p_j,s_z\rangle$ then in quantum state $|p_i, s'_z\rangle$, the contribution of that energy-momentum eigenstate to $\langle n|n\rangle$ remains unaffected. That is:

$$|c(p_1, \tfrac{1}{2}; ...; p_i, -\tfrac{1}{2}...; p_j, \tfrac{1}{2}; ...; p_n, -\tfrac{1}{2})|^2 = |c(p_1, \tfrac{1}{2}; ...; p_j, \tfrac{1}{2}; ...; p_i, -\tfrac{1}{2}; ...; p_n, -\tfrac{1}{2})|^2 \quad (7.III)$$

for spin-$\tfrac{1}{2}$ fields,

$$|c(p_1, ...p_i, ...p_j, ..., p_n)|^2 = |c(p_1, ...p_j, ...p_i, ..., p_n)|^2 \qquad (7.IV)$$

for spin-0 fields and

$$|c(p_1, 1; ...; p_i, -1...; p_j, 1; ...; p_n, -1|^2 = |c(p_1, 1; ...; p_j, 1; ...; p_i, -1; ...; p_n, -1|^2 \quad (7.V)$$

for spin-1 fields.

Thus, the coefficients in a linear combination which yields an arbitrary particle state $|n\rangle$ of any quantum field, allow for two distinct possibilities. Either:

$$c(p_1, \tfrac{1}{2}; ...; p_i, -\tfrac{1}{2}...; p_j, \tfrac{1}{2}; ...; p_n, -\tfrac{1}{2}) = +c(p_1, \tfrac{1}{2}; ...; p_j, \tfrac{1}{2}; ...; p_i, -\tfrac{1}{2}; ...; p_n, -\tfrac{1}{2}) \quad (7.VI)$$

or:

$$c(p_1, \tfrac{1}{2}; ...; p_i, -\tfrac{1}{2}...; p_j, \tfrac{1}{2}; ...; p_n, -\tfrac{1}{2}) = -c(p_1, \tfrac{1}{2}; ...; p_j, \tfrac{1}{2}; ...; p_i, -\tfrac{1}{2}; ...; p_n, -\tfrac{1}{2}) \quad (7.VII)$$

(accordingly for (7.IV) and (7.V)).

These two distinct possibilities reflect two distinct physical situations. If the interchange between two quantum states (such as $|p_i, -\tfrac{1}{2}\rangle$ and $|p_j, \tfrac{1}{2}\rangle$) does not change the sign of the relevant coefficient, a situation evident in (7.VI), then the formalism which we developed in the previous sections applies identically. Each plane wave of probability in Fig.21 may be occupied by an arbitrary number of quantum particles of the same eigenvalue s_z. This property of unrestricted occupation numbers for the quantum states signifies the *Bose-Einstein statistics*. Fields of this property are known as *boson fields*. Clearly, boson fields express themselves in terms of *bosons*, particles which adhere to the Bose-Einstein statistics. Since the Bose-Einstein statistics allows for unrestricted occupation numbers, no quantum state of a boson field is ever full. Such a state is, for that matter, the foremost expression of egalitarianism! It places no restrictions to those excitations of the underlying quantum vacuum which wish to enter it and, once a member, each such excitation has the same momentum, energy and quantum numbers s and s_z as any other! We shall see that it is precisely this property which allows bosons to build the smooth classical potentials and this property which causes the boson fields to mediate interactions.

The tacit assumption of Bose-Einstein statistics associated with zero spin which underlies the formalism in the previous section is manifest in energy-momentum eigenstates such as $|2_{p_1}\rangle|0_{p_2}\rangle|0_{p_3}\rangle$.... The real *Klein-Gordon* field $\hat{\phi}(x)$, as such a boson field is formally known, is the most simple of the free quantum fields and serves for this reason as the most convenient illustration of the most fundamental aspects of their dynamical behavior. This does not extend to the centrally important statistical behavior. The fact that identical particles invariably express the same underlying quantum field has enforced the division of all fields into two major categories. One of them is the category of boson fields. The other, emerges from (7.VII).

If the interchange between two quantum states changes the sign of the relevant coefficient then the occupation numbers are severely restricted. This is readily understood by merely considering the eventuality that quantum state of momentum magnitude p_i and $s_z = \frac{1}{2}$ be occupied by two identical particles:

$$c(p_1, \frac{1}{2}; ...; p_i, \frac{1}{2}; p_i, \frac{1}{2}; ...; p_n, -\frac{1}{2}) = -c(p_1, \frac{1}{2}; ...; p_i, \frac{1}{2}; p_i, \frac{1}{2}; ...; p_n, -\frac{1}{2}) = 0$$

which signifies a zero contribution of energy-momentum eigenstate

$$|1_{p_1}, \frac{1}{2}\rangle ... |2_{p_i}, \frac{1}{2}\rangle ... |1_{p_n}, -\frac{1}{2}\rangle ...$$

to $\langle n|n \rangle$ - on qualitative but non-rigorous terms, zero probability that this energy-momentum eigenstate be realized. Thus, no quantum state in fields of the second category may be occupied by more than one quantum particle. For such fields there are only two possible occupation numbers for each quantum state, the number zero which signifies a plane wave of probability void of particles and the number one. This restrictive behavior signifies the *Fermi-Dirac statistics*. Fields of this statistical behavior are known as *fermion fields*. Clearly, fermion fields express themselves in terms of *fermions*, particles which adhere to the Fermi-Dirac statistics. The central property of fermion fields, for that matter, is the *Pauli exclusion principle*, a statement to the effect that no two identical fermions may simultaneously occupy the same quantum state. Each quantum state of a fermion field describes at most one particle which categorically refuses to share power! Within such a state a single fermion reigns supreme! We shall see that it is precisely this property which conforms with our intuition to the effect that no two particles can simultaneously occupy the same point in space and which, for that matter, builds the semi-classical physical systems which we perceive as "ordinary matter" around us.

Such considerations do not by themselves determine whether the proton field which we used as the vehicle of our analysis in this section has the statistical behavior which (7.VI) signifies and is thus a boson field, or the statistical behavior which (7.VII) signifies and is thus a fermion field. Specifically, how do we determine whether a single particle is a boson or a fermion? If all quantum fields fall in these two categories, there has to be a way of distinguishing between them without recourse to their statistical behavior. In 1940 Wolfgang Pauli[57] resolved this field-theoretical issue in a deep theorem which links the statistical behavior of a particle to that particle's spin. More rigorously, the *spin-and-statistics theorem* links the statistical behavior of a quantum field to the spin of that field. Quantum fields of integer spin are characterized by the Bose-Einstein statistics and their particles are, thus, bosons, whereas quantum fields of half-integer spin are characterized by the Fermi-Dirac statistics and their particles are, thus, fermions. For

---

[57]Wolfgang E. Pauli (1900 - 1958). One of the pioneers of quantum theory and Nobel-prize winner for his exclusion principle.

that matter, (7.VI) is precluded for the proton field. The proton field adheres to (7.VII) and is thus a fermion field.

In order to illustrate the fundamental distinction between boson fields and fermion fields, it is instructive to give a comprehensive list of examples in each category before we proceed.

In the category of bosons:

(i) The π^0, π^+, π^- *mesons* of nuclear interactions all of whose spin is equal to 0. The π^0 meson is, in fact, an example of the real Klein-Gordon field which has served as the basis of our general formalism. As already stated in the first section, under Lorentz transformations the Klein-Gordon field receives an invariant numerical value at each space-time point and is, thus, a *scalar field*. The π^+ and π^- mesons are examples of the charged Klein-Gordon field. In the first section we saw that at each point $\vec{x}$ of a three-dimensional surface of equal time x^0 a classical wave field signifies a finite number of degrees of freedom. The charged Klein-Gordon field is an example of a field which comprises more than one *field degrees of freedom*. It is, in fact, a complex scalar field, as we shall see in the next chapter.

(ii) The photon of the electromagnetic interactions, the *gluons* of the strong interactions and the W^+, W^- and Z of the weak interactions, all of whose spin is equal to 1. They are all examples of a four-dimensional *vector field* of invariant magnitude under Lorentz transformations.

(iii) The theoretically predicted *graviton* of the gravitational interaction whose spin is equal to 2. A thoroughly involved field lacking, until now, a rigorous formulation as a consistent theory of quantum gravity has yet to be attained.

In the category of fermions:

(i) The *baryons* and respective anti-baryons. They are *spinor fields* of spin equal to either $\frac{3}{2}$, or $\frac{1}{2}$. The baryons involved in the composition of "ordinary matter" are the proton and the neutron. The nuclei are states of protons and neutrons bound by nuclear interactions mediated by mesons.

(ii) The *leptons*, the *quarks* and their respective anti-particles. They are spinor fields of spin equal to $\frac{1}{2}$. The leptons comprise the electron and its anti-particle, the *muon* and its anti-particle, the *tau-particle* and its anti-particle, as well as all associated types of *neutrinos* and *anti-neutrinos*. The quarks come in three "colours" and six "flavours" (descriptive references to quantum numbers). The mesons and the baryons are bound states of quarks, with the former being bound states of a quark and an anti-quark and the latter being bound states of three quarks.

(iii) The theoretically predicted supersymmetric counterpart to graviton, the *gravitino*, an example of a spinor field of spin equal to $\frac{3}{2}$.

Pursuant to the third premise in the Introduction to the effect that all classical systems emerge as the semi-classical limit of the dynamical behavior of corresponding quantum fields, our analysis is not complete until it reveals the manner in which quantum fields eventuate in both, the classical (electromagnetic and gravitational) fields and the classical objects which "compose" the world as we perceive it. As stated, the former objective concerns the semi-classical limit of boson fields and will be pursued in the next chapter.

The latter objective concerns the semi-classical limit of fermion fields, a limit which centrally involves the Pauli exclusion principle. As this particularly intricate semi-classical limit requires an elaborate analysis of each specific state of fermionic matter in which it eventuates, rather than formally analyze it in the sixth chapter it is more consistent with our conceptual approach to qualitatively discuss that limit's salient features in each such state now.

Quantum fields are in unremitting interaction with each other. The strong interaction carried by the gluon field binds the quarks into hadrons (baryons and mesons) over a distance range of $\sim 10^{-16}m$. The residual strong interaction which "leaks out" from each hadron over a distance range of $\sim 10^{-15}m$ binds the hadrons in positively-charged atomic nuclei. The electromagnetic interaction carried by the photon field binds the nuclei and the negatively-charged electrons into atoms over a distance range of $\sim 10^{-8}m$. In a manner analogous to the strong interaction, the residual electromagnetic interaction which extends beyond the atoms over a distance range of $\sim 10^{-7}m$ primarily binds the latter into molecules and has a further binding effect between molecules. The distance range of this additional molecule-binding effect depends on the strength of the residual electromagnetic interaction and on the effect which the Pauli exclusion principle has, as we shall see immediately below. Residual forces of adequately high magnitude bind the molecules into the macroscopic structures which we perceive as solid objects over an average distance range of $\sim 10^{-5}m$. Weaker residual forces result in macroscopic liquid and gas structures respectively. We can qualitatively understand the reason that objects in the solid, liquid and gas state which compose the world of "every-day experience"[58] typically appear to be classical objects by first considering a single molecule in a static attractive potential. Such a quantum system has the typically discrete energy spectrum which we studied in the first chapter. If, instead, this molecule is part of a crystal lattice then it is effectively subjected to all forces from all other molecules. Since each molecule generates its own attractive potential, the relative separation $\frac{\Delta E_n}{E_n}$ between successive energy levels in the effective potential which generates the net force on the molecule in consideration is effectively vanishing. In turn, the energy spectrum E_n in that potential is effectively continuous! In the seventh section of the first chapter we saw that an effectively continuous energy spectrum necessarily implies a highly "localized" wavefunction over half a wavelength and the concomitant emergence of a classical particle! This is the first aspect of that particularly involved semi-classical limit ("$\hbar \to 0$") of interacting quantum fields which eventuates in the objects around us. We see that it spans a distance range of eleven orders of magnitude and it involves not only the fermion fields, but also the boson fields on equal footing.

The most striking attribute of the objects which we perceive as classical is their volume, the fact that they "take space". For objects in the solid, liquid, or gas state, this property can be traced to the fact that not all electrons of an atom can be in the same quantum state, thereby not all electrons can be in the ground state, the state of least energy, as would be the case absent the Fermi-Dirac statistics. The Pauli exclusion principle results, for that matter, in an effective volume for the atom. In turn, the Pauli exclusion

---

[58]"Macroscopic" does not necessarily mean "classical". The superconductor, for example, is an *extended quantum object.*

principle enforces an effective minimum distance between the molecules of an "ordinary" body which precludes those molecules from being confined to arbitrarily small volumes with their outer electrons "merging" in a single quantum state. Common knowledge such as that to the effect that no two solid objects can occupy the same place at the same moment or pass through each other is a direct consequence of the Pauli exclusion principle. Of course, this is not the case with liquids and gases. Bodies in the liquid or gas state can mix with each other. This, however, is the exclusive consequence of the fact that the effective minimum distance between molecules in the liquid state and, all the more so, in the gas state is much bigger than the effective minimum distance in the solid state. Clearly, the much larger effective distance between molecules allows for such a behavior while upholding the Pauli exclusion principle.

The "fourth" relativistic-plasma state is the matter-state of the stars. The stars of the main sequence owe their stability to a state of equilibrium between the inward pull of gravity and the outward pressure of radiation produced by exothermic nuclear reactions at the core of the star. As such, the stars of the main sequence are complicated physical systems which render a discussion as to the relevance of the Pauli exclusion principle to their stability non-trivial. We shall only mention in this respect that the latter is amply upheld in the relativistic-plasma state in view of the large number of quantum states available to the electrons, protons and all fermions involved. We shall, however, stress that the Pauli exclusion principle is exclusively responsible for the stability of exotic objects of "degenerate matter" such as white dwarfs and neutron stars. Neither of these two objects produces outward radiation in its core. Instead, the factor which precludes total gravitational collapse is the Pauli exclusion principle. In white dwarfs the pull of gravity has "compressed" all electron quantum states to the ultimate extent of full occupation: Each electron quantum state is occupied so that there are no available electron quantum states to allow for further gravitational collapse. Nevertheless, if the mass of the object exceeds a certain limit (the *Chandrashekar limit*) then the object's gravity overwhelms the effect of the Pauli exclusion principle, the outward *electron degeneracy pressure*, forcing the electrons to combine with the object's protons and produce neutrons. The result is a neutron star, an object consisting almost entirely of neutrons. In a neutron star each neutron quantum state is likewise occupied so that there are no available neutron quantum states to allow for further gravitational collapse. Unless, of course, the mass exceeds the new limit which the Pauli exclusion principle sets (the *Tolman-Oppenheimer-Volkoff limit*). Objects of even higher mass than that which the *neutron degeneracy pressure* can sustain collapse to a black hole.

Thus, gravity always wins in the end! At least, this is the statement in the popular narratives of physics. What this statement, as it stands, dissembles is that gravity does not win over the Pauli exclusion principle. If the object's inward gravitational pull is bigger than the outward neutron degeneracy pressure, it becomes energetically favorable for neutrons to form two-neutron bound states of total spin equal to 0 or 1 (analogous to the well-understood *Cooper pairs* of superconductivity). The latter, being composite bosons, may occupy the same quantum state in arbitrary numbers. In turn, such *boson condensation* effectively terminates all outward pressure and resistance to total gravitational collapse. It is clear, for that matter, that such boson condensation[59] renders the Pauli exclusion principle irrelevant to the formation of a black hole.

We deem such qualitative remarks as we have made on "terrestrial" and "extraterrestrial matter" a sufficient statement on the semi-classical limit of fermion quantum fields.

We finally turn to the implications which the preceding analysis has for the physically non-rigorous, at-best approximate, theory of quantum mechanics. In order to arrive at the quantum-mechanical perspective we consider again the two-proton state in (7.I). Since, as we have already stressed, the semblance of quantum mechanics emerges at the non-relativistic limit of very low energies, contact with quantum mechanics can be made if at that limit we <u>arbitrarily</u> disregard all infinitely many void degrees of freedom. For that matter, if the two protons respectively occupy quantum states $|1_{p_j}, \frac{1}{2}\rangle$ and $|1_{p_i}, -\frac{1}{2}\rangle$ then the quantum-mechanical state which describes them is

$$|2\rangle = c_{i\downarrow,j\uparrow}|1_{p_i}, -\frac{1}{2}\rangle|1_{p_j}, \frac{1}{2}\rangle + c_{j\uparrow,i\downarrow}|1_{p_j}, \frac{1}{2}\rangle|1_{p_i}, -\frac{1}{2}\rangle$$

Since for fermion fields it is $c_{i\downarrow,j\uparrow} = -c_{j\uparrow,i\downarrow}$, this implies

$$|2\rangle = c_{i\downarrow,j\uparrow}\left(|1_{p_i}, -\frac{1}{2}\rangle|1_{p_j}, \frac{1}{2}\rangle - |1_{p_j}, \frac{1}{2}\rangle|1_{p_i}, -\frac{1}{2}\rangle\right)$$

in which, clearly, $c_{i\downarrow,j\uparrow}$ has been reduced from a factor signifying the contribution of an energy-momentum eigenstate to $\langle 2|2\rangle$ to a mere multiplicative factor, which for that matter, can be absorbed in the expression it multiplies. Consequently, the quantum-mechanical state of the two protons is described by the *anti-symmetric wavefunction*

$$|2\rangle = |1_{p_i}, -\frac{1}{2}\rangle|1_{p_j}, \frac{1}{2}\rangle - |1_{p_j}, \frac{1}{2}\rangle|1_{p_i}, -\frac{1}{2}\rangle$$

The quantum-mechanical state of two bosons is likewise obtained from the two-particle state of the corresponding <u>free</u> boson field at the non-relativistic limit of very low energies. For example, in the case of two spin-0 bosons (7.IV) implies the *symmetric wavefunction*

$$|1_{p_i}\rangle|1_{p_j}\rangle + |1_{p_j}\rangle|1_{p_i}\rangle$$

Accordingly, the symmetric wavefunction which at the non-relativistic limit describes the quantum state of two spin-1 bosons respectively occupying quantum states $|1_{p_i}, -1\rangle$ and $|1_{p_j}, 1\rangle$ is[60]

$$|1_{p_i}, -1\rangle|1_{p_j}, 1\rangle + |1_{p_j}, 1\rangle|1_{p_i}, -1\rangle$$

---

[59]This is also the case for the degenerate quark state which possibly replaces the degenerate neutron state, as certain current theories hold.

[60]Of course, we are not referring to photons which are identically relativistic! The reference is to massive spin-1 bosons such as the ρ-meson and to composite spin-1 bosons.

These expressions have been derived for non-interacting quantum-mechanical particles. Their significance, however, is general: Fermions and bosons in whichever quantum-mechanical states are respectively described by antisymmetric and symmetric wavefunctions. This can be qualitatively understood by the fact that, in general, quantum-mechanical states dissemble systems of interacting quantum fields with fermion fields at non-relativistic low-energy limit and boson fields at semi-classical limit. As the antisymmetry in the quantum-mechanical $|2\rangle$ stated above concerns non-relativistic plane waves, it also concerns the superposition of non-relativistic plane waves. Whence, all fermion quantum-mechanical states are antisymmetric! For example, from the perspective of quantum mechanics, the quantum states in atoms express the evolution of electrons against the classical electromagnetic potential of the nucleus. Yet, underlying such semblance of quantum mechanics at the non-relativistic low-energy limit is the interaction between a system of relativistic quantum fields, the photon field and the electron field with its characteristic antisymmetry under reversal of order concerning excitations in distinct quantum states.

Using the Klein-Goron field as an illustrative device, we have studied the most basic aspects of the dynamical behavior of quantum fields. We have further seen how the introduction of the spin degree of freedom enlarges the Fock space and enforces a fundamental distinction of quantum fields into two categories according to their statistical behavior. Yet, the physical essence of quantum fields is infinitely more complex than the basic level at which our conceptual analysis has hitherto been limited. In particular, we have stressed that a theory of free quantum fields has no physical content. Fields reveal themselves only through their interactions. In order to study the manner in which quantum fields interact a further leap in abstract thought is necessary, a leap which will allow for a substantial advance toward the essence of quantum fields precisely because it is founded upon the mathematical formalism of quantum field theory. We shall, in the next chapter, introduce that formalism and the formal quantum field theory itself without compromising our conceptual approach to the latter. In anticipation of it, we proceed now to the formal examination of one of quantum field theory's most fundamental aspects.

8. The Feynman Propagator

Let us turn again to the superposition (6.I) which yields the one-particle state $|1\rangle$. If we concretize the abstract ket $|1\rangle$ to wavefunction $\psi_1(x^0, \vec{x})$ then such a superposition is formally identical to the superposition (2.II) in the first chapter, albeit in Minkowski space-time: The wavefunction as a function of position - regardless of whether the position is a dynamical variable, or merely a label - is a superposition of infinitely many plane waves of probability. Mathematically and physically however, the two superpositions are very different. Whereas each term in the quantum-mechanical superposition (2.II) is a single plane wave, each term in (6.I) is a product of the infinitely many momentum states which the free quantum field comprises. Whereas the quantum-mechanical wavefunction for a given number of particles signifies a finite number of degrees of freedom, the wave-function for a given number of particles in quantum field theory signifies the infinitely many degrees of freedom of the underlying quantum field.

Thus, we really have no right to speak of a particle wavefunction in dissociation from the quantum field itself. The quantum-mechanical particle wavefunction in (2.II) of the first chapter expresses a finite number of degrees of freedom and the position $\vec{x}$ as a dynamical variable. The wavefunction of the one-particle state $\psi_1(x^0, \vec{x})$ in (6.I), or of any particle state of a quantum field for that matter, expresses the infinitely many degrees of freedom of the underlying quantum field and the position $\vec{x}$ as a label. The precise statement is that the n-particle wavefunction of a free quantum field describes the given n particles in terms of which the free quantum field expresses itself. Such a particle interpretation emerges as the exclusive consequence of that formal procedure which singles out the excited momentum states and acknowledges them as distinct from the infinitely many void momentum states of the vacuum state. The exclusive reality is that of the quantum field with its infinitely many degrees of freedom. The finite number of excited momentum states, exclusively expresses the underlying quantum field. It is for that matter that quantum mechanics emerges, at best, as an approximate description of Nature: It is only when we disregard the infinitely many void momentum states in each term of the superposition in (6.I) - an arbitrary and unjustified operation - that the latter reduces to the superposition in (2.II) and to the quantum-mechanical wave-function of the first chapter. In fact, not even this statement is correct! For, the plane waves which emerge upon dismissing the infinitely many void degrees of freedom are relativistic and do not, for that matter, satisfy the non-relativistic Schrödinger equation with its inherent interpretation of $|\psi(t, \vec{x})|^2$ as probability density for position. Quantum mechanics emerges only when we arbitrarily disregard the infinitely many void momentum states at the non-relativistic limit of very low energies! This aspect of quantum field theory will receive further elucidation in the seventh section of the next chapter. The point here is that quantum mechanics signifies, at best, an approximate description of Nature. Particles are only expressions of underlying quantum fields. No procedure can elevate them to fundamental physical entities. A free quantum field expresses itself in terms of actual quantum particles. Actual quantum particles reveal the underlying quantum field through their resolution to the infinitely many degrees of freedom in each energy-momentum eigenstate in (6.I), (6.V) and (6.VIII).

The mathematical formalism which expresses actual particles as excitations of the vacuum state (Fig.22) is necessary in order to give precise meaning to the above qualitative remarks. This will be introduced in the next chapter. Nevertheless, we can arrive at a sufficiently rigorous understanding of the emergence of particle states and their relation to the infinitely many degrees of freedom through the simple observation that for all quantum fields that emergence necessarily involves the action of corresponding field operator $\hat{\phi}(x)$ on corresponding vacuum state $|0\rangle$. The infinitely many degrees of freedom are inherent in $\hat{\phi}(x)$ and manifest themselves in the operation

$$\hat{\phi}(x^0, \vec{x})|0\rangle = \psi_1(x^0, \vec{x}) \tag{8.I}$$

which yields the one-particle state $|1\rangle$ in (6.I) expressed as a state "analogous" to the quantum-mechanical wavefunction for position. In fact, (8.I) yields the precise significance of $|1\rangle$ as a wavefunction: *We interpret the action of operator $\hat{\phi}(x)$ on the vacuum state as the emergence of a one-particle state out of the latter.* This is, indeed, the case! In quantum field theory the wavefunction is exclusively a matter of interpretation, precisely as the particle itself is. Higher number particle states accordingly interpreted as wavefunctions are obtained through successive applications of $\hat{\phi}(x)$ on $|0\rangle$. Thus, the physical content of (8.I) involves two aspects, that of superposition and that of emergence and annihilation. They can be understood on the following lines.

The quantization of the classical field $\phi(x)$ will necessarily preserve the superposition (2.II) in the second section of the present chapter. At a more fundamental level, the superposition inherent in (8.I) represents, in fact, all one-particle states in (6.I) which conform with (6.II). The precise reason will be given in the next chapter where the mathematical formalism will elucidate the difference between a particle interpreted as an elevated momentum state and a particle interpreted as a transition of probability over finite space-time separations. For the purposes of the present analysis let it suffice to state the following. Momentum states exist as plane waves of probability throughout Minkowski space-time. Over finite space-time separations no information as to the actual momentum eigenvalue $p \equiv |\vec{p}|$ of a particle exists. Pursuant to the fundamental premise in the third section of the first chapter, lack of information as to the actual momentum eigenvalue implies that the particle is in a superposition of all infinitely many momentum states. Of course, this premise is valid all the same also for each of the infinitely many momentum states which compose the vacuum state. This then is the first conclusion concerning the physical content of (8.I): Over finite space-time separations particles emerge from the vacuum state in a superposition of all momentum states, in the sense that the quantum field itself is in a superposition of all energy-momentum eigenstates characterized by total occupation number equal to one.

In order to understand in what sense over finite space-time separations the stated superposition signifies emergence from the vacuum state, we must examine the consequences which the interpretation accorded to (8.I) has for the amplitude $\langle x_2^0, \vec{x}_2 | x_1^0, \vec{x}_1 \rangle$ that a quantum particle transit from space-time point $x_1 = (x_1^0, \vec{x}_1)$ to space-time point $x_2 = (x_2^0, \vec{x}_2)$. In line with (3.I) in the third chapter, the evolution of the one-particle state $\psi_1(x^0, \vec{x})$ as registered in an inertial frame from moment x_1^0 to moment x_2^0 is (in this general discussion we make the simplifying assumption of a real quantum field, $\hat{\phi}^\dagger(x) = \hat{\phi}(x)$):

$$\langle \vec{x}_2, x_2^0 | \psi_1 \rangle = \int d^3x_1 \langle \vec{x}_2, x_2^0 | \vec{x}_1, x_1^0 \rangle \langle \vec{x}_1, x_1^0 | \psi_1 \rangle = \int d^3x_1 \langle 0|\hat{\phi}(x_2^0, \vec{x}_2)\hat{\phi}(x_1^0, \vec{x}_1)|0\rangle \langle 0|\hat{\phi}(x_1^0, \vec{x}_1)$$

with the direct implication that *by dint of interpretation, the transition amplitude between two space-time points is*

$$\langle x_2^0, \vec{x}_2 | x_1^0, \vec{x}_1 \rangle \equiv \langle 0|\hat{\phi}(x_2^0, \vec{x}_2)\hat{\phi}(x_1^0, \vec{x}_1)|0\rangle \qquad (8.II)$$

In conformity with $\langle x', t' | x, t \rangle$ in the third chapter, this expression is the *propagator* for the quantum field $\hat{\phi}(x)$.

Thus, a particle picture emerges from (8.I) and (8.II): Each point $\vec{x}_1$ in space as it is at moment x_1^0 is interpreted as a position eigenstate now signifying a possible emergence of a particle out of the vacuum state (as in quantum mechanics each position eigenstate signifies a possible position of a particle according to (3.VI) in the first chapter). In a manner "analogous" to (2.III) in the third chapter, the evolution of this position eigenstate yields the transition amplitude in (8.II) with the implication that the particle which emerges from the vacuum state at point $\vec{x}_1$ and moment x_1^0, propagates in space-time and vanishes into the vacuum state at point $\vec{x}_2$ and moment x_2^0 in the position eigenstate which $\vec{x}_2$ by dint of interpretation signifies. This annihilation of the quantum particle into $|0\rangle$ at space-time point x_2 is the inevitable consequence of the particle's emergence at space-time point x_1. As over finite space-time separations particles emerge out of the vacuum state as a superposition of momentum states according to (8.I), so they vanish into the vacuum state as such. This can also be independently understood as a consequence of energy conservation: If that were not the case, an inertial observer would register the spontaneous creation of particles out of the vacuum. For that matter, the propagator in (8.II) is interpreted as *the amplitude that a particle emerge from the vacuum state at space-time point x_1, propagate to space-time point x_2 and vanish at that point into the vacuum state.*

It follows, for that matter, that a field operator also annihilates particles into the vacuum state. On non-rigorous terms it comprises two components, the creation-component $\hat{\phi}^{(+)}(x)$ and the annihilation-component $\hat{\phi}^{(-)}(x)$. We shall examine the specific expression of these components for the scalar, fermion and photon quantum field respectively. However, in the context of the simplifying assumption $\hat{\phi}^\dagger(x) = \hat{\phi}(x)$ which serves our purposes in this section, either of these is adjoint to the other ($\hat{\phi}^{(-)\dagger}(x) = \hat{\phi}^{(+)}(x)$; $\hat{\phi}^{(+)\dagger}(x) = \hat{\phi}^{(-)}(x)$), so that: $\hat{\phi}(x) = \hat{\phi}^{(-)}(x) + \hat{\phi}^{(+)}(x) = \hat{\phi}^\dagger(x)$. They are both evident in (8.I) in which the obvious annihilating effect $\hat{\phi}^{(-)}(x)$ has on the vacuum state ($\hat{\phi}^{(-)}(x)|0\rangle = 0$) implies that $\hat{\phi}(x)|0\rangle = \hat{\phi}^{(-)}(x)|0\rangle + \hat{\phi}^{(+)}(x)|0\rangle = \hat{\phi}^{(+)}(x)|0\rangle = \psi_1(x)$. Accordingly,

$$\hat{\phi}(x_2^0, \vec{x}_2)\hat{\phi}(x_1^0, \vec{x}_1)|0\rangle = \hat{\phi}^{(-)}(x_2^0, \vec{x}_2)\hat{\phi}^{(+)}(x_1^0, \vec{x}_1)|0\rangle + \hat{\phi}^{(+)}(x_2^0, \vec{x}_2)\hat{\phi}^{(+)}(x_1^0, \vec{x}_1)|0\rangle =$$

$$\int \frac{d^3p}{(2\pi)^3 2p^0} e^{-ip.(x_2 - x_1)}|0\rangle + \hat{\phi}^{(+)}(x_2^0, \vec{x}_2)\hat{\phi}^{(+)}(x_1^0, \vec{x}_1)|0\rangle \qquad (8.III)$$

The first term in the second equality we shall derive in the next chapter and will not concern us at this stage.

As the interpretation of particle creation from and particle-annihilation into the vacuum state is crucially contingent upon a non-vanishing propagator, we would like to explore the implication which (8.III) has for the transition amplitude in (8.II). In the fifth (refer to (5.IX)) and eleventh section of the first chapter, we saw that in a Dirac bra-ket the action of an operator on the quantum state to its right is identical to the action of the adjoint operator on the quantum state to its left. Since, for that matter, in (8.II) $\hat{\phi}^{(+)}(x_2)$ operates on $\langle 0|$ to its left as $\hat{\phi}^{(+)\dagger}(x_2)|0\rangle = \hat{\phi}^{(-)}(x_2)|0\rangle = 0$, it is[61]

$$\langle 0|\hat{\phi}^{(+)}(x_2^0, \vec{x}_2)\hat{\phi}^{(+)}(x_1^0, \vec{x}_1)|0\rangle = 0$$

The first term on the right side of (8.III) does not involve any operators. For a given space-time separation it is just a number of no consequence to $|0\rangle$ and $\langle 0|$. We, thus, conclude that

$$\langle 0|\hat{\phi}(x_2^0, \vec{x}_2)\hat{\phi}(x_1^0, \vec{x}_1)|0\rangle = \langle 0| \int \frac{d^3 p}{(2\pi)^3 2p^0} e^{-ip.(x_2 - x_1)}|0\rangle = \int \frac{d^3 p}{(2\pi)^3 2p^0} e^{-ip.(x_2 - x_1)} \langle 0|0\rangle$$

$$= \int \frac{d^3 p}{(2\pi)^3 2p^0} e^{-ip.(x_2 - x_1)} \tag{8.IV}$$

on the understanding that as it is certain the undisturbed vacuum state of the free quantum field throughout space will advance from past infinity to future infinity, the *vacuum-to-vacuum transition amplitude* is $\langle 0|0\rangle = 1$. This statement expresses the stability of the vacuum state in Minkowski space-time. We have already hinted at it in the third section of the third chapter and we shall further examine it in this and the next section.

Careful observation of the result in (8.IV) reveals that the propagator is a consequence of the stated superpositions of energy-momentum eigenstates at x_1 and x_2, with the exclusive dependence on the difference $x_2 - x_1$ (in contradistinction to a dependence on the specific x_1 and x_2) being the direct consequence of the homogeneity of Minkowski space-time. Although we shall deal with the formal derivation of (8.IV) in the next chapter, we could have elicited this result merely as a consequence of space-time homogeneity without the above brief outline which eventuated in it. However, the primary significance which (8.IV) has for this analysis is the support which it lends to the interpretation of the propagator in (8.II) in terms of particle-emergence from and particle-annihilation into the vacuum state over finite space-time separations. We reiterate that each point $\vec{x}_1$ in space as it is at moment x_1^0 is interpreted as a position eigenstate in which a particle emerges from the vacuum state. The non-vanishing result for the propagator in (8.IV) directly implies that each point $\vec{x}_2$ in space as it is at moment x_2^0 is necessarily interpreted as a position eigenstate in which a particle is annihilated into the vacuum state. This interpretation is in perfect conformity with physical expectations: The vacuum state in Minkowski space-time is stable. It does not decay, giving rise to particles in arbitrary numbers.

Before we turn to a crucial issue inherent in the preceding considerations, something must be stated about the relation of the transition amplitude in (8.II) to the quantum-mechanical path integral. We saw in the third chapter that although mathematically the quantum-mechanical path integral is not - yet, at least - rigorously defined, it has a concrete physical content based upon the concept of particle-trajectory. The latter is,

---

[61]For the mathematically-inclined reader: Equivalently, $\hat{\phi}^{(+)}(x_2)$ operates on $\hat{\phi}^{(+)}(x_1)|0\rangle$ to its right giving rise to an expression which involves $\langle 0|2\rangle = 0$ (in Fock space, particle states of different occupation numbers are *orthogonal* to each other).

in turn, rigorously derived from the action functional for a non-relativistic particle. On the contrary, as we already remarked earlier in this chapter and shall rigorously show in the fifth chapter, the action functional for a field concerns field configurations and is as such utterly irrelevant to particle-trajectories. For that matter, no path integral over particle-trajectories is defined in quantum field theory, even by dint of interpretation. In fact, this can be directly deduced from the nature of a field itself: As in quantum field theory the position is only a label, the path in the concrete sense of trajectory is altogether meaningless!

To the afore-mentioned issue now which lies at the foundation of quantum field theory: Causality! In the opening section of the present chapter I remarked that field theory cannot be understood without a thorough understanding of special relativity and recommended a careful study of the Appendix. In the third chapter, we saw that (2.V) allows for superluminal propagation of a quantum-mechanical particle. This is, of course, the case also in quantum field theory. A simple calculation reveals that the propagator $\langle 0|\hat{\phi}(x_2)\hat{\phi}(x_1)|0\rangle$ in (8.II) does not vanish for space-like separations

$$(x_2 - x_1)^2 = (x_2^0 - x_1^0)^2 - |\vec{x}_2 - \vec{x}_1|^2 < 0$$

It does decay rapidly out of the light cone, but it is nonetheless non-zero. It is not difficult to sense the underlying reason. First, take note of the elementary relativistic premise to the effect that a particle of rest mass m has mass equal to $\frac{m}{\sqrt{1-\frac{v^2}{c^2}}}$ in a frame with respect to which the particle's speed is v. A free quantum field contains momentum states of arbitrarily high momenta $p = \hbar k$ as $v \to c$ (Fig.21). As to the speeds themselves, although all momentum states, being <u>observable</u> (actual particles), are characterized by the classical constraint $v < c$ if $m \neq 0$; $v = c$ if $m = 0$, nothing prevents the <u>unobservable</u> propagating particle from completing a transition over a space-like interval.

This situation presents an apparent inconsistency. Pursuant to the analysis associated with Fig.A5 in the Appendix, two inertial frames which register two events, P and Q, either of which lies out of the other's light cone may never agree as to which event precedes the other. For space-like separations there is no Lorentz invariant way of ordering events. Equivalently, if event P and event Q are connected by a signal which propagates at a speed higher than that of light, no causal relation between them exists. Infinitely many inertial observers agree that event P precedes event Q and infinitely many others agree that event Q precedes event P. A quantum particle can propagate between two points in space in arbitrarily short time, thereby violating causality. A physically consistent theory must preclude such an eventuality.

In order to understand how quantum field theory resolves the issue of causality, we must first place that issue in its proper physical context. The theory of relativity precludes propagation of classical particles at superluminal speeds. On the contrary, in quantum field theory particles can, by rights, propagate at superluminal speeds! This, however, is not the physically relevant issue. We reiterate that since a particle which by dint of interpretation emerges at x_1 and vanishes at x_2 is observable only at these two space-time points, that particle may propagate between them even if $(x_2 - x_1)^2 < 0$. Since no causal relation between space-time points in space-like separations is admissible, the physically relevant issue is whether the observation of field-strength at x_1 can affect the observation of field-strength at x_2 in the event that $(x_2 - x_1)^2 < 0$. On rigorous terms, in order that quantum field theory be consistent with the physical demand for causality, it must establish that a measurement of field-strength made at $x_1 = (x_1^0, \vec{x}_1)$ does not affect a measurement of field-strength made at $x_2 = (x_2^0, \vec{x}_2)$ in the event that $(x_2 - x_1)^2 < 0$. Recall, in this respect, the analysis in the third section of the first chapter. If the order in which two physical attributes are measured is irrelevant, if they can both be simultaneously measured with absolute certainty, the corresponding operators commute. The direct implication is that the field operators $\hat{\phi}(x_1)$ and $\hat{\phi}(x_2)$ which correspond to the values of the quantum field at space-time points x_1 and x_2 must commute for space-like separations in all inertial frames:

$$\hat{\phi}(x_1)\hat{\phi}(x_2) = \hat{\phi}(x_2)\hat{\phi}(x_1) \ ; \quad if \ \ (x_2 - x_1)^2 < 0 \qquad (8.V)$$

if quantum field theory is to be physically consistent. We can show that this condition is satisfied in a few simple steps.

First, recall that the procedure of equal-time canonical quantization has quantized the classical field on a three-dimensional surface of equal time x^0 with respect to an inertial frame. The separation between space-time points $x_1 = (x^0, \vec{x}_1)$ and $x_2 = (x^0, \vec{x}_2)$ on this surface is clearly space-like:

$$(x_2 - x_1)^2 = (x^0 - x^0)^2 - |\vec{x}_2 - \vec{x}_1|^2 < 0 \qquad (8.VI)$$

and remains space-like under Lorentz transformations since the latter leave the magnitude of space-time vectors invariant. Next, consider that on this surface (8.V) is identically satisfied on account of the simultaneous measurement of the quantum field $\hat{\phi}(x)$ at $x_1 = (x^0, \vec{x}_1)$ and $x_2 = (x^0, \vec{x}_2)$:

$$\hat{\phi}(x^0, \vec{x}_1)\hat{\phi}(x^0, \vec{x}_2) - \hat{\phi}(x^0, \vec{x}_2)\hat{\phi}(x^0, \vec{x}_1) = 0 \qquad (8.VII)$$

How does this quantity behave under Lorentz transformations? Transformations do not generate something out of nothing! Since this quantity is equal to zero in the inertial frame which registers the stated three-dimensional surface of equal time, it necessarily remains equal to zero under all Lorentz transformations. Let us choose the frame associated with coordinates $x' = (x'^0, \vec{x}')$. In these new coordinates the same two space-time points are respectively labelled as x_1', x_2' and satisfy $(x_2' - x_1')^2 < 0$. Accordingly, the operators $\hat{\phi}'(x_1')$ and $\hat{\phi}'(x_2')$ (for the same field, now observed in this new frame) satisfy

$$\hat{\phi}'(x_1'^0, \vec{x}_1')\hat{\phi}'(x_2'^0, \vec{x}_2') - \hat{\phi}'(x_2'^0, \vec{x}_2')\hat{\phi}'(x_1'^0, \vec{x}_1') = 0 \qquad (8.VIII)$$

which is identical to (8.V).

We have thus shown that (8.V) is valid in all inertial frames and that, for that matter, causality is upheld. However, (8.V) is not as simple as it appears and we have yet to reach the physical essence of the matter. If, in simple quantum mechanics, two quantities A and B can be simultaneously and precisely measured then an initial measurement of A followed by a measurement of B is equivalent to an initial measurement of B followed by a measurement of A. There is never an issue of temporal ordering because in three-dimensional Euclidean space the speed of light is effectively infinite! Quantum field theory, on the contrary, is formulated in four-dimensional Minkowski space-time, a central aspect of which is the invariant light cone and the concomitant distinction between time-like, null and space-like separations. For space-like separations, an initial measurement of field-strength $\hat{\phi}(x_1)$ at moment x_1^0 followed by a measurement of field-strength $\hat{\phi}(x_2)$ at moment x_2^0 ($x_2^0 > x_1^0$) is equivalent to an initial measurement of field-strength $\hat{\phi}(x_2)$ at moment x_2^0 followed by a measurement of field-strength $\hat{\phi}(x_1)$ at moment x_1^0 ($x_1^0 > x_2^0$). Yet, the unique, frame-independent arrow of time - invariably pointing from past infinity to future infinity - precludes two opposite temporal orders in the same inertial frame. Let us, for that matter, also establish causality for space-like separations in an independent manner which illustrates the underlying physical essence of (8.V).

We start directly with the inertial frame associated with coordinates x' which we now denote simply by x. In this inertial frame the arrow of time determines the unique temporal order $x_2^0 > x_1^0$ between events $P \equiv x_1 = (x_1^0, \vec{x}_1)$ and $Q \equiv x_2 = (x_2^0, \vec{x}_2)$ in Fig.A5. Clearly, in the expression $\hat{\phi}(x_1)\hat{\phi}(x_2) - \hat{\phi}(x_2)\hat{\phi}(x_1)$ only the second term is consistent with this temporal order. The first term is consistent with $x_1^0 > x_2^0$ and refers, for that matter, to the observations in another inertial frame associated with coordinates $x' = (x'^0, \vec{x}')$ and temporal order $x_2'^0 > x_1'^0$ between events $Q \equiv x_1' = (x_1'^0, \vec{x}_1')$ and $P \equiv x_2' = (x_2'^0, \vec{x}_2')$. If causality is to be established in the arbitrary but fixed inertial frame associated with coordinates x, the first term must be reduced to that frame and to the inherent order $x_2^0 > x_1^0$ between the two measurements. That is, we must establish that $\hat{\phi}(x_1)\hat{\phi}(x_2) = \hat{\phi}(x_2)\hat{\phi}(x_1)$. We explicitly consider at this point the simplest case of a real Klein-Godon field (which, as we shall see, remains invariant under Lorentz transformations), the better to illustrate the essence of the matter. Although not all fields are Lorentz invariant, the gist of the argument remains the same, as we shall see. Note in this respect that each individual term in $\hat{\phi}(x_1)\hat{\phi}(x_2) - \hat{\phi}(x_2)\hat{\phi}(x_1)$, being the product of two Lorentz-invariant terms, is itself Lorentz invariant. Different inertial frames may disagree on the temporal order between events x_1 and x_2, but they always agree on the numerical value of $\hat{\phi}(x_1)\hat{\phi}(x_2)$ and independently on that of $\hat{\phi}(x_2)\hat{\phi}(x_1)$. Let us, for that matter, consider a Lorentz transformation which takes the space-time vector $x_2 - x_1$ defined by the two measurements as observed in the inertial frame associated with the x-coordinates to the space-time vector $x_2' - x_1' = x_1 - x_2$ defined by the two measurements as observed in the inertial frame associated with the x'-coordinates[62]. This is most easily accomplished by taking x_1 at the origin of the invariant light cone ($x_1 = (0, \vec{0})$) so that the transformation $x_2 - x_1 \to x_1 - x_2$ implies $x_2 \to -x_2$. Fig.23(a) illustrates this procedure. The two hyperbolae $XY = -1$ in the two-dimensional Minkowski diagram in Fig.A4 correspond to the three-dimensional hyperboloid in the actual four-dimensional Minkowski space-time. The future and past light cones are likewise three-dimensional.

---

[62]Both sides of $x_2' - x_1' = x_1 - x_2$ signify the same vector. In x-coordinates it is labelled $x_1 - x_2$, in x'-coordinates it is labelled $x_2' - x_1'$. Note that $x_2 - x_1 \to x_1 - x_2$ reverses both, the temporal and the spatial order.

A continuous Lorentz transformation from $x_2 - x_1$ to $x_1 - x_2$ in the event that $(x_2 - x_1)^2 < 0$ is possible precisely because of the three-dimensional hyperboloid. Since in $\hat{\phi}(x_1)\hat{\phi}(x_2) - \hat{\phi}(x_2)\hat{\phi}(x_1)$ each individual term is Lorentz invariant, performing such a Lorentz transformation on the second term yields $\hat{\phi}(x_2)\hat{\phi}(x_1) = \hat{\phi}(x_1)\hat{\phi}(x_2)$. The two terms are equal *in the same* inertial frame, precisely as required. Moreover, as that frame is utterly arbitrary, (8.V) is valid in all inertial frames and causality for space-like separations is upheld!

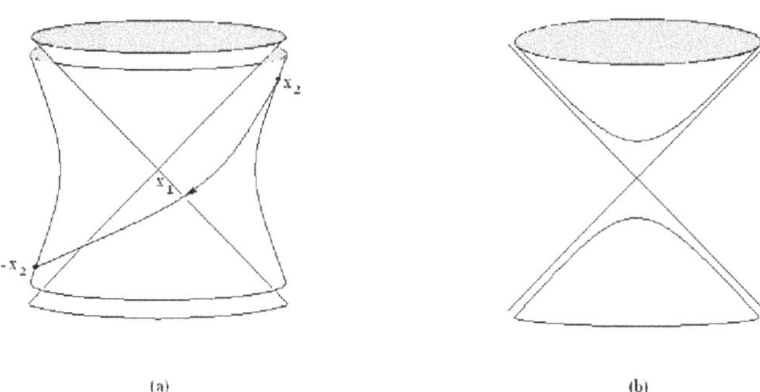

FIGURE 23. The three-dimensional hyperboloid which yields the unit of length in all inertial frames renders a continuous Lorentz transformation from $x_2 - x_1$ to $x_1 - x_2$ possible (a). The two disconnected hyperboloid which yield the unit of time in all inertial frames render a continuous Lorentz transformation from $x_2 - x_1$ to $x_1 - x_2$ impossible with the implication that for time-like separations a measurement at x_1 does affect a measurement at x_2 (b).

Thus, the physical demand for causality determines this: *At any two space-time points x_1 and x_2 in space-like separation - that is, which lie out of each other's light cone and cannot, for that matter, be connected by a light signal (much less by any other signal) - the field operators $\hat{\phi}(x_1)$ and $\hat{\phi}(x_2)$ commute and the field-strengths which these operators respectively express can, for that matter, be measured precisely and independently of each other.* In making this statement, it must be kept in mind that the inherent interpretation of field-strengths $\hat{\phi}(x_1)$ and $\hat{\phi}(x_2)$ as observable physical quantities is rather rhetorical! Fields reveal themselves to us through their interactions. Yet, interactions signify the negation of free quantum fields. The theory of free quantum fields is a linear theory in whose context individual particles exist in their isolated quantum states, in the complete absence of interactions amongst them. Such a theory allows for no observations whatsoever. Nevertheless, the interpretation of field-strengths as observable quantities is physically consistent if only because the theory of free quantum fields constitutes the basis of the formalism which underlies the theory of interacting quantum fields. If actual particles which emerge as a result of interactions between quantum fields are observable then, logically, the actual particles which express free quantum fields are also observable. For ultimately, the observable character of actual particles is independent of the strength of interactions. However indirect and difficult observation may be in the context of weakly interacting fields, actual particles are ultimately observable. Equivalently, a progressive decrease of the strength of interactions between quantum fields leaves the actual particles invariably observable. It is physically consistent for that matter, to interpret the actual particles of the free quantum fields as observable entities at the limit at which all interactions vanish and the fields de-couple from each other. Of course, at that ideal limit both the observable actual quantum particles and the free quantum fields themselves emerge as idealized entities and it is as such that they also allow for the physically consistent interpretation of field-strength as an observable quantity: If the interpretation of idealized actual particles as observable entities is physically consistent, the interpretation of field-strength as an observable quantity also is! We reiterate that we have arrived at the above-stated central conclusion regarding the commutativity of field operators for space-like separations in the context of the real scalar Klein-Gordon field which we use as a prototype of all quantum fields on account of its comparative simplicity. We shall see that this conclusion is directly relevant to all quantum fields, even though, formally, it is only valid for boson fields.

We see now the implication which causality has for particle propagation. Considered in isolation, the propagator introduced in (8.II) violates causality for space-like separations and is not, for that matter, physically acceptable. However, the fundamental relation (8.V) directly implies that

$$\langle 0|\hat{\phi}(x_2)\hat{\phi}(x_1)|0\rangle - \langle 0|\hat{\phi}(x_1)\hat{\phi}(x_2)|0\rangle = 0 \; ; \quad if \; (x_2 - x_1)^2 < 0 \qquad (8.IX)$$

Let's look at this fundamental relation somewhat more carefully than as a mere implication of (8.V). For convenience, we first introduce the *commutator* $[\hat{\phi}(x_2), \hat{\phi}(x_1)]$ for field operators $\hat{\phi}(x_1)$ and $\hat{\phi}(x_2)$:

$$[\hat{\phi}(x_2), \hat{\phi}(x_1)] = \hat{\phi}(x_2)\hat{\phi}(x_1) - \hat{\phi}(x_1)\hat{\phi}(x_2)$$

It is now a straightforward matter to deduce that

$$[\hat{\phi}(x_2), \hat{\phi}(x_1)] = \langle 0|[\hat{\phi}(x_2), \hat{\phi}(x_1)]|0\rangle = \langle 0|\hat{\phi}(x_2)\hat{\phi}(x_1)|0\rangle - \langle 0|\hat{\phi}(x_1)\hat{\phi}(x_2)|0\rangle \qquad (8.X)$$

for all space-time separations. The second equality in (8.X) is quite obvious. The first equality follows for space-like separations directly from (8.V). However, the first equality is also valid for time-like separations. That this is the case follows from this simple argument. Since the commutator $[\hat{\phi}(x_2), \hat{\phi}(x_1)]$ is equal to zero for $(x_2 - x_1)^2 < 0$, we expect it to be equal to a number $c \neq 0$ for $(x_2 - x_1)^2 > 0$. Numbers are trivial operators! Their action on quantum states merely signifies multiplication of the specific quantum state by the specific number. Thus, $\langle 0|c|0\rangle = c\langle 0|0\rangle = 1$, with the second equality being a direct consequence of $\langle 0|0\rangle = 1$. This establishes that the first equality in (8.X) is also valid for time-like separations. It is, in fact, the validity of (8.X) for all space-time separations which prompted this brief derivation of (8.IX).

This then is the implication in (8.IX). Although $\langle 0|\hat{\phi}(x_2)\hat{\phi}(x_1)|0\rangle$ allows for superluminal propagation, (8.V) upholds causality by offsetting the effect with $\langle 0|\hat{\phi}(x_1)\hat{\phi}(x_2)|0\rangle$. Again, for space-like separations there is no Lorentz-invariant way of ordering events. As observed from inertial frame K, event $P = x_1$ of particle emergence from $|0\rangle$ precedes event $Q = x_2$ of particle annihilation into $|0\rangle$. As observed from inertial frame K', event $Q = x_1'$ of particle emergence from $|0\rangle$ precedes event $P = x_2'$ of particle annihilation into $|0\rangle$. If P and Q are separated by a space-like interval, a particle propagates from P to Q with respect to K, but from Q to P with respect to K'. Propagation between two events in space-like separation involves two particles, one associated with inertial frame K and the other associated with inertial frame K'. In any measurement, the two transition amplitudes cancel against each other. In fact, we already know that they cancel because they are reducible to the same inertial frame. The reason is always the three-dimensional hyperboloid which determines the unit of length in all inertial frames and renders a continuous Lorentz transformation from $x_2 - x_1$ to $x_1 - x_2$ possible for space-like separations (Fig.23(a)).

There is a lot more which must be stated about this cancellation. We have arrived at this conclusion in the context of the real Klein-Gordon field which is only a special aspect of the general complex Klein-Gordon field. When we study the latter, we shall see that for space-like separations inertial frame K registers the propagation of a particle, whereas inertial frame K' registers the propagation of an antiparticle. We shall see that in quantum field theory conformity with causality requires that to each particle there exist a corresponding antiparticle with the same mass and opposite quantum numbers, such as the electric charge. In the case of the real Klein-Gordon field, which has in this section served as a vehicle for the illustration of the issues associated with causality and of their resolution, the particle is its own antiparticle.

Having established causality for space-like separations, we note in passing the physical consistency of the theory for time-like separations. For $(x_2 - x_1)^2 > 0$, there is no issue of causality. All inertial frames agree that the event of emergence x_1 precedes the event of annihilation x_2. In conformity with causality (event x_1 causes event x_2), the commutator $[\hat{\phi}(x_2), \hat{\phi}(x_1)]$ inherent in (8.V) does not vanish. Nor, for that matter, are the two terms $\langle 0|\hat{\phi}(x_2)\hat{\phi}(x_1)|0\rangle$ and $\langle 0|\hat{\phi}(x_1)\hat{\phi}(x_2)|0\rangle$ in (8.X) equal. The underlying reason, of course, is the absence of a continuous Lorentz transformation which takes $x_2 - x_1 \to x_1 - x_2$ for $(x_2 - x_1)^2 > 0$ (Fig.23(b)).

The preceding considerations render transparent the fact that taken in isolation the transition amplitude in (8.II) which served as the incipient point of our analysis on particle propagation is inadequate because it is predicated on a specific temporal order between space-time points x_1 and x_2. For space-like separations the commutator $[\hat{\phi}(x_2), \hat{\phi}(x_1)]$ which is central to the issue of causality also enforces the transition amplitude predicated on the opposite temporal order. The two transition amplitudes are observed in two distinct inertial frames. A physically consistent transition amplitude must treat both temporal orders on equal footing. It follows that the physically consistent transition amplitude is exclusively the sum-total[63]

$$i\Delta_F(x_2-x_1) = \langle 0|\hat{\phi}(x_2)\hat{\phi}(x_1)|0\rangle\theta(x_2^0-x_1^0)+\langle 0|\hat{\phi}(x_1)\hat{\phi}(x_2)|0\rangle\theta(x_1^0-x_2^0) \qquad (8.XI)$$

of the two transition amplitudes in the two inertial frames. For space-like separations both temporal orders, $x_2^0 > x_1^0$ and $x_1^0 > x_2^0$, between events x_1 and x_2 exist. We demand, for that matter, that the first term in (8.XI) vanish for $x_1^0 > x_2^0$ and that the second term vanish for $x_2^0 > x_1^0$. Formally, these two demands are respectively inherent in the *Heaviside function* featured in each term and defined as

$$\theta(x_2^0 - x_1^0) = \left\{ \begin{array}{l} 0 \; ; \; if \; x_2^0 < x_1^0 \\ 1 \; ; \; if \; x_2^0 \geq x_1^0 \end{array} \right.$$

(accordingly for $\theta(x_1^0 - x_2^0)$). It is, of course, hardly necessary to stress that the arrow of time is fixed from past infinity to future infinity in all frames. Accordingly, for time-like separations the temporal order $x_2^0 > x_1^0$ is unique. As an immediate consequence, the second term vanishes in all frames.

The transition amplitude $\Delta_F(x_2 - x_1)$ in (8.XI) is thus Lorentz invariant and valid *for all* space-time separations. It is known as the *Feynman propagator* and constitutes a central aspect of quantum field theory. In defining the Feynman propagator the following aspects of the preceding analysis must be kept into perspective: (i) Causality is upheld in the commutator $[\phi(x_2), \phi(x_1)]$. The latter vanishes strictly for space-like separations and receives non-vanishing values for time-like separations. For all separations in space-time the stated commutator comprises the two terms in (8.V) and, equivalently, in (8.X). (ii) The transition amplitude between two space-time points is determined by the Feynman propagator which comprises *only one* term in each specified inertial frame. Note, in this respect, the resemblance between the Feynman propagator $\Delta_F(x_2 - x_1)$ and the propagator defined in (8.II) and featured in $\Delta_F(x_2 - x_1)$ respectively as $\langle 0|\hat{\phi}(x_2)\hat{\phi}(x_1)|0\rangle$ and $\langle 0|\hat{\phi}(x_1)\hat{\phi}(x_2)|0\rangle$: Both propagators are manifestly Lorentz invariant (although each term in $\Delta_F(x_2 - x_1)$ is not since the Heaviside function causes it to vanish upon a Lorentz transformation to a coordinate system of reverse temporal order between x_1 and x_2). Both propagators comprise only one term in each specific frame. Both propagators are consistent with causality! There is essentially no contradiction between the present statement and the preceding statement to the effect that taken in isolation the transition amplitude in (8.II) is inadequate because here we do not consider the latter in isolation, tied to a specific inertial frame and associated coordinate system, but in its general behavior under Lorentz transformations between inertial frames. From the perspective of its behavior under Lorentz transformations, $\langle 0|\hat{\phi}(x_2)\hat{\phi}(x_1)|0\rangle$ necessarily implies $\langle 0|\hat{\phi}(x_1)\hat{\phi}(x_2)|0\rangle$ for space-like separations between x_1 and x_2. The two transition amplitudes are equal and, pursuant to (8.X) and (8.IX), cancel against each other in order to uphold causality. Perhaps in our endeavour to arrive at the physically consistent Feynman propagator we have been somewhat too severe with the propagator defined in

---

[63]The multiplicative factor i appears for mathematical convenience and is physically irrelevant. The observable quantity is the *positive-definite* transition probability $|\Delta_F(x_2 - x_1)|^2$.

(8.II). Physically, there is no essential difference between the Feynman propagator and the propagator on condition that we consider the latter from the general perspective of all inertial frames, not just from the particular perspective of a specific inertial frame. The physically consistent statement, for that matter, is that although both propagators are consistent with the demand for causality, causality is manifest only in the Feynman propagator precisely because only the latter manifestly features $\langle 0|\hat{\phi}(x_2)\hat{\phi}(x_1)|0\rangle$ and $\langle 0|\hat{\phi}(x_1)\hat{\phi}(x_2)|0\rangle$.

The Feynman propagator $\Delta_F(x_2 - x_1)$ receives an extremely useful diagrammatic representation in Minkowski space-time through its association with a line connecting the two space-time points x_1 and x_2. We have already seen, of course, that each of these two points represents Minkowski space-time in its entirety. As it is necessary that each category of quantum fields receive a distinct diagrammatic representation, we associate the Feynman propagator for a scalar field (the *scalar propagator*) with a syncopated straight line, the Feynman propagator for a fermion field (the *fermion propagator*) with a smooth straight line and the Feynman propagator for a vector field such as the Feynman propagator for the photon field which is of immediate interest to us (the *photon propagator*) with an undulating line (Fig.24). The Feynman propagator and its respective diagrammatic representations lie at the foundation of *perturbative field theory* which describes interacting - and thus, *actual* - quantum fields through successive approximations to the non-linear dynamical behavior of the interacting system.

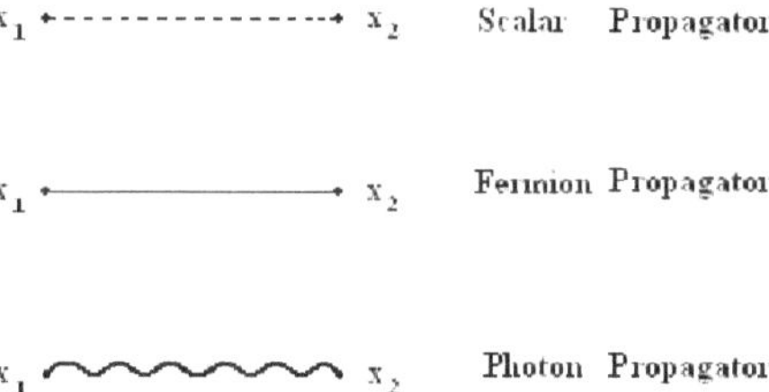

FIGURE 24. Minkowski space-time representation of the Feynman propagator for the scalar, fermion and photon field respectively.

We conclude with a few remarks on the amplitude that the vacuum state at past infinity transit to itself at future infinity. This is tantamount to the vacuum state on a three-dimensional surface of equal time advancing from past infinity to future infinity. That is, the vacuum-to-vacuum transition amplitude is defined in the entirety of Minkowski space-time. Since, as already stated, it is stable and would not, for that matter, evolve to another state if left to itself, it clearly satisfies

$$\langle 0|0 \rangle = 1 \qquad\qquad (8.XII)$$

Relation (8.XII) can be viewed as a generalization of (3.III) in the third chapter to the infinitely many degrees of freedom of quantum field theory. It applies to all free quantum fields and expresses a simple but fundamental property of the vacuum state in Minkowski space-time. It is probably worth adding that, in general, (8.XII) is not as trivial as it may impress upon. It certainly has a concrete and universal physical significance for the preferred set of inertial observers in Minkowski space-time since for them the vacuum state is the only Lorentz-invariant state of a quantum field. However, it ceases to be valid for accelerating observers in Minkowski space-time (as they describe the vacuum state differently from inertial observers - they register particles) and for static but curved (not to mention dynamic) space-times (as in them different observers define the vacuum state differently according to their state of motion). Although we study quantum field theory from the perspective of inertial observers in Minkowski space-time and shall not extend our analysis beyond it, we made these brief remarks in order to call attention to the fact that in the presence of gravity quantum field theory is decisively more intricate.

We have only arrived at the basic physical content and most general properties of the Feynman propagator in Minkowski space-time. We are still far from uncovering its central implications of *CPT* invariance and antiparticles. We advance in this direction now. We turn to the powerful mathematical formalism of quantum field theory and to the immeasurably deeper insight which that formalism avails.

PART III

FREE QUANTUM FIELDS - ADVANCED THEORY

CHAPTER 5.

THE KLEIN-GORDON FIELD

1. The Real Klein-Gordon Field

By way of initiating the formal approach to quantum field theory recall that in the fourth section of the preceding chapter we introduced a system of units in which $c = \hbar = 1$. We shall, as of this chapter, use this expedient system when mathematical simplicity demands it and avoid its use when conceptual clarity is necessary. As of this chapter we shall, in addition, replace the descriptive term "continuous summation" which, allowing for the third chapter, we have hitherto favored, by the formal term *integration*. The formal term *integral* will signify the sum itself - that is, the result of integration.

By now we are substantially acquainted with the real Klein-Gordon field $\hat{\phi}(x)$ which comprises only one field degree of freedom. Being the simplest of all fields, this spin-0 scalar quantum field has served as the device for the qualitative analysis of the most fundamental aspects of canonical field quantization and of the dynamical behavior of all free quantum fields. In order to advance past this qualitative level and make contact with the actual mathematical formalism which rigorously conveys the essence of the real quantum Klein-Gordon field, we return to the real classical Klein-Gordon field. The latter is the superposition of plane waves of energy in (2.VII) of the fourth chapter with each wave number $k \equiv |\vec{k}|$ labelling two terms, $e^{-i(w_k t - \vec{k}.\vec{x})} = e^{-ik.x}$ and $e^{i(w_k t - \vec{k}.\vec{x})} = e^{ik.x}$, either of them being the complex conjugate of the other $\left((e^{\pm ik.x})^* = e^{\mp ik.x}\right)$. At this stage, we limit ourselves to merely stating that the reason is inherent in the character of the Klein-Gordon equation as a second-order differential equation with respect to each space-time variable $x^\mu = (x^0, x^1, x^2, x^3)$, a fact which has the implication that if $e^{-ik.x}$ is a solution then $e^{ik.x}$ also is. This is not the case with the quantum-mechanical Schrödinger equation which is first-order with respect to time. We shall be able to appreciate this difference in the seventh section of the present chapter where we shall state the two equations and discuss their respective implications. Until then it is important to remember that *all classical wave fields are linear combinations of plane waves $e^{-ik.x}$ and their complex conjugates $e^{ik.x}$*. For that matter, the significance which the generic expression (2.VII) has for the classical Klein-Gordon wave field is $(c = 1)$:

$$\phi(t, \vec{x}) = \int d^3k \frac{1}{\sqrt{(2\pi)^3 2w_k}} [a(k)e^{-i(w_k t - \vec{k}.\vec{x})} + a^*(k)e^{i(w_k t - \vec{k}.\vec{x})}] \tag{1.Ia}$$

or, concisely:

$$\phi(x) = \int d^3k \frac{1}{\sqrt{(2\pi)^3 2k^0}} [a(k)e^{-ik.x} + a^*(k)e^{ik.x}] \tag{1.Ib}$$

Two physical properties of the Klein-Gordon field in (1.I) are immediately obvious. The first is its real character: $\phi^*(x) = \phi(x)$. The real Klein-Gordon field is a particular aspect of the complex Klein-Gordon field which we shall examine in the next section. The second is its scalar character. This symmetry is a particular expression of the Klein-Gordon field's Lorentz covariance which we extensively discussed in the preceding chapter: Under a Lorentz transformation $x \to x'$ - that is, which takes the expression of a given point in the unprimed Cartesian coordinates to the expression of the same point in the primed Cartesian coordinates associated with a frame in relative uniform motion (refer to the Appendix) - the Klein-Gordon field remains invariant. That is

$$\phi'(x') = \phi(x) \tag{1.II}$$

Before we proceed to quantization, we must deal with an issue the resolution of which has been pending from the outset. We saw in the second section of the fourth chapter that the plane wave of energy $a(k)e^{-i(w_k t - \vec{k}.\vec{x})}$ involves, by demand, exclusively the positive frequency $w_k = +\sqrt{(kc)^2 + m^2 c^4}$. We must, for that matter, account for the reason that in $w_k^2 = (kc)^2 + m^2 c^4 \implies w_k = \pm\sqrt{(kc)^2 + m^2 c^4}$ the negative square root is precluded. Naively, it would appear that since complex conjugation has the effect of converting positive signs to negative and negative signs to positive, the complex conjugate $a^*(k)e^{i(w_k t - \vec{k}.\vec{x})}$ involves the negative value $w_k = -\sqrt{(kc)^2 + m^2 c^4}$. That this is not the case follows from the simple relation $(e^{i\theta})^* = e^{-i\theta} = e^{i(-\theta)} = cos(-\theta) + isin(-\theta) = cos\theta - isin\theta$ which makes a statement to the effect that the y-projection of the position vector tilted at $-\theta$ is identical to the reverse y-projection of the position vector tilted at $+\theta$ with the direct implication that θ is invariably positive (Fig.25). Since $e^{-i(w_k t - \vec{k}.\vec{x})}$ and $e^{i(w_k t - \vec{k}.\vec{x})}$ are sinusoidal waveforms in each direction of space at a given moment t and in the course of time at each point $\vec{x}$, the argument $w_k t \equiv \theta$ is, for an arbitrary but fixed point $\vec{x}$, invariably positive. Thus, as long as the arrow of time points from past infinity to future infinity $(t > 0)$, the frequency in both, $e^{-i(w_k t - \vec{k}.\vec{x})}$ and $e^{i(w_k t - \vec{k}.\vec{x})}$, is invariably positive: $w_k = +\sqrt{|\vec{k}|^2 c^2 + m^2 c^4} \equiv +\sqrt{k^2 c^2 + m^2 c^4}$. The apparent negative value of w_k has disappeared. The implications inherent in the absence of negative values of frequency in the classical theory point to the resolution of the apparent contradiction between the theory of relativity and the quantum theory and to a triumph of quantum field theory itself, as we shall soon see.

The magic of field quantization unfolds now... Fields as particles! The discrete spectrum of amplitudes which each plane wave of probability admits and the associated particle occupation numbers (Fig.22) are the direct consequence of operators $\hat{a}^\dagger(k)$ and $\hat{a}(k)$. In replacing the plane waves of energy by plane waves of probability, field quantization elevates the classical amplitudes to operators! Amplitude $a(k)$ is elevated to operator $\hat{a}(k)$ and amplitude $a^*(k)$ is elevated to the adjoined operator $\hat{a}^\dagger(k)$. To each momentum magnitude $p = \hbar k$ correspond operators $\hat{a}^\dagger(k) \equiv \hat{a}^\dagger(p)$ and $\hat{a}(k) \equiv \hat{a}(p)$; $(\hbar = 1)$. The former excites the simple quantum harmonic oscillator defined by p. For example, the action of $\hat{a}^\dagger(p)$ on the vacuum state $|0\rangle$ of the free quantum Klein-Gordon field excites the amplitude of the void plane wave of probability which propagates momentum p by an increment of one. Equivalently, upon operating on $|0\rangle$, $\hat{a}^\dagger(p)$ creates one particle in momentum state (p):

$$\hat{a}^\dagger(p)|0\rangle = |1_p\rangle \tag{1.III}$$

on the understanding that $|1_p\rangle$ has the exclusive significance stated in (5.VIII) of the preceding chapter. The operation of $\hat{a}^\dagger(p)$ on $|0\rangle$ yields the energy-momentum eigenstate $|1_p\rangle \equiv |0_k\rangle|0_l\rangle...|1_p\rangle...|0_r\rangle...$ characterized by occupation number $n_p = 1$. We interpret

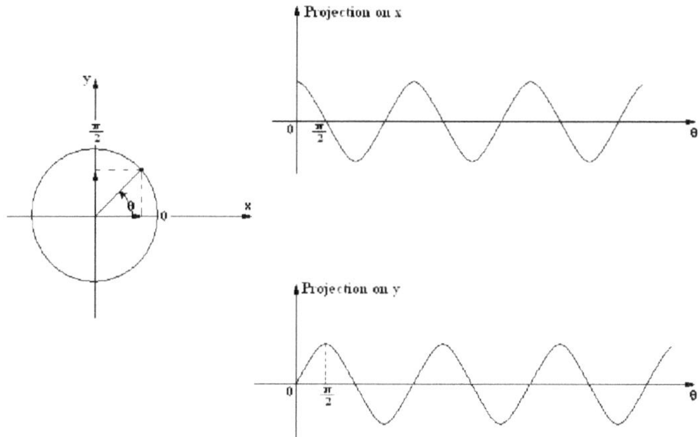

FIGURE 25. The complex function $e^{i\theta} = cos\theta + isin\theta$ signifies a unit circle ($|e^{i\theta}| = 1$) in the complex plane with $x = Re\,e^{i\theta} = cos\theta$ and $y = Im\,e^{i\theta} = sin\theta$. Since $cos(-\theta) = cos\theta$ and $sin(-\theta) = -sin\theta$, the clockwise motion of the position vector ($\theta < 0$) is always reducible to its counterclockwise motion ($\theta > 0$): The angle θ is invariably positive.

this energy-momentum eigenstate as a one-particle state in the sense that we single out the excited momentum state *without* disregarding the infinitely many void degrees of freedom which this energy-momentum state comprises. At once, it is because $\hat{a}^{\dagger}(p)$ creates this particle out of the vacuum state that we interpret $\hat{a}^{\dagger}(p)$ as a *creation operator*. Repeated application of $\hat{a}^{\dagger}(p)$ on $|0\rangle$ excites the oscillator defined by p to its n-th energy eigenstate. Equivalently, repeated application of $\hat{a}^{\dagger}(p)$ on $|0\rangle$ yields n particles in momentum state (p) - that is, the n-particle state $|n_p\rangle$:

$$(\hat{a}^{\dagger}(p))^{n}|0\rangle = \frac{1}{\sqrt{n!}}|n_p\rangle \qquad (1.IV)$$

with $n! = 1 \times 2 \times ... \times n$ (parenthetically, the multiplicative factor $\frac{1}{\sqrt{n!}}$ enforces consistency with the *Bose-Einstein statistics*). The n-particle state $|n_p\rangle$ is always the product of momentum state (p) and the infinitely many void momentum states which $|0\rangle$ comprises, precisely as stated above for $|1_p\rangle$. Clearly now, the operation of $\hat{a}^{\dagger}(p)$ on $|n_p\rangle$ yields the $(n+1)$-particle state $|(n+1)_p\rangle$:

$$\hat{a}^{\dagger}(p)|n_p\rangle = \sqrt{n+1}|(n+1)_p\rangle \qquad (1.V)$$

Accordingly, the operator $\hat{a}(p)$ lowers the corresponding simple quantum harmonic oscillator defined by p from its n-th energy eigenstate to its $(n-1)$-th energy eigenstate, thus annihilating a particle in momentum state (p):

$$\hat{a}(p)|n_p\rangle = \sqrt{n}|(n-1)_p\rangle \qquad (1.VI)$$

It is for that matter that $\hat{a}(p)$ is an *annihilation operator*. The immediate consequence which this has for the no-particle vacuum state is

$$\hat{a}(p)|0\rangle = 0 \qquad (1.VII)$$

- that is, *the vacuum state is annihilated by all operators $\hat{a}(p)$*. Alongside $|0_{p_1}\rangle|0_{p_2}\rangle|0_{p_3}\rangle$ ... $|0_{p_n}\rangle... \equiv |0\rangle$ and $\hat{N}|0\rangle = 0|0\rangle$, this property serves as yet another equivalent definition of the vacuum state in Minkowski space-time. Note also that, as an immediate consequence of (1.V) and (1.VI), it is

$$\hat{a}^{\dagger}(p)\hat{a}(p)|n_p\rangle = n|n_p\rangle$$

which reveals that

$$\hat{N}_p = \hat{a}^{\dagger}(p)\hat{a}(p) \qquad (1.VIII)$$

is the occupation-number operator introduced in the fifth section of the preceding chapter.

The formalism in terms of creation and annihilation operators is the language of canonical field quantization. The entire qualitative analysis in the preceding fourth chapter finds its rigorous expression in this formalism. All particle states and the entire Fock space of a free quantum field emerge as the result of the action of creation and annihilation operators. For example, the state

$$|1_{p_1}\rangle|1_{p_2}\rangle...|1_{p_n}\rangle \equiv |1_{p_1}, 1_{p_2}, ..., 1_{p_n}\rangle = \hat{a}^{\dagger}(p_1)\hat{a}^{\dagger}(p_2)...\hat{a}^{\dagger}(p_n)|0\rangle \qquad (1.IX)$$

generated by the successive action of n creation operators on the vacuum state is a specific energy-momentum eigenstate of total number of particles equal to n (it is hardly necessary to state again that in this notation the infinitely many void momentum states are implicit). We have seen in the sixth section of the preceding fourth chapter that the n-particle state $|n\rangle$ is a linear combination of all possible distributions of n-particles in the infinitely many momentum states. The collection of all particle states obtained by acting on the vacuum state with all possible combinations of creation operators

$$|0\rangle, \hat{a}^{\dagger}(p)|0\rangle, \hat{a}^{\dagger}(p)\hat{a}^{\dagger}(q)|0\rangle, \hat{a}^{\dagger}(p)\hat{a}^{\dagger}(q)\hat{a}^{\dagger}(r)|0\rangle, ... \qquad (1.X)$$

spans the Fock space of the theory.

The free quantum Klein-Gordon field which emerges from the quantization of the solution (1.I) to the equation for a real classical Klein-Gordon wave field is a spin-0 field expressed by the operator ($c = \hbar = 1 \implies k = p$; $w_k = E_p$):

$$\hat{\phi}(t,\vec{x}) = \int d^3p \frac{1}{\sqrt{(2\pi)^3 2E_p}} [\hat{a}(p)e^{-i(E_p t - \vec{p}.\vec{x})} + \hat{a}^\dagger(p)e^{i(E_p t - \vec{p}.\vec{x})}] \qquad (1.XIa)$$

or, concisely:

$$\hat{\phi}(x) = \int d^3p \frac{1}{\sqrt{(2\pi)^3 2p^0}} [\hat{a}(p)e^{-ip.x} + \hat{a}^\dagger(p)e^{ip.x}] \qquad (1.XIb)$$

which acts on the Fock space in the manner specified for $\hat{a}^\dagger(p)$ and $\hat{a}(p)$. Only the second sector in (1.XI) associated with $\hat{a}^\dagger(p)$ yields a non-vanishing result in the operation $\hat{\phi}(x^0,\vec{x})|0\rangle$. Inherent in this result are all the fundamental conclusions to which we arrived in the preceding chapter. It is, in fact, difficult to overstate the importance of our discussion in the eighth section regarding the intimate relation between particles and fields. The superposition

$$\psi_1(x^0,\vec{x}) = \hat{\phi}(x^0,\vec{x})|0\rangle = \int d^3p \frac{1}{\sqrt{(2\pi)^3 2E_p}} [\hat{a}^\dagger(p)e^{i(E_p t - \vec{p}.\vec{x})}]|0\rangle$$

$$= \int d^3p \frac{1}{\sqrt{(2\pi)^3 2E_p}} e^{i(E_p t - \vec{p}.\vec{x})}|1_p\rangle \qquad (1.XII)$$

is the one-particle state $|1\rangle$ announced in (8.I) expressed as a state "analogous" to the wavefunction for position, that superposition of one-particle energy-momentum eigen-states $|1_p\rangle$ which signifies the expression of the real Klein-Gordon field in terms of one particle in the event that its actual energy-momentum eigenvalue is unknown. The plane waves $e^{-i(E_p t - \vec{p}.\vec{x})}$ and $e^{i(E_p t - \vec{p}.\vec{x})}$ in the operator - more rigorously, in the *operator-valued function* $\hat{\phi}(x^0,\vec{x})$ - make a statement to the effect that the free quantum field which emerges in the context of equal-time canonical quantization is a superposition of them in the sense that these plane waves of probability are actually inherent in the energy-momentum eigenstates of $\hat{\phi}(x^0,\vec{x})$ in general and in the vacuum state $|0\rangle$ in particular. Specifically, $|1_p\rangle = \hat{a}^\dagger(p)|0\rangle$ in (1.XII) is merely the maximum amplitude of plane wave $e^{i(E_p t - \vec{p}.\vec{x})}$ already present in $|0\rangle$. Thus, (1.XII) is formally identical to (2.II) in the first chapter with $|1_p\rangle$ corresponding to the magnitude (the maximum value) $\phi(p)$ of $\phi(t,p) = \phi(p)e^{iE_p t}$ and $e^{-i\vec{p}.\vec{x}}$ corresponding to $\psi_p(x)$. The reason we are reluctant to separate *energy-momentum eigenstate* $|1_p\rangle$ in a space-dependent factor analogous to $\psi_p^*(x)$ and a time-dependent $\phi^*(t,p)$ is because - unlike non-relativistic $\psi(t,x)$ in (2.II) - $\psi_1(x^0,\vec{x})$ in (1.XII) is Lorentz invariant. Such Lorenz invariance would cease to be manifest upon the stated separation. This point need not concern us further now. We shall return to it in the third section of the present chapter. The essence of the matter here is that the identity between (1.XII) and (2.II) is only formal! As we thoroughly understand by now, in sharp contrast to quantum mechanics, each term in the super-position (1.XII) is a product of infinitely many momentum states and each momentum state is an independent degree of freedom. The formal identity between (1.XII) and (2.II) reinforces our interpretation of the excited energy-momentum eigenstates as par-ticles as well as our interpretation of $\hat{\phi}(x^0,\vec{x})|0\rangle \equiv \psi_1(x^0,\vec{x})$ as the wavefunction of a particle which emerges from $|0\rangle$ throughout space at moment x^0: *We interpret the action of operator $\hat{\phi}(x^0, \vec{x})$ on $|0\rangle$ as the emergence at moment x^0 of a particle at point $\vec{x}$.* This has the same significance as that in quantum mechanics: As in $\psi(t, \vec{x}) \equiv \langle \vec{x}, t | \psi \rangle$ each position eigenstate $|\vec{x}, t\rangle$ evolves independently of all the others (refer to (3.I) in the third chapter), so in the "wavefunction" $\hat{\phi}(x^0, \vec{x})|0\rangle$ each "position eigenstate" $|x^0, \vec{x}\rangle$ which emerges out of $|0\rangle$ evolves independently of all the others throughout infinite space at moment x^0. Nevertheless, we reiterate that $\psi_1(x^0, \vec{x})$ does not signify probabilities for position. We would not, in fact, overstate the essence of the matter if we stressed again that this particle picture is only formal, merely a matter of interpretation: "Experimentally we are concerned with particles", yet wavefunction $\psi_1(x^0, \vec{x})$ expresses essentially the underlying quantum Klein-Gordon field and its infinitely many degrees of freedom!

So what do we actually observe, if not the position? In principle, we could answer this subtle question in the intuitive context which we delineated in the opening paragraph of the preceding chapter's third section. If, for the moment, we return to the classical perspective from which fields are products of sources and consider - for convenience - the electromagnetic field as the product of an electron then upon quantization the probabilistic character of the latter's position implies that at each space-time point the strength of the electromagnetic field which the electron produces has itself a probabilistic character. At each space-time point it consists in a continuous set of values each of which is realized with a predictable probability. In this intuitive context, such values represent the infinitely many amplitudes of the infinitely many plane waves of probability to which we formally arrived in the fifth section of the preceding chapter[64]. Let us, for that matter, also consider a second electron at a "certain distance" from the first, through which we aspire to observe the stated probability distribution of photon field-strength. That second electron will, at each moment, be effectively subjected to infinitely many possible values of a repulsive force. Each such value is, in principle, observable and identical to the photon field-strength produced by the first electron. Of course, the two electrons interact with each other through "virtual" photons (the precise nature of which we shall examine soon) transiting between any possible position of the first and any possible position of the second. Thus, the observation of photon field-strength through either electron, is at once formally equivalent to the observation of the end-points of virtual propagation: *At any space-time point, the observation of photon field-strength as a predictable probability distribution is formally interpreted as observation of an emergent (accordingly, vanishing) photon.* Such a formal interpretation is clearly consistent with (1.XII): Absence of knowledge as to the value of repulsive force and, thereby, of photon field-strength prior to observation implies that the emergent (accordingly, vanishing) virtual photon is in a one-particle state $|x^0, \vec{x}\rangle$ - in the superposition $\hat{\phi}(x)|0\rangle$, as it were.

We see, for that matter, that in quantum field theory the observation of position is invariably a matter of interpretation, never a matter of fact! In turn, this further elucidates the relation of propagation to causality. The entire analysis on this matter in the eighth section of the preceding chapter identically applies to the Klein-Gordon

---

[64]Recall our discussion on the nature of observation on free quantum fields in the eighth section of the preceding chapter: The larger the "distance from the source" is, the more accurate this statement becomes.

field, which we used as a prototype. Central to that analysis was our statement to the effect that for space-like separations, a measurement of field-strength made at space-time point x_1 does not affect a measurement of field-strength made at space-time point x_2. In view of the preceding remarks, *we formally interpret the measurement of field-strength at a space-time point as the observation of the emergent particle's position $\vec{x}$ at moment x^0.* This is, of course, in perfect conformity with the formal interpretation of $\langle x_2^0, \vec{x}_2 | x_1^0, \vec{x}_1 \rangle$ as a transition amplitude for a particle. This is also in perfect conformity with the language of field configurations. Repeated applications of $\hat{\phi}(x^0, \vec{x})$ on $|0\rangle$ $(...\phi(x^{0''}, \vec{x}'')\phi(x^{0'}, \vec{x}')\phi(x^0, \vec{x})|0\rangle)$ yield accordingly the general n-particle state $|n\rangle$; $n = 1, 2, 3, ...$ At a given moment x^0 each point $\vec{x}$ labels infinitely many possible values of field-strength in each energy-momentum eigenstate inherent in the superposition which yields $|n\rangle$. They are always the amplitudes - now treated as values of field-strength - of that eigenstate's infinitely many plane waves of probability at $x = (x^0, \vec{x})$. For that matter, the equivalence between the language of energy-momentum eigenstates and the language of field configurations, between the language of particles and the language of fields, is inherent in (1.XI) and (1.XII). This renders transparent the precise meaning of the qualitative expression "values of field-strength $\hat{\phi}(x^0, \vec{x})$" which we used in the preceding chapter.

Onward to the statistical behavior of the quantum Klein-Gordon field. In view of its scalar behavior under Lorentz transformations (refer to (1.II)), $\hat{\phi}(x)$ is a boson field. Thus, it satisfies

$$|1_{p_i}, 1_{p_j}\rangle = |1_{p_j}, 1_{p_i}\rangle \qquad (1.XIII)$$

as we have seen in the seventh section of the preceding chapter. In view of (1.IX), this relation is enforced only if

$$\hat{a}^\dagger(p_i)\hat{a}^\dagger(p_j) = \hat{a}^\dagger(p_j)\hat{a}^\dagger(p_i) \iff [\hat{a}^\dagger(p_i), \hat{a}^\dagger(p_j)] = 0 \qquad (1.XIVa)$$

which is indeed a central aspect of quantum field theory: *The creation operators of a boson field commute amongst themselves.* Clearly, as a consequence of (1.XIII), the annihilation operators also commute amongst themselves:

$$\hat{a}(p_i)\hat{a}(p_j) = \hat{a}(p_j)\hat{a}(p_i) \iff [\hat{a}(p_i), \hat{a}(p_j)] = 0 \qquad (1.XIVb)$$

The relation between creation and annihilation operators is somewhat more involved. Since we did not introduce the canonical momentum of the Klein-Gordon field, we shall not advance its rigorous derivation. We shall, instead, arrive at it through the observation that although

$$\hat{a}(p_i)\hat{a}^\dagger(p_j)|0\rangle = \hat{a}^\dagger(p_j)\hat{a}(p_i)|0\rangle = 0 \; ; \; if \; p_i \neq p_j$$

it is

$$\hat{a}(p_i)\hat{a}^\dagger(p_j)|0\rangle \neq \hat{a}^\dagger(p_j)\hat{a}(p_i)|0\rangle = 0 \; ; \; if \; p_i = p_j$$

Since, for that matter, $\hat{a}^\dagger(p_j)$ and $\hat{a}(p_i)$ do not commute for $p_i = p_j$, the relation between them necessarily is:

$$[\hat{a}(p_i), \hat{a}^\dagger(p_j)] = \delta^{(3)}(\vec{p}_i - \vec{p}_j) \qquad (1.XIVc)$$

We conclude with the resolution of an outstanding issue. The quantization of the classical Klein-Gordon wave field translated the issue of negative frequencies to an issue of negative energies. Since the momentum magnitude p and the energy E_p of a momentum state satisfy the relativistic relation $E_p^2 = (pc)^2 + (mc^2)^2$, we have to account for the negative value of energy $E_p = -\sqrt{(pc)^2 + (mc^2)^2}$ and associated infinitely many negative-energy momentum states which are absent in the analysis of the preceding chapter. By now the resolution of this issue is trivial. All energies $E_p = \hbar w_p$ associated with the infinitely many momentum states in (1.XI) are positive for the same reason that all frequencies w_k in the classical theory are positive! This resolves the issue which was briefly raised in the fourth section of the preceding chapter: *As long as the arrow of time points from past infinity to future infinity, no negative-energy states are admissible!*

The profound significance of this statement is not yet obvious. The relation $E = \pm\sqrt{(pc)^2 + (mc^2)^2}$ also applies to a relativistic classical particle without cause for concern. We trivially discard the negative value on the grounds that negative energies are physically unacceptable. Why then is it imperative that we preclude all negative-energy values at quantum level? Why is "relativistic quantum mechanics" with its characteristic negative-energy states physically unacceptable? Moreover, why stress the observable arrow of time? Could it ever reverse its direction? The fact is that neither are negative energies at all as easy to dismiss in the quantum theory as they are in the classical theory, nor is the observable arrow of time as unique and universal as it appears in a world which accommodates antiparticles. In anticipation of the rigorous justification of such claims, let it suffice to state for now that the distinction between particles and antiparticles is, in the case of the real quantum Klein-Gordon field, trivial. The real quantum Klein-Gordon field expresses itself in terms of spin-0 electrically-neutral bosons. Each such boson is the antiparticle of itself. A typical example is the π^0-meson.

We may now generalize these results.

2. The Complex Klein-Gordon Field

An extension which yields two field degrees of freedom is the *charged Klein-Gordon field*. The corresponding classical relativistic wave field is described by the expression

$$\phi(x) = \frac{1}{\sqrt{2}}[\phi_1(x) + i\phi_2(x)] \qquad (2.I)$$

The components $\phi_1(x)$ and $\phi_2(x)$ are real Klein-Gordon fields which we assume to be of equal rest mass. The charged Klein-Gordon field $\phi(x)$ is thus a complex scalar field with $\phi_1(x)$ and $\phi_2(x)$ being its two field degrees of freedom. Since $\phi_1(x)$ and $\phi_2(x)$ define as much the *complex conjugate field*

$$\phi^*(x) = \frac{1}{\sqrt{2}}[\phi_1(x) - i\phi_2(x)] \qquad (2.II)$$

as they define the complex field $\phi(x)$, the complex Klein-Gordon field is represented as much by the two field degrees of freedom $\phi(x)$ and $\phi^*(x)$ as it is represented by the two field degrees of freedom $\phi_1(x)$ and $\phi_2(x)$. These two equivalent representations yield the same interpretation of the quantum complex Klein-Gordon field $\hat{\phi}(x)$ as an electrically-charged quantum scalar field whose two field degrees of freedom

$$\hat{\phi}(x) = \frac{1}{\sqrt{2}}[\hat{\phi}_1(x) + i\hat{\phi}_2(x)] \qquad (2.III)$$

and

$$\hat{\phi}^\dagger(x) = \frac{1}{\sqrt{2}}[\hat{\phi}_1(x) - i\hat{\phi}_2(x)] \qquad (2.IV)$$

are associated with positively-charged scalar particles and negatively-charged scalar particles of the same rest mass. We shall see that the latter are the antiparticles of the former and that the former are the antiparticles of the latter, a fact which justifies our assumption of equal rest mass. Pursuant to (1.XI) in the preceding section, the quantum complex Klein-Gordon field is formally a superposition of plane waves. Now, however, the classical theory involves two distinct classes of infinitely many plane waves of energy, one of which defines $\phi_1(x)$ according to (1.I) and the other, likewise, defines $\phi_2(x)$. The general solution $\phi(x)$ to the classical field equation, the solution which describes the classical complex Klein-Gordon field, is a combination of these two superpositions in conformity with $\phi(x) \sim \phi_1(x) + i\phi_2(x)$. Accordingly, the complex conjugate $\phi^*(x)$ to the general solution $\phi(x)$ is a combination of the same two superpositions in conformity with $\phi^*(x) \sim \phi_1(x) - i\phi_2(x)$. In turn, upon quantization, $\hat{\phi}(x)$ and $\hat{\phi}^\dagger(x)$ respectively emerge as a superposition of plane waves in each class according to (1.XI) and a combination of these two superpositions in conformity with $\hat{\phi}_1(x) \pm i\hat{\phi}_2(x)$. The superposition which yields $\hat{\phi}_1(x)$ is predicated on $\hat{a}_1(p)$ and $\hat{a}_1^\dagger(p)$, whereas that which yields $\hat{\phi}_2(x)$ is predicated on $\hat{a}_2(p)$ and $\hat{a}_2^\dagger(p)$. Inherent in the combination of the two superpositions is the expression of the quantum complex Klein-Gordon field in terms of positively-charged and negatively-charged particles. Formally, field operator $\hat{\phi}(x)$ annihilates the former and creates the latter. Conversely, field operator $\hat{\phi}^\dagger(x)$ creates the former and annihilates the latter. This follows directly if in (2.III) and (2.IV) the real fields $\hat{\phi}_1(x)$ and $\hat{\phi}_2(x)$ are respectively expressed through (1.XI) with the subsequent identifications

$$\frac{1}{\sqrt{2}}[\hat{a}_1(p) + i\hat{a}_2(p)] = \hat{a}_+(p) \; ; \; \frac{1}{\sqrt{2}}[\hat{a}_1^\dagger(p) - i\hat{a}_2^\dagger(p)] = \hat{a}_+^\dagger(p)$$

$$(2.V)$$

$$\frac{1}{\sqrt{2}}[\hat{a}_1(p) - i\hat{a}_2(p)] = \hat{a}_-(p) \; ; \; \frac{1}{\sqrt{2}}[\hat{a}_1^\dagger(p) + i\hat{a}_2^\dagger(p)] = \hat{a}_-^\dagger(p)$$

242

For then it is an elementary exercise to show that

$$\hat{\phi}(x^0,\vec{x}) = \int \frac{d^3p}{\sqrt{(2\pi)^3 2E_p}}[\hat{a}_+(p)e^{-i(E_p t - \vec{p}.\vec{x})} + \hat{a}_-^\dagger(p)e^{i(E_p t - \vec{p}.\vec{x})}] \qquad (2.VIa)$$

or, concisely

$$\hat{\phi}(x^0,\vec{x}) = \int \frac{d^3p}{\sqrt{(2\pi)^3 2E_p}}[\hat{a}_+(p)e^{-ip.x} + \hat{a}_-^\dagger(p)e^{ip.x}] \qquad (2.VIb)$$

and

$$\hat{\phi}^\dagger(x^0,\vec{x}) = \int \frac{d^3p}{\sqrt{(2\pi)^3 2E_p}}[\hat{a}_+^\dagger(p)e^{i(E_p t - \vec{p}.\vec{x})} + \hat{a}_-(p)e^{-i(E_p t - \vec{p}.\vec{x})}] \qquad (2.VIIa)$$

or, concisely

$$\hat{\phi}^\dagger(x^0,\vec{x}) = \int \frac{d^3p}{\sqrt{(2\pi)^3 2E_p}}[\hat{a}_+^\dagger(p)e^{ip.x} + \hat{a}_-(p)e^{-ip.x}] \qquad (2.VIIb)$$

with the creation and annihilation operators satisfying the commutation relations

$$[\hat{a}_+(p_i),\hat{a}_+^\dagger(p_j)] = [\hat{a}_-(p_i),\hat{a}_-^\dagger(p_j)] = \delta^{(3)}(\vec{p}_i - \vec{p}_j) \qquad (2.VIIIa)$$

$$[\hat{a}_+(p_i),\hat{a}_-^\dagger(p_j)] = [\hat{a}_-(p_i),\hat{a}_+^\dagger(p_j)] = 0 \qquad (2.VIIIb)$$

$$[\hat{a}_\pm(p_i),\hat{a}_\pm(p_j)] = [\hat{a}_\pm^\dagger(p_i),\hat{a}_\pm^\dagger(p_j)] = 0 \qquad (2.VIIIc)$$

which directly follow from (1.XIV) in view of the fact that the momentum state on which $\hat{a}_+(p)$ acts is distinct from the momentum state on which $\hat{a}_-^\dagger(p)$ acts (likewise for $\hat{a}_-(p)$ and $\hat{a}_+^\dagger(p)$).

The implication which (2.VI) has for the no-particle vacuum state $|0\rangle$ of the complex Klein-Gordon field is that $\hat{a}_+(p)|0\rangle = 0$ and $\hat{a}_-^\dagger(p)|0\rangle = |1_p^-\rangle$, for all values of p. This is consistent with the following interpretation. Operator $\hat{a}_+(p)$ annihilates positively-charged particles and operator $\hat{a}_-^\dagger(p)$ creates negatively-charged particles. Pursuant to our discussion in the preceding chapter's eighth section, as well as to the formal interpretation which we introduced in the present chapter's preceding section, the operation of $\hat{\phi}(x)$ on $|0\rangle$ yields the negatively-charged one-particle state $|1^-\rangle$ expressed as a state "analogous" to the wavefunction for position. Specifically, (1.XII) in the previous section implies that

$$\hat{\phi}(x^0,\vec{x})|0\rangle = \psi_1^-(x^0,\vec{x}) \qquad (2.IX)$$

Likewise, the implication which (2.VII) has for $|0\rangle$ is that $\hat{a}_-(p)|0\rangle = 0$ and $\hat{a}_+^\dagger(p)|0\rangle = |1_p^+\rangle$, for all values of p. Thereby, the interpretation is that operator $\hat{a}_-(p)$ annihilates negatively-charged particles and operator $\hat{a}_+^\dagger(p)$ creates positively-charged particles. It follows that the operation of $\hat{\phi}^\dagger(x)$ on $|0\rangle$ yields the positively-charged one-particle state $|1^+\rangle$ expressed as a state "analogous" to the wavefunction for position:

$$\hat{\phi}^\dagger(x^0, \vec{x})|0\rangle = \psi_1^+(x^0, \vec{x}) \qquad (2.X)$$

Let us now consider the $\pi^\pm$-meson field as a specific example of the charged quantum Klein-Gordon field and inquire as to the transition amplitude $\langle +; x_2^0, \vec{x}_2 | x_1^0, \vec{x}_1; + \rangle$ for a positively-charged excitation π^+. As we saw in the preceding chapter's eighth section, π^+ emerges from $|0\rangle$ in position eigenstate $|x_1^0, \vec{x}_1; +\rangle$, propagates in space-time and vanishes into $|0\rangle$ in position eigenstate $|x_2^0, \vec{x}_2; +\rangle$. The transition amplitude for this process is

$$\langle +; x_2^0, \vec{x}_2 | x_1^0, \vec{x}_1; + \rangle = \langle 0|\hat{\phi}(x_2^0, \vec{x}_2)\hat{\phi}^\dagger(x_1^0, \vec{x}_1)|0\rangle \; ; \; x_2^0 > x_1^0 \qquad (2.XI)$$

where we have used (8.II) in the preceding chapter in the context of (2.VI), (2.VII) and (2.X) above. Note in particular, that the creation of a π^- "alongside" the π^+ through the operation $\hat{\phi}(x_2^0, \vec{x}_2)\hat{\phi}^\dagger(x_1^0, \vec{x}_1)|0\rangle$ yields a vanishing (zero) contribution to $\langle +; x_2^0, \vec{x}_2 | x_1^0, \vec{x}_1; + \rangle$ for the reason stated as a passing remark in the eighth section of the preceding chapter ($\langle 0|2\rangle = 0$). Equivalently, we are at liberty to use (5.IX) in the first chapter. In that case, $\hat{\phi}(x_2^0, \vec{x}_2)$ operates on $\langle 0|$ to its left as $\hat{\phi}^\dagger(x_2^0, \vec{x}_2)|0\rangle$ on the understanding that this yields $\psi_1^{+*}(x_2^0, \vec{x}_2)$ which, in turn, signifies the annihilation of a π^+ at $x_2 = (x_2^0, \vec{x}_2)$ in line with the transition amplitude in (2.XI). It should be obvious, of course, that in this operation the sector of $\hat{\phi}^\dagger(x_2)$ associated with the annihilation of a π^- yields again a vanishing contribution in perfect conformity with physical expectations: As total charge is conserved in the course of time, a positively-charged particle does not transform to a negatively-charged one.

Relation (2.XI) has been established for $x_2^0 > x_1^0$. Since for space-like separations $(x_2 - x_1)^2 < 0$ between the end-points of propagation $x_1 = (x_1^0, \vec{x}_1)$ and $x_2 = (x_2^0, \vec{x}_2)$ the temporal order $x_2^0 > x_1^0$ is not Lorentz invariant, the physical demand for causality - stated for a real Klein-Gordon field in (8.V) of the previous chapter - implies that for a complex Klein-Gordon field it is

$$\hat{\phi}(x_2)\hat{\phi}^\dagger(x_1) = \hat{\phi}^\dagger(x_1)\hat{\phi}(x_2) \; ; \qquad if \; (x_2 - x_1)^2 < 0$$

or, more elegantly

$$[\hat{\phi}(x_2), \hat{\phi}^\dagger(x_1)] = 0 \; ; \qquad\qquad if \; (x_2 - x_1)^2 < 0 \qquad (2.XII)$$

In order to understand the reason that (8.V) for a real Klein-Gordon field generalizes to (2.XII) we must answer the following question. In how many ways can a unit of positive electric charge $+1$ be carried between space-time points $x_1 = (x_1^0, \vec{x}_1)$ and $x_2 = (x_2^0, \vec{x}_2)$ given the fact that the complex Klein-Gordon field expresses itself in terms of the π^+ which carries one unit of positive electric charge $+1$ and the π^- which carries one unit of negative electric charge -1? There are obviously only two ways: One way is by propagating a π^+ from x_1 to x_2, thus increasing the charge by $+1$ unit at x_1 and reducing the charge by $+1$ unit at x_2. The other is by propagating a π^- from x_2 to x_1, thus reducing the charge by $+1$ unit at x_2 and increasing the charge by $+1$ unit at x_1. Thus, upon measuring the strength of the complex Klein-Gordon field at space-time points x_1 and x_2 in space-like separation, inertial frames in which $x_2^0 > x_1^0$ register the emergence of π^+ at x_1 and its annihilation at x_2, whereas inertial frames in which $x_1^0 > x_2^0$ register the emergence of π^- at x_2 and its annihilation at x_1. It is clear, for that matter, that the consistent statement of causality is that in (2.XII). The latter reduces to (8.V) for a real, electrically neutral Klein-Gordon field. In turn, by dint of (8.IX) in the previous chapter, (2.XII) implies that[65]

$$\langle 0|\hat{\phi}(x_2)\hat{\phi}^\dagger(x_1)|0\rangle - \langle 0|\hat{\phi}^\dagger(x_1)\hat{\phi}(x_2)|0\rangle = 0 \quad if \ (x_2 - x_1)^2 < 0 \qquad (2.XIII)$$

The first term in (2.XIII) expresses the emergence of a π^+ at x_1, its propagation to x_2 in space-like separation from x_1 and its annihilation at x_2. Such a transition is registered in inertial frames in which the temporal order is $x_2^0 > x_1^0$. The second term in (2.XIII) expresses the emergence of a π^- at x_2, its propagation to x_1 in space-like separation from x_2 and its annihilation at x_1. Such a transition is accordingly registered in inertial frames in which the temporal order is $x_1^0 > x_2^0$. Although $\langle +; x_2^0, \vec{x}_2 | x_1^0, \vec{x}_1; + \rangle$ and independently $\langle -; x_2^0, \vec{x}_2 | x_1^0, \vec{x}_1; - \rangle$ allow for superluminal propagation of a π^+ and a π^- respectively in a specified inertial frame, causality is upheld because, by dint of (2.XIII), the transition amplitude $\langle -; x_2^0, \vec{x}_2 | x_1^0, \vec{x}_1; - \rangle$ offsets the transition amplitude $\langle +; x_2^0, \vec{x}_2 | x_1^0, \vec{x}_1; + \rangle$ and conversely the latter offsets the former through a Lorentz transformation to an inertial frame of reverse temporal order between space-time points x_1 and x_2. As observed from inertial frame K, event $P = x_1$ concerning the emergence of a π^+ out of $|0\rangle$ precedes event $Q = x_2$ concerning the annihilation of that π^+ into $|0\rangle$. As observed from inertial frame K', event $Q = x_1'$ concerning the emergence of a π^- out of $|0\rangle$ precedes event $P = x_2'$ concerning the annihilation of that π^- into $|0\rangle$. If P and Q are separated by a space-like interval then the propagation of a π^+ from P to Q with respect to K implies the propagation of a π^- from Q to P with respect to K'. Note, in this respect, the already announced result: In order that either propagation offset the other in (2.XIII) both, the π^+ and the π^- must have the same rest mass.

We, thus, arrive at this central conclusion: *Causality implies the existence of antiparticles*! Propagation between two events in space-like separation involves two particles of equal rest mass, the particle associated with inertial frame K and the antiparticle associated with inertial frame K'. The propagation of the former is indistinguishable from the propagation of the latter. In any measurement, the two transition amplitudes cancel against each other. For the Klein-Gordon field, the π^- is the antiparticle to the π^+ and the π^+ is the antiparticle to the π^-. In fact, this interpretation is just as valid for the real Klein-Gordon field. Underlying the entire analysis in the eighth section of

---

[65]Note in this respect that for a complex Klein-Gordon field it is $\langle 0|\hat{\phi}(x_2^0, \vec{x}_2)\hat{\phi}(x_1^0, \vec{x}_1)|0\rangle = 0$ and $\langle 0|\hat{\phi}^\dagger(x_2^0, \vec{x}_2)\hat{\phi}^\dagger(x_1^0, \vec{x}_1)|0\rangle = 0$.

the preceding chapter is the fact that for a real Klein-Gordon field the particle π^0 is in identity with its own antiparticle.

Of course, for time-like separations between x_1 and x_2 there is no issue of causality. Since all inertial frames agree that the event of emergence of a π^+ at x_1 precedes the event of annihilation of that π^+ at x_2, the two terms on the right side of the general relation

$$[\hat{\phi}(x_2), \hat{\phi}^\dagger(x_1)] = \langle 0|\hat{\phi}(x_2)\hat{\phi}^\dagger(x_1)|0\rangle - \langle 0|\hat{\phi}^\dagger(x_1)\hat{\phi}(x_2)|0\rangle \qquad (2.XIV)$$

are not equal and the commutator on the left side does not vanish. Likewise, since all inertial frames agree that the event of emergence of a π^- at x_1 precedes the event of annihilation of that π^- at x_2, the two terms on the right side of the general relation

$$[\hat{\phi}^\dagger(x_2), \hat{\phi}(x_1)] = \langle 0|\hat{\phi}^\dagger(x_2)\hat{\phi}(x_1)|0\rangle - \langle 0|\hat{\phi}(x_1)\hat{\phi}^\dagger(x_2)|0\rangle \qquad (2.XV)$$

are not equal and the commutator on the left side does not vanish.

Thus, the transition amplitude for the quantum Klein-Gordon field which is manifestly consistent with the demand for causality is the Feynman propagator

$$i\Delta_F(x_2-x_1) = \langle 0|\hat{\phi}(x_2)\hat{\phi}^\dagger(x_1)|0\rangle\theta(x_2^0-x_1^0)+\langle 0|\hat{\phi}^\dagger(x_1)\hat{\phi}(x_2)|0\rangle\theta(x_1^0-x_2^0) \qquad (2.XVI)$$

which is also known as the *Klein-Gordon propagator*. It is valid for all space-time separations, it is Lorentz invariant and such that the first term vanishes for $x_1^0 > x_2^0$ and the second term vanishes for $x_2^0 > x_1^0$. Its diagrammatic representation in Minkowski space-time has already been given in Fig.24. It is obvious, of course, that the Klein-Gordon propagator in (2.XVI) reduces to the Klein-Gordon propagator in (8.XI) for the special case of the real Klein-Gordon field.

Our analysis as of the eighth section of the preceding chapter has established that the transition amplitude in quantum field theory is radically different from that in quantum mechanics. The quantum-mechanical transition amplitude concerns a single particle and the finite number of degrees of freedom associated with it. In quantum field theory, on the contrary, the transition concerns the underlying field itself. A particle in transition between two space-time points emerges out of, and vanishes into, the quantum vacuum, thereby reflecting an underlying evolution of the infinitely many degrees of freedom of the quantum field. In turn, the relativistic nature of quantum fields and the concomitant demand for causality render the transition of a quantum particle inseparable from the transition of its antiparticle. The existence of antiparticles is yet another expression of the fact that particles are not fundamental physical entities and can neither be described in isolation from the underlying quantum field, nor in isolation from associated particles which the underlying quantum field implies. Despite its rigor, our analysis on the transition amplitude in quantum field theory has yet to reach the essence of the matter. For, the relation between particles and antiparticles in Minkowski space-time points to the existence of an underlying duality between these two categories. The reversal of fundamental attributes of space-time reverses the antiparticles to particles! In order to reveal that duality which will finally establish the precise relation between particles and antiparticles it is necessary to study the Feynman propagator also in momentum space, the definition of which as the collection of all four-dimensional momenta $p^\mu = (p^0, p^1, p^2, p^3)$ has been given in the fourth section of the fourth chapter.

3. Scalar Propagation in Momentum Space

For the purposes of the ensuing analysis, it is expedient to work in the system of units $c = \hbar = 1$ which, as we have seen in the fourth section of the preceding chapter, allows for the reduction of the classical relativistic relation $E_p^2 = (|\vec{p}|c)^2 + (mc^2)^2$ to merely $p^2 = m^2$, with $p^2 = (p^0)^2 - |\vec{p}|^2 = (p^0)^2 - (p^1)^2 - (p^2)^2 - (p^3)^2$.

Let us directly advance the following claim: The Minkowski space-time Feynman propagator in (2.XVI) receives the expression

$$\Delta_F(x_2 - x_1) = \int \frac{d^4 p}{(2\pi)^4} \frac{1}{p^2 - m^2 + i\epsilon} e^{-ip.(x_2 - x_1)} \tag{3.I}$$

with the momentum-space Feynman propagator being

$$\tilde{\Delta}_F(p) = \frac{1}{p^2 - m^2 + i\epsilon} \tag{3.II}$$

(the explanation concerning the "$i\epsilon$-term" in the denominator of both expressions is pending).

Before we establish the validity of this claim, it is necessary to make a few observations. Note first that since

$$p.(x_2 - x_1) = p^0(x_2^0 - x_1^0) - p^1(x_2^1 - x_1^1) - p^2(x_2^2 - x_1^2) - p^3(x_2^3 - x_1^3) =$$

$$p^0(x_2^0 - x_1^0) - \vec{p}.(\vec{x}_2 - \vec{x}_1) \equiv \eta_{\mu\nu} p^\mu (x_2 - x_1)^\nu$$

(3.I) is manifestly Lorentz-invariant. Inherent in this Lorentz-invariant expression which relates the Feynman propagator $\Delta_F(x_2 - x_1)$ in Minkowski space-time to the Feynman propagator $\tilde{\Delta}_F(p)$ in momentum space is the following implication: Since $\Delta_F(x_2 - x_1)$ expresses the probability amplitude that a particle propagate between space-time points x_1 and x_2, *the momentum-space Feynman propagator $\tilde{\Delta}_F(p)$ expresses the probability amplitude that a particle of rest mass m merely propagate with four-momentum magnitude p.*

Note further that this is the first time we continuously sum over the entire four-dimensional momentum space. In the superpositions (1.XI), (2.VI), (2.VII) the energy $\frac{E}{c} \equiv p^0$ was, for each value of three-momentum $|\vec{p}|$, fixed by the classical relativistic relation $E_p^2 = (|\vec{p}|c)^2 + (mc^2)^2$. In (3.I) no physical condition applies. We merely expressed the mathematical function $\Delta_F(x_2 - x_1)$ as an integral over the new four-dimensional variable[66] p^μ. Only in retrospect did we physically identify $p^\mu = (p^0, \vec{p})$ with the four-momentum and the new function $\tilde{\Delta}_F(p)$ in (3.II) with the Feynman propagator in momentum space, a physical attribute irrelevant to space-time separations with their inherent superpositions over energy-momentum eigenstates. We stress that $\tilde{\Delta}_F(p)$ is not the probability amplitude that a particle of rest mass m propagate with four-momentum magnitude p over space-time separation $x_2 - x_1$. It is, instead, the probability amplitude that a particle of rest mass m merely propagate in Minkowski space-time with four-momentum magnitude p - that is, regardless of the specific value of $x_2 - x_1$.

Now to the rigorous proof of (3.I). In view of the prominent importance which integration over the energy has, we temporarily disregard the $i\epsilon$-term in (3.II) and recast $\tilde{\Delta}_F(p)$ in the form

$$\tilde{\Delta}_F(p) = \frac{1}{(p^0 - \sqrt{|\vec{p}|^2 + m^2})(p^0 + \sqrt{|\vec{p}|^2 + m^2})}$$

The value $|\vec{p}|$ which enters $\tilde{\Delta}_F(p) \equiv \tilde{\Delta}_F(p^0, |\vec{p}|)$ is fixed (once chosen) but arbitrary. Clearly, for any such value of three-momentum, the value of energy p^0 is just as arbitrary. Therefore, for any fixed value of $|\vec{p}|$, $\tilde{\Delta}_F(p)$ is defined for all values of p^0 except for the two classical values $p^0 = \pm\sqrt{|\vec{p}|^2 + m^2}$. Before we tackle the crucial issue which these two classical values pose, note that as the integral over p^0 in (3.I) extends from $-\infty$ to $+\infty$, the negative values of energy are as crucially important as the positive values. This stands in stark contrast to the classical theory in which the energy value $p^0 = -\sqrt{|\vec{p}|^2 + m^2}$ of a relativistic particle is trivially discarded on the grounds that negative energies are unphysical (not to mention the contrast with the invariably positive energy $E = \frac{p^2}{2m}$ of a free non-relativistic particle in classical and quantum mechanics).

Thus, integration over the energy in

$$\int \frac{dp^4}{(2\pi)^4} \frac{1}{p^2 - m^2} e^{-ip.(x_2 - x_1)} =$$

$$\frac{1}{(2\pi)^4} \int d^3p\, e^{i\vec{p}.(\vec{x}_2 - \vec{x}_1)} \int_{-\infty}^{+\infty} dp^0 \frac{e^{-ip^0(x_2^0 - x_1^0)}}{(p^0 - \sqrt{|\vec{p}|^2 + m^2})(p^0 + \sqrt{|\vec{p}|^2 + m^2})}$$

is meaningless since in summing over p^0 for any given $\vec{p}$ we inevitably encounter the problematic two classical values. $\Delta_F(x_2 - x_1)$ in (3.I) has not been defined until we state what exactly we mean by

$$\int_{-\infty}^{+\infty} dp^0 \frac{e^{-ip^0(x_2^0 - x_1^0)}}{(p^0 - \sqrt{|\vec{p}|^2 + m^2})(p^0 + \sqrt{|\vec{p}|^2 + m^2})}$$

In order to give concrete meaning to this integral, we *analytically extend* the real-variable p^0 to a complex-variable p^0. We accomplish this by rendering the initial real-variable p^0 the real part Rep^0 of the complex $p^0 = Rep^0 + iImp^0$ which, for that matter, extends

---

[66]Refer to the *Fourier transform* of a function mentioned in passing in connection to (2.II) in the first chapter.

throughout the complex plane. Specifically (refer to the information in the Mathematical Preliminaries), we set

$$p^0 = |p^0|e^{i\theta} = |p^0|cos\theta + i|p^0|sin\theta$$

so that for $\theta = 0$ ($\Longrightarrow cos\theta = 1$; $sin\theta = 0$) and $\theta = \pi$ ($\Longrightarrow cos\theta = -1$; $sin\theta = 0$) respectively, we recover the initial real-variable $p^0 \equiv Rep^0$ which extends from $-\infty$ to $+\infty$. Of course, such a mathematical operation cannot be arbitrary. The final result must identically conform with (3.I). We shall see how this is accomplished in what immediately follows.

The propagator $\tilde{\Delta}_F(p)$ as a function of this new analytically-extended complex p^0 is characterized by the same ambiguity. It is not defined at the two classical values of energy. Such values of a complex variable as render a function meaningless are known as the *poles of the function*. The poles of $\tilde{\Delta}_F(p)$ are, for that matter, $p^0 = \pm\sqrt{|\vec{p}|^2 + m^2}$ and clearly necessitate a *prescription* which will elucidate what exactly we mean when integrating over p^0. It is at this point that the enormous advantage of the analytical extension of the real p^0 to the entire complex plane becomes obvious. In continuously summing over complex variable p^0 from $-\infty$ to $+\infty$, we can now arbitrarily approach $p^0 = -\sqrt{|\vec{p}|^2 + m^2}$ without ever reaching it! Instead, we avoid it by deforming the path of integration over p^0 to a small semicircle in the complex plane as in Fig.26. Accordingly, we avoid the positive pole $p^0 = +\sqrt{|\vec{p}|^2 + m^2}$ by a small semicircle. This procedure is justified by the fact that the eventual limit of vanishing radius imposed on the two small semicircles will yield the integral over p^0 in (3.I) now rigorously defined in view of the fact that at no stage of the integration did we encounter a pole!

The fundamental question in this respect is the orientation of the two semicircles. For clearly, as we advance from negative infinity to positive infinity along Rep^0, we have the choice of avoiding each pole either by a small semicircle in the upper-half complex plane, or by a small semicircle in the lower-half complex plane. In order to appreciate the crucial importance of such a choice to the procedure of integration, we must first establish the equivalence between the stated integration along the Rep^0-axis (excluding the two poles) and the integration along a closed contour of infinite radius. The idea is the following: If, at some very big value of energy $p^0 = P^0 >> 1$, we digress from the straight path of integration along the Rep^0-axis to a large semicircle of radius $|P^0|$ and form a closed contour of integration by arriving at $-P^0 << -1$, as in either of the two contours in Fig.26, then the eventual limit of infinite radius $|P^0| \to \infty$ for that semicircle will exclusively yield the integral along the entire Rep^0-axis provided that we are careful enough to ensure that $\frac{e^{-ip^0(x_2^0 - x_1^0)}}{p^2 - m^2}$ vanishes on the infinite semicircle. On rigorous terms, the integral along either of the two closed contours in Fig.26 - the *contour integral* $\oint dp^0$, as it is formally known - involves integration over two variables in four steps: (i) Keeping θ fixed at $\theta = 0$, integrate over p^0 from zero to a large number $P^0 >> 1$. (ii) Keeping $|p^0|$ fixed at $P^0 >> 1$, integrate over θ counterclockwise from 0 to π for the large semicircle in the upper half-plane, or clockwise from 0, to $-\pi$ for the large semicircle in the lower half-plane. (iii) Keeping θ fixed at $+\pi$ or $-\pi$ respectively, integrate over p^0 from $-P^0$ to zero. These three steps accomplish the procedure of *contour integration* along either contour in Fig.26. Finally, (iv) since P^0 is arbitrary, elevate it to a variable and let it go to infinity: $P^0 \to +\infty$. If the integral along the relevant infinite semicircle vanishes, the contour integral

$$\oint_{|p^0|\to\infty} dp^0 \frac{e^{-ip^0(x_2^0-x_1^0)}}{(p^0 - \sqrt{|\vec{p}|^2 + m^2})(p^0 + \sqrt{|\vec{p}|^2 + m^2})}$$

reduces to the integral along the $Re p^0$-axis and, for that matter, to the above-stated integral over p^0.

Let us put this concept into practice. In order to integrate along either large semicircle of radius $|p^0| = P^0 \gg 1$ we need to use the elementary rule of differentiation which states that the derivative of function e^{cx} with respect to its independent variable x, is merely the function itself multiplied by the constant c:

$$\frac{de^{cx}}{dx} = ce^{cx} \implies de^{cx} = ce^{cx}dx$$

We shall introduce the concept of differentiation later in this chapter. If, however, you are not already familiar with it, you can merely accept this mathematical statement as a heuristic rule. For that matter, with the angle θ between complex variable p^0 and the $Re p^0$-axis spanning either the entire large semicircle in the upper-half p^0-plane (from $\theta = 0$ to $\theta = \pi$), or the entire large semicircle in the lower-half p^0-plane (from $\theta = 0$ to $\theta = -\pi$), it is:

$$dp^0 = d(|p^0|e^{i\theta}) = |p^0|de^{i\theta} = |p^0|ie^{i\theta}d\theta = ip^0 d\theta$$

so that with the fixed radius of either semicircle being $|p^0| = P^0 \gg 1$, contour-integration in the upper-half p^0-plane for example, is

$$\oint dp^0 \frac{e^{-ip^0(x_2^0-x_1^0)}}{(p^0 - \sqrt{|\vec{p}|^2 + m^2})(p^0 + \sqrt{|\vec{p}|^2 + m^2})} =$$

$$i\int_0^\pi d\theta p^0 \frac{e^{-iP^0(x_2^0-x_1^0)e^{i\theta}}}{(p^0 - \sqrt{|\vec{p}|^2 + m^2})(p^0 + \sqrt{|\vec{p}|^2 + m^2})} +$$

$$\int_{-P^0}^{+P^0} dp^0 \frac{e^{-ip^0(x_2^0-x_1^0)}}{(p^0 - \sqrt{|\vec{p}|^2 + m^2})(p^0 + \sqrt{|\vec{p}|^2 + m^2})} =$$

$$i\int_0^\pi d\theta p^0 \frac{e^{-iP^0(x_2^0-x_1^0)cos\theta}e^{P^0(x_2^0-x_1^0)sin\theta}}{(p^0 - \sqrt{|\vec{p}|^2 + m^2})(p^0 + \sqrt{|\vec{p}|^2 + m^2})} +$$

$$\int_{-P^0}^{+P^0} dp^0 \frac{e^{-ip^0(x_2^0-x_1^0)}}{(p^0 - \sqrt{|\vec{p}|^2 + m^2})(p^0 + \sqrt{|\vec{p}|^2 + m^2})}$$

where, of course, we have used the identities $e^{i\theta} = cos\theta + isin\theta$ and $i^2 = -1$.

We note now that if $x_2^0 > x_1^0 \implies (x_2^0 - x_1^0) > 0$ then the limit $P^0 \to +\infty$ implies that $e^{P^0(x_2^0-x_1^0)sin\theta} \to +\infty$ and, for that matter, that the above contour integral itself diverges (tends to infinity). On the contrary, contour-integration in the lower-half p^0-plane entails integration over θ from 0 to $-\pi$ which, in view of $sin(-\theta) = -sin\theta$, replaces $e^{P^0(x_2^0-x_1^0)sin\theta}$ by its inverse $e^{-P^0(x_2^0-x_1^0)sin|\theta|}$. The latter clearly vanishes on the infinite semicircle. Thus, we conclude that for $x_2^0 > x_1^0$ the required vanishing contribution stems from the infinite semicircle in the lower-half p^0-plane. By the same reasoning, it immediately follows that for $x_2^0 < x_1^0$ the required vanishing contribution stems from the infinite semicircle in the upper-half p^0-plane. We state, for that matter, that

$$\int_{-\infty}^{+\infty} dp^0\, \frac{e^{-ip^0(x_2^0-x_1^0)}}{p^2 - m^2 + i\epsilon} = \oint_{|p^0|\to+\infty} dp^0\, \frac{e^{-ip^0(x_2^0-x_1^0)}}{p^2 - m^2 + i\epsilon} \begin{cases} -\pi \le \theta \le 0 \;;\; if\; x_2^0 > x_1^0 & (3.IIIa) \\[2ex] 0 \le \theta \le \pi \;;\; if\; x_2^0 < x_1^0 & (3.IIIb) \end{cases}$$

where the $i\epsilon$-factor suggests the, as of yet unspecified, manner in which the two poles are to be avoided.

The result in (3.III) renders the crucial importance of the manner in which the two poles are avoided manifest. Each of the two large semicircles respectively in the upper-half and lower-half complex plane yields four distinct contours depending on whether each of the two poles is avoided by a small semicircle in the upper-half complex plane, or by a small semicircle in the lower-half complex plane. Once a choice of contour has been made, the corresponding contour integral is evaluated by dint of a powerful theorem in the domain of complex functions, the *residue theorem*. The latter applies to functions which are *analytic* in a region of the complex plane bounded by a closed contour. The definition of an analytic function necessitates the concept of derivative which we have not yet introduced. For our purposes, it suffices that we interpret the analytic function at a point z_0 in the complex plane as that complex function $f(z)$ which receives a well-defined value $f(z_0)$ at that point. If a function $f(z)$ is not analytic at z_0 then, as already stated, z_0 is a pole of $f(z)$. The *residue* $Res f(z_0)$ of $f(z)$ at $z = z_0$ is the value of the product $(z - z_0)f(z)$ at $z = z_0$. For example, the residue of our function $\frac{e^{-ip^0(x_2^0-x_1^0)}}{p^2-m^2}$ at $p^0 = +\sqrt{|\vec{p}|^2 + m^2}$ is

$$Res\, \frac{e^{-ip^0(x_2^0-x_1^0)}}{(p^2 - m^2)}\Big|_{p^0=\sqrt{|\vec{p}|^2+m^2}} = \qquad\qquad (3.IVa)$$

$$[\frac{e^{-ip^0(x_2^0-x_1^0)}}{(p^0 - \sqrt{|\vec{p}|^2 + m^2})(p^0 + \sqrt{|\vec{p}|^2 + m^2})}(p^0 - \sqrt{|\vec{p}|^2 + m^2})]\Big|_{p^0=\sqrt{|\vec{p}|^2+m^2}} = \frac{1}{2}\frac{e^{-i\sqrt{|\vec{p}|^2+m^2}(x_2^0-x_1^0)}}{\sqrt{|\vec{p}|^2 + m^2}}$$

Likewise, the residue of our function at $p^0 = -\sqrt{|\vec{p}|^2 + m^2}$ is

$$Res\, \frac{e^{-ip^0(x_2^0-x_1^0)}}{p^2 - m^2}\Big|_{p^0=-\sqrt{|\vec{p}|^2+m^2}} = -\frac{1}{2}\frac{e^{i\sqrt{|\vec{p}|^2+m^2}(x_2^0-x_1^0)}}{\sqrt{|\vec{p}|^2 + m^2}} \qquad\qquad (3.IVb)$$

To the residue theorem now which formally states this: *If a function $f(z)$ is analytic in a region of the complex plane bounded by a closed counterclockwise-oriented contour C except for a finite number of poles $z = z_k \;;\; k = 1, 2, ...n$, the contour integral of $f(z)$ along C is the sum of the residues $Res f(z_k)$ multiplied by $2\pi i$:*

$$\oint_C dz f(z) = 2\pi i \Sigma_{k=1}^{n} Res f(z_k) \qquad (3.V)$$

This is a statement of enormous mathematical power! For it determines that *the integral along a closed contour is exclusively determined by the poles which that contour encompasses.* Integration along a closed contour in the complex plane was the reason we digressed from the Rep^0-axis to the infinitely big semicircle in the complex plane. The residue theorem applies now to each of the four contours for which each of the two infinitely big semicircles allows. For example, if $x_2^0 > x_1^0$, the contour integral, and thereby the integral over the Rep^0-axis, is determined exclusively by $+\sqrt{|\vec{p}|^2 + m^2}$ for the choice of small semicircle depicted in Fig.26. On the contrary, if $x_1^0 > x_2^0$, the contour integral, and thereby the integral over the Rep^0-axis, is determined exclusively by $-\sqrt{|\vec{p}|^2 + m^2}$ for the choice of small semicircle depicted in Fig.26.

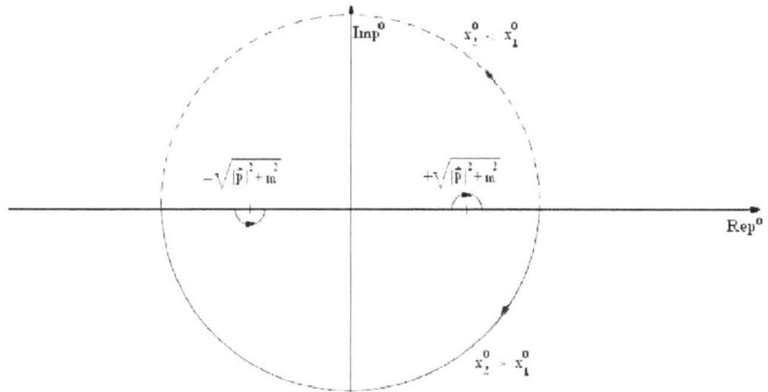

FIGURE 26. *Contour Integration in the Complex Plane*: For each value of $|\vec{p}|$ there are two poles $\pm\sqrt{|\vec{p}|^2 + m^2}$ along the path of integration Rep^0 obviated as shown according to the $i\epsilon$-prescription. For $x_2^0 > x_1^0$, the integral in (3.I) over the analytically extended p^0 vanishes along the infinite semicircle in the lower-half complex plane. Thus, for forward-time propagation the infinite contour extends clockwise and includes only the positive pole. Accordingly, for backward-time propagation $x_1^0 > x_2^0$, the infinite contour extends counterclockwise and includes only the negative pole.

This is as far as mathematical operations take us. Which choice of contour we should make is determined exclusively by the physical content of (3.I). Of course, we have noticed from the outset that the two conditions $x_2^0 > x_1^0$ and $x_1^0 > x_2^0$ are none other than those respectively associated with the propagation of a particle and its antiparticle in Minkowski space-time as stated in (2.XVI) of the previous section. Nevertheless, pursuant to (3.I) and (3.II), the equivalent formulation of propagation in momentum space, an abstract mathematical space irrelevant to space-time points and to temporal ordering, yields an equivalent but altogether distinct physical interpretation of these two conditions. In order to arrive at it we note that only positive energies propagating from past to future exist in Minkowski space-time. Negative energies and backward-time propagation are physically unacceptable. In momentum space, on the contrary, the equivalent formulation of propagation renders the unobservable negative energies as physically valid as the observable positive energies. At once, for that matter, it renders, the unobserved reverse arrow of time pointing from future infinity to past infinity as physically valid as the observed arrow of time pointing from past infinity to future infinity. The two conditions $x_2^0 > x_1^0$ and $x_1^0 > x_2^0$ which in momentum space respectively enforce a vanishing contribution to integration over the energy along the infinitely-big semicircles define precisely the two distinct arrows of time! As positive energies, being invariably observable, are associated with the observed arrow of time pointing from past infinity to future infinity and conveyed by the condition $x_2^0 > x_1^0$, so negative energies, being invariably unobservable, are associated with the reverse unobservable arrow of time pointing from future infinity to past infinity and conveyed by the condition $x_1^0 > x_2^0$. This then is the prescription - or, more rigorously, the *boundary condition* - which we have set out to determine in order to give meaning to the integral over the energy p^0 in (3.I):

Positive energies propagate exclusively forward in time with the direct implication that the positive pole $+\sqrt{|\vec{p}|^2 + m^2}$ be obviated in the upper-half complex plane so as to be included in the closed contour determined by $x_2^0 > x_1^0$ in the lower-half complex plane. Negative energies propagate exclusively backward in time with the direct implication that the negative pole $-\sqrt{|\vec{p}|^2 + m^2}$ be obviated in the lower-half complex plane so as to be included in the closed contour determined by $x_2^0 < x_1^0$ in the upper-half complex plane.

If the physical content of this statement is not immediately obvious, consider again that the integral along each of the two infinitely big semicircles vanishes, reducing whichever contour integral to the integral over the real variable p^0. If for forward-time propagation $x_2^0 > x_1^0$ a negative value of energy contributed to the integral over Rep^0, that value would be in identical contradiction to the above-stated boundary condition. Accordingly, if for backward-time propagation $x_1^0 > x_2^0$ a positive value of energy contributed to the integral over Rep^0, that value would be in identical contradiction to the above-stated boundary condition.

Note now the crucial implication inherent in this statement. If negative energies were available to a physical system, nothing would prevent that system from achieving minimum energy by collapsing to a state of infinite negative energy! It is for that matter that of the two values $p^0 = \pm\sqrt{|\vec{p}|^2 + m^2}$ which the relation $p^2 = m^2$ implies for a relativistic classical particle we dismiss the negative value on the grounds that "negative energies are unphysical". By this we mean that negative energies are unphysical because in Minkowski space-time the arrow of time points unconditionally from past infinity to future infinity. On the contrary, the negative energies in momentum space evident in (3.I) propagate in a course of time determined by an unobservable inverse arrow which points from future infinity to past infinity. With respect to that inverse arrow of time physical systems can only achieve a state of maximum energy if left to themselves and, for that matter, nothing would prevent them from achieving maximum energy by advancing to a state of infinite positive energy! The absence of negative energies in the observed course of time and the absence of positive energies in the unobservable course of time determined by the inverse arrow ensures that quantum field theory is physically consistent!

We may now apply the stated boundary condition to the two contour integrals. In order to obviate the positive pole with a small semicircle in the upper-half complex plane, we add an imaginary $i\epsilon$-term, ϵ being a very small positive real number, to the expression $|\vec{p}|^2 + m^2$ underneath the square root and advance the following mathematical operation:

$$+\sqrt{|\vec{p}|^2 + m^2 + i\epsilon} = \sqrt{(|\vec{p}|^2 + m^2)(1 + i\frac{\epsilon}{|\vec{p}|^2 + m^2})} = \sqrt{|\vec{p}|^2 + m^2}\sqrt{1 + i\frac{\epsilon}{|\vec{p}|^2 + m^2}} =$$

$$\sqrt{|\vec{p}|^2 + m^2}(1 + \frac{i}{2}\frac{\epsilon}{|\vec{p}|^2 + m^2} + \frac{1}{8}\frac{\epsilon^2}{(|\vec{p}|^2 + m^2)^2} + ...) \sim$$

$$+\sqrt{|\vec{p}|^2 + m^2} + \frac{i}{2}\frac{\epsilon}{\sqrt{|\vec{p}|^2 + m^2}} \tag{3.VIa}$$

where in the third equality we used the elementary identity in differential calculus

$$\sqrt{1 + x} = 1 + \frac{1}{2}x - \frac{1}{8}x^2 + \frac{1}{16}x^3 - ... \; ; \; if \; |x| < 1$$

Accordingly, we obviate the negative pole with a small semicircle in the lower-half complex plane:

$$-\sqrt{|\vec{p}|^2 + m^2 + i\epsilon} \sim -\sqrt{|\vec{p}|^2 + m^2} - \frac{i}{2}\frac{\epsilon}{\sqrt{|\vec{p}|^2 + m^2}} \tag{3.VIb}$$

The required contour of integration for $x_2^0 > x_1^0$ is now specified by the infinite semicircle in the lower-half complex plane and (3.VI) which determines that it include the pole $p^0 = +\sqrt{|\vec{p}|^2 + m^2}$, but not the pole $p^0 = -\sqrt{|\vec{p}|^2 + m^2}$. The residue theorem in (3.V) applied to the contour integral in (3.IIIa) "picks", for that matter, the residue in (3.IVa). Note that as an immediate consequence of clockwise contour integration in (3.IIIa), the result is necessarily the opposite to that stated in (3.V) and (3.IVa):

$$\int_{-\infty}^{+\infty} dp^0 \frac{e^{-ip^0(x_2^0 - x_1^0)}}{p^2 - m^2 + i\epsilon} = -\pi i \frac{e^{-i\sqrt{|\vec{p}|^2 + m^2 + i\epsilon}(x_2^0 - x_1^0)}}{\sqrt{|\vec{p}|^2 + m^2 + i\epsilon}} \; ; \; if \; x_2^0 > x_1^0 \tag{3.VIIa}$$

Accordingly, the required contour of integration for $x_1^0 > x_2^0$ is now specified by the infinite semicircle in the upper-half complex plane and (3.VI) which determines that it include the pole $p^0 = -\sqrt{|\vec{p}|^2 + m^2}$, but not the pole $p^0 = +\sqrt{|\vec{p}|^2 + m^2}$. The residue theorem in (3.V) applied to the contour integral in (3.IIIb) "picks", for that matter, the residue in (3.IVb):

$$\int_{-\infty}^{+\infty} dp^0 \frac{e^{-ip^0(x_2^0 - x_1^0)}}{p^2 - m^2 + i\epsilon} = -\pi i \frac{e^{i\sqrt{|\vec{p}|^2 + m^2 + i\epsilon}(x_2^0 - x_1^0)}}{\sqrt{|\vec{p}|^2 + m^2 + i\epsilon}} \; ; \; if \; x_1^0 > x_2^0 \tag{3.VIIb}$$

We may now take the limit $\epsilon \to 0$ and combine the two independent contributions in (3.VII) to a single integral over p^0:

$$\int_{-\infty}^{+\infty} dp^0 \, \frac{e^{-ip^0(x_2^0 - x_1^0)}}{p^2 - m^2 + i\epsilon} = \qquad (3.VIII)$$

$$\frac{\pi}{i} \frac{1}{\sqrt{|\vec{p}|^2 + m^2}} \left[e^{-i\sqrt{|\vec{p}|^2 + m^2}(x_2^0 - x_1^0)} \theta(x_2^0 - x_1^0) + e^{i\sqrt{|\vec{p}|^2 + m^2}(x_2^0 - x_1^0)} \theta(x_1^0 - x_2^0) \right]$$

Thus, the integral over the energy p^0 is now rigorously defined by virtue of the stated boundary condition which we imposed on the propagator $\Delta_F(x_2 - x_1)$. Replacing (3.VIII) in (3.I), we finally obtain:

$$i\Delta_F(x_2 - x_1) =$$

$$\frac{\pi}{(2\pi)^4} \int \frac{d^3 p}{\sqrt{|\vec{p}|^2 + m^2}} \left[e^{-i\sqrt{|\vec{p}|^2 + m^2}(x_2^0 - x_1^0)} e^{i\vec{p}\cdot(\vec{x}_2 - \vec{x}_1)} \theta(x_2^0 - x_1^0) + e^{i\sqrt{|\vec{p}|^2 + m^2}(x_2^0 - x_1^0)} e^{-i\vec{p}\cdot(\vec{x}_2 - \vec{x}_1)} \theta(x_1^0 - x_2^0) \right]$$

or (with $c = \hbar = 1$):

$$i\Delta_F(x_2 - x_1) = \int \frac{d^3 p}{(2\pi)^3 2E_p} \left[e^{-i(E_p(t_2 - t_1) - \vec{p}\cdot(\vec{x}_2 - \vec{x}_1))} \theta(t_2 - t_1) + e^{i(E_p(t_2 - t_1) - \vec{p}\cdot(\vec{x}_2 - \vec{x}_1))} \theta(t_1 - t_2) \right] =$$

$$\int \frac{d^3 p}{(2\pi)^3 2E_p} \left[e^{-ip\cdot(x_2 - x_1)} \theta(t_2 - t_1) + e^{ip\cdot(x_2 - x_1)} \theta(t_1 - t_2) \right] \qquad (3.IX)$$

on the understanding that Lorentz invariance directly implies that

$$x_2^0 - x_1^0 \to x_1^0 - x_2^0 \implies \vec{x}_2 - \vec{x}_1 \to \vec{x}_1 - \vec{x}_2$$

$\Delta_F(x_2 - x_1)$ in (3.IX) is the expression which we declared as the Feynman propagator in Minkowski space-time at the outset in (3.I). In order to prove the veracity of our claim, we show now that the established Feynman propagator in (2.XVI) of the preceding section reduces indeed to (3.IX) above.

Upon replacing (2.VI) and (2.VII) in (2.XVI), we obtain

$$i\Delta_F(x_2 - x_1) =$$

$$\int \frac{d^3 p}{\sqrt{(2\pi)^3 2E_p}} \int \frac{d^3 p'}{\sqrt{(2\pi)^3 2E_{p'}}} \left[\langle 0 | \hat{a}_+(p) e^{-i(E_p t_2 - \vec{p}\cdot\vec{x}_2)} \hat{a}_+^\dagger(p') e^{i(E_{p'} t_1 - \vec{p}'\cdot\vec{x}_1)} | 0 \rangle \theta(t_2 - t_1) \right] +$$

$$\int \frac{d^3 p}{\sqrt{(2\pi)^3 2E_p}} \int \frac{d^3 p'}{\sqrt{(2\pi)^3 2E_{p'}}} \left[\langle 0 | \hat{a}_-(p) e^{-i(E_p t_1 - \vec{p}\cdot\vec{x}_1)} \hat{a}_-^\dagger(p') e^{i(E_{p'} t_2 - \vec{p}'\cdot\vec{x}_2)} | 0 \rangle \theta(t_1 - t_2) \right] =$$

$$\int \frac{d^3 p}{\sqrt{(2\pi)^3 2E_p}} \int \frac{d^3 p'}{\sqrt{(2\pi)^3 2E_{p'}}} \left[\langle 0 | \hat{a}_+(p) \hat{a}_+^\dagger(p') | 0 \rangle e^{-i(E_p t_2 - \vec{p}\cdot\vec{x}_2)} e^{i(E_{p'} t_1 - \vec{p}'\cdot\vec{x}_1)} \theta(t_2 - t_1) \right] +$$

$$\int \frac{d^3p}{\sqrt{(2\pi)^3 2E_p}} \int \frac{d^3p'}{\sqrt{(2\pi)^3 2E_{p'}}} \left[\langle 0|\hat{a}_-(p)\hat{a}_-^\dagger(p')|0\rangle e^{-i(E_p t_1 - \vec{p}.\vec{x}_1)} e^{i(E_{p'} t_2 - \vec{p}'.\vec{x}_2)} \theta(t_1 - t_2) \right]$$

At this point we exploit the fact that $\hat{a}_\pm|0\rangle = 0$ in order to recast the above relation as

$$i\Delta_F(x_2 - x_1) =$$

$$\int \frac{d^3p}{\sqrt{(2\pi)^3 2E_p}} \int \frac{d^3p'}{\sqrt{(2\pi)^3 2E_{p'}}} \left[\langle 0|\hat{a}_+(p)\hat{a}_+^\dagger(p') - \hat{a}_+^\dagger(p')\hat{a}_+(p)|0\rangle e^{-i(E_p t_2 - \vec{p}.\vec{x}_2)} e^{i(E_{p'} t_1 - \vec{p}'.\vec{x}_1)} \theta(t_2 - t_1) \right] +$$

$$\int \frac{d^3p}{\sqrt{(2\pi)^3 2E_p}} \int \frac{d^3p'}{\sqrt{(2\pi)^3 2E_{p'}}} \left[\langle 0|\hat{a}_-(p)\hat{a}_-^\dagger(p') - \hat{a}_-^\dagger(p')\hat{a}_-(p)|0\rangle e^{-i(E_p t_1 - \vec{p}.\vec{x}_1)} e^{i(E_{p'} t_2 - \vec{p}'.\vec{x}_2)} \theta(t_1 - t_2) \right] =$$

$$\int \frac{d^3p}{\sqrt{(2\pi)^3 2E_p}} \int \frac{d^3p'}{\sqrt{(2\pi)^3 2E_{p'}}} \left[\delta^{(3)}(\vec{p} - \vec{p}\,')\langle 0|0\rangle e^{-i(E_p t_2 - \vec{p}.\vec{x}_2)} e^{i(E_{p'} t_1 - \vec{p}'.\vec{x}_1)} \theta(t_2 - t_1) \right] +$$

$$\int \frac{d^3p}{\sqrt{(2\pi)^3 2E_p}} \int \frac{d^3p'}{\sqrt{(2\pi)^3 2E_{p'}}} \left[\delta^{(3)}(\vec{p} - \vec{p}\,')\langle 0|0\rangle e^{-i(E_p t_1 - \vec{p}.\vec{x}_1)} e^{i(E_{p'} t_2 - \vec{p}'.\vec{x}_2)} \theta(t_1 - t_2) \right]$$

where we have used (2.VIIIa). Considering now that $\langle 0|0\rangle = 1$ and that on account of $\delta^{(3)}(\vec{p} - \vec{p}\,')$, the integral over $\vec{p}\,'$ yields the integrated function itself with $\vec{p}\,' = \vec{p}$, we obtain:

$$i\Delta_F(x_2 - x_1) =$$

$$\int \frac{d^3p}{(2\pi)^3 2E_p} \left[e^{-i(E_p(t_2 - t_1) - \vec{p}.(\vec{x}_2 - \vec{x}_1))} \theta(t_2 - t_1) \right] + \int \frac{d^3p}{(2\pi)^3 2E_p} \left[e^{i(E_p(t_2 - t_1) - \vec{p}.(\vec{x}_2 - \vec{x}_1))} \theta(t_1 - t_2) \right] =$$

$$\int \frac{d^3p}{(2\pi)^3 2E_p} \left[e^{-ip.(x_2 - x_1)} \theta(t_2 - t_1) + e^{ip.(x_2 - x_1)} \theta(t_1 - t_2) \right]$$

which is the above-stated (3.IX), as announced!

Thus, the operation of contour integration which the "$i\epsilon$-prescription" in (3.I) signifies yields the concrete expression (3.IX) for the Feynman propagator $\Delta_F(x_2 - x_1)$ in Minkowski space-time, the two integrals themselves in (3.IX) being respectively the concrete expression of $\langle 0|\hat{\phi}(x_2)\hat{\phi}^\dagger(x_1)|0\rangle$ and $\langle 0|\hat{\phi}^\dagger(x_1)\hat{\phi}(x_2)|0\rangle$ in (2.XVI). Of course, the same expression is obtained through direct replacement of (2.VI) and (2.VII) in the latter. However, the enormous value of (3.I) is that it points to the identification of the Feynman propagator $\tilde{\Delta}_F(p)$ in momentum space and states its precise relation to $\Delta_F(x_2 - x_1)$. We see that whereas in Minkowski space-time the two conditions $x_2^0 > x_1^0$ and $x_1^0 > x_2^0$ are respectively associated with the propagation of a positive-energy particle and its positive-energy antiparticle in two distinct frames, in momentum space they are respectively associated with the positive and negative energies which characterize the propagation of a single particle. Whereas in Minkowski space-time with its unique observable arrow of time they express two opposite temporal orders between two events in space-like separation, in momentum space they express two opposite arrows of time. In physical Minkowski space-time inertial frames in which $x_2^0 > x_1^0$ register the propagation of a positive-energy π^+-meson and inertial frames in which $x_1^0 > x_2^0$ register the propagation of a positive-energy π^--meson. In abstract momentum space exclusively the propagation of the π^+-meson (or exclusively the propagation of the π^--meson) exists in forward-time propagation of positive energies and backward-time propagation of negative energies. Our procedure has established that the former is, in Minkowski space-time, identical to the observed positive-energy π^+-meson and the latter is identical to the observed positive-energy π^--meson. All abstract negative energies of a propagating particle are equivalent to the positive energies of the observed propagating antiparticle. These results are clearly in perfect conformity with the conclusion in the opening section of the present chapter to the effect that as long as the arrow of time points from past infinity to future infinity, no negative-energy states are admissible. We have thus revealed the underlying duality between particles and antiparticles and established the precise relation between the two categories to the effect that:

(i) *No negative energies appear in the observed course of time. The theoretical emergence of negative-energy states is inextricably linked to propagation with respect to a reverse arrow of time in a manner equivalent to and indistinguishable from the observed propagation of positive-energy states from past to future.*

(ii) *The positive-energy states describe observable particles. The unobservable negative-energy states in backward-time propagation describe observable positive-energy antiparticles. To each particle, there corresponds an antiparticle of opposite charge but identical to the particle in all other respects. The Feynman propagator in Minkowski space-time and in momentum space describes the propagation of a particle and its antiparticle at once.*

These conclusions are valid for all fields. As a direct consequence, the fundamental premise to the effect that quantum fields express themselves in terms of particles includes the antiparticles on equal footing. This finds its formal expression in *CT invariance: A quantum state remains invariant under the simultaneous reversal of charge and time.* Again, a negative-energy state which describes a π^+-meson in backward-time propagation is invariant under the simultaneous charge reversal C which produces a π^--meson and time reversal T $(t \to -t)$ which produces forward-time propagation of a positive-energy state. Invariance under CT transformations is, for that matter, the fundamental conclusion to which we have arrived. It characterizes all free quantum fields and the vast majority of interacting contexts between quantum fields. It must be stated, however, that even this invariance is inadequate. As it turns out, Nature discreetly favors one direction in space over the opposite direction. Succinctly put, there are physical contexts in which the mirror image of a quantum state is not physically equivalent to the quantum state itself. Such *parity violations* have been observed in the *weak interactions* which slightly favor one direction in space over the other. The exact symmetry, for that matter, is that of *CPT* invariance. In addition to the simultaneous *CT* reversal, the simultaneous reversal of *parity P*, signifying the reversal of all three spatial coordinates $(x \to -x \,,\; y \to -y \,,\; z \to -z)$ and the concomitant reversal of spin orientation, is also required: *A quantum state remains invariant under CPT reversal in all physical contexts.*

We may now complete the list of Feynman propagators cited in Fig.24. To each Feynman propagator in the Minkowski space-time representation, corresponds the Feynman propagator in the momentum-space representation. Fig.27 cites both representations.

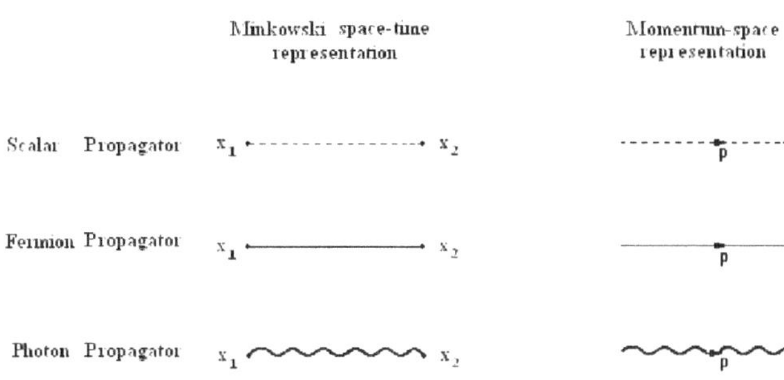

FIGURE 27. Minkowski space-time representation and corresponding momentum-space representation of the Feynman propagator for the scalar, fermion and photon field respectively. The former features an exclusive space-time dependence. The latter features an exclusive momentum dependence.

Before we conclude, we merely mention that all conclusions to which we arrived in this section identically apply also to the propagator for a real scalar field $\phi^\dagger(x) = \phi(x)$ and to its momentum-space counterpart. In particular, in deriving (3.IX) through the detailed two independent procedures we also confirmed our claim as to the first term in (8.III) in the last section of the preceding chapter. For, clearly, the ensuing (8.IV) which this term yields, is the first term in (3.IX).

We have already stressed the central importance of the vacuum state as that energy-momentum eigenstate of a free quantum field in which all oscillators are in their ground state and commented that as the vacuum state is not the zero-field state it is characterized by the absence of actual quantum particles, but not by the absence of probabilities for field-values. We now turn to the formal examination of the Klein-Gordon quantum vacuum. As always, the conclusions to which we shall arrive in examining this prototype of a vacuum state will be relevant to all quantum fields.

258

4. The Quantum Vacuum - I. Vacuum Fluctuations

Let us first briefly summarize the knowledge which we have hitherto gained about the vacuum state and put it in its proper physical perspective. Pursuant to the analysis associated with Fig.21 and Fig.22, the wavefunctions $|0_{p_1}\rangle$, $|0_{p_2}\rangle$, $|0_{p_3}\rangle$, ... in (5.VI) and the non-vanishing probabilities which they signify describe momentum states whose occupation numbers are equal to zero. Those void momentum states respectively receive the abstract representation of a simple quantum harmonic oscillator in its ground state. The ground state of a simple quantum harmonic oscillator is the "Gaussian" "bell-shaped" curve in Fig.4 of course. In turn, the abstract representation of $|0_{p_1}\rangle$, $|0_{p_2}\rangle$, $|0_{p_3}\rangle$, ... is the succession of ground-state oscillator wavefunctions in Fig.28. The striking feature of this succession is the progressively decreasing uncertainty for position Δx. As we climb to ever higher values of p and thus to ever steeper parabolic potentials, the ground-state energy E_0 accordingly rises and Δx, the region within which the probability for position is highest, grows progressively smaller and vanishes as the parabolic potential becomes infinitely strong ($\Delta x \to 0$ as $p \to \infty$). This feature is readily understood. The more attractive (steeper) the parabolic potential is, the more "localized" the quantum oscillator about that potential's bottom point also is.

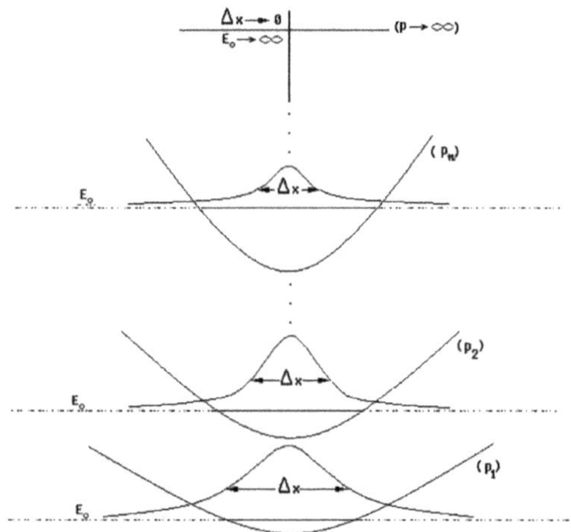

FIGURE 28. The infinitely many classical parabolic potentials against which the simple quantum harmonic oscillators in Fig.21(b) evolve. The uncertainty Δx of the ground state is infinite for $p = 0$, progressively decreases as the momentum $p_1 < p_2 < ... < p_n < ...$ of the corresponding plane wave progressively increases and vanishes for a plane wave of infinite momentum.

We may now turn to the implications which these considerations have for all quantum fields in general and for the quantum Klein-Gordon field $\hat{\phi}(x)$ in particular. Recall that each point $\vec{x}$ on a three-dimensional surface of equal time x^0 labels infinitely many field-values (Fig.20). In arriving at this central result, we tacitly disregarded the arbitrarily high field-values. Arbitrarily high field-values also occur in the classical theory. Formally, classical wave fields may receive arbitrarily high values at certain space-time points just as classical plane waves of energy may have arbitrarily high amplitudes. This is not much of a problem in the classical theory since infinite energies are required in order to excite a classical field to arbitrarily high values. We expect, for that matter, a corresponding situation in quantum field theory. Indeed, we shall soon see in detail that arbitrarily high field-values have arbitrarily small probabilities for occuring. Yet in its relation to arbitrarily high values, divergences and infinities, quantum field theory is drastically different from classical theories and quantum mechanics. The divergences, the stated hallmark of quantum field theory, are all ultimately reducible to the dynamical behavior of the vacuum state at vanishing space-time separations. Inherent in the infinitely many degrees of freedom represented as abstract simple quantum harmonic oscillators in Fig.28 is the implication of a new, non-trivial divergence. For the vacuum state manifests its character as the state of no-particles only throughout Minkowski space-time. Such defining properties of the quantum vacuum as zero momentum, zero energy and zero occupation numbers are manifest only over infinite space-time separations. Over finite space-time separations they become ambiguous. Over finite space-time separation $x_2 - x_1$ there are effectively non-zero values of momentum and energy in $|0\rangle$. Formally, this is explained as an expression of the *uncertainty in field-values*:

$$\langle 0|\hat{\phi}(x_2)\hat{\phi}(x_1)|0\rangle - \langle 0|\hat{\phi}(x_2)|0\rangle\langle 0|\hat{\phi}(x_1)|0\rangle$$

which is, of course, (4.VII) in the first chapter concretized as $\hat{A} \equiv \hat{\phi}(x)$ when $\hat{\phi}(x)$ is in its vacuum state $|0\rangle$, the analogue of (5.III) in the context of infinitely many degrees of freedom in which the dynamical variable is the field $\hat{\phi}(x)$ and x is only a label. However, the concrete physical significance of this relation which accounts for the stated non-zero values of momentum and energy, for the dynamical behavior of quantum fields over finite space-time separations, is not obvious. It is "buried" under layers of mathematical formalism (see, for example, [27], p. 35). In this and the next section I advance, instead, a rather qualitative but sufficiently rigorous line of thought which reveals the essence of the matter without compromizing the necessary analytical process.

Let us state again that such defining properties of the quantum vacuum as zero momentum, zero energy and zero occupation numbers signify are manifest only over infinite space-time separations. Over finite space-time separations they become ambiguous. This is inevitable because the vacuum state has been defined in the entirety of Minkowski space-time as a collection of void plane waves of probability with the direct consequence that over finite space-time separations there is an inherent lack of information as to that particular state. Following always the fundamental premise to the effect that lack of information as to a specific eigenvalue is equivalent to a superposition of all possible quantum states, we understand that the ambiguity which such non-zero values signify is formally interpreted as probability amplitude for the emergence of a particle from the vacuum at x_1, its propagation to x_2 and its annihilation into the vacuum at that space-time point pursuant to the entire analysis on the nature of the propagator

$\langle x_2^0, \vec{x}_2 | x_1^0, \vec{x}_1 \rangle$ in the eighth section of the preceding chapter and of the Feynman propagator $\Delta_F(x_2 - x_1)$. Such transition probabilities and the ambiguity which they signify express the fluctuations of the quantum vacuum over finite space-time separations and are aptly known as *vacuum fluctuations* or *vacuum activity*. Since such vacuum fluctuations over finite space-time separations "raise" the values of momentum and energy above their zero values, it is reasonable to expect that at vanishing space-time separations they yield a divergent result. As we shall see now, the dynamical behavior of the vacuum state confirms such an intuitive expectation, yielding an infinity the precise nature of which is the objective of our analysis in this and the next section.

We shall examine the precise nature of the propagating particle which expresses the fluctuations of the vacuum state in the next section. Whatever the case, however, such a transition is described by the propagator $\langle 0|\hat{\phi}(x_2)\hat{\phi}(x_1)|0\rangle$. The activity of a free vacuum state itself is described, for that matter, by the relevant propagator which for the sake of simplicity we represent here by the propagator of the real Klein-Godon field (the extension to the complex Klein-Gordon field is straightforward). In anticipation of the rigorous confirmation of the divergence inherent in the propagator, let us reiterate the intuitive expectation that since finite space-time separations signify a detraction from the zero values of momentum and energy of $|0\rangle$, those values necessarily grow progressively larger at progressively smaller space-time separations and become infinite at vanishing space-time separations. This is indicative of the fact that the activity of the vacuum state itself diverges at the coincidence space-time limit $x_2 \to x_1$. Indeed, (3.IX) in the preceding section directly implies that at $x_2 \to x_1 \equiv x$ it is:

$$lim_{x_2 \to x_1} \Delta_F(x_2 - x_1) = \int \frac{d^3p}{(2\pi)^3 2p^0}$$

and that consequently:

$$\langle 0|\hat{\phi}^2(x)|0\rangle = lim_{x_2 \to x_1} \langle 0|\hat{\phi}(x_2)\hat{\phi}(x_1)|0\rangle = \infty \tag{4.I}$$

since the integral features three powers of the four-momentum p^μ in the numerator, but only one power in the denominator. If this is not immediately obvious, consider that integration over $\frac{1}{p^0}$ occurs successively with respect to p^1, p^2 and p^3. As integration with respect to each of these three variables extends to arbitrarily high values, d^3p in the numerator tends to infinity much faster than $p^0 = +\sqrt{|\vec{p}|^2 + m^2} = +\sqrt{(p^1)^2 + (p^2)^2 + (p^3)^2 + m^2}$ in the denominator. The result at the limit $p \to \infty$ is the *quadratic divergence* $(3 - 1 = 2)$ in (4.I).

The character of the infinity in (4.I) as a quadratic divergence becomes more transparent if, instead of (3.IX), we consider the coincidence space-time limit $x_2 \to x_1$ on (3.I):

$$lim_{x_2 \to x_1} \Delta_F(x_2 - x_1) = \int d^4p \, \frac{1}{p^2 - m^2} \tag{4.II}$$

In this mathematical statement the integral on the right side features four powers of p in the numerator, but only two powers of p in the denominator. Integration over $\frac{1}{p^2 - m^2}$

occurs successively with respect to p^0, p^1, p^2 and p^3. As integration with respect to each of these four variables extends to arbitrarily high values, d^4p in the numerator tends to infinity much faster than $p^2 - m^2$ in the denominator. Again, the result at the limit $p \to \infty$ is the quadratic divergence $(4 - 2 = 2)$ in (4.I):

$$lim_{x_2 \to x_1} \Delta_F(x_2 - x_1) = \infty \qquad (4.III)$$

The result which (4.I) expresses is of universal significance. It is relevant to all quantum fields in whichever space-time they may evolve. Products of quantum fields considered at the same space-time point yield a divergent result![67] In the preceding chapter we renormalized the divergent energy of the vacuum state to zero value by subtracting an infinite constant from that formally divergent but unobservable absolute energy. This new divergence cannot be eliminated! The reason will become clear as we examine the implications inherent in (4.I). Be that as it may, this divergence is a central aspect of vacuum activity. In order to obtain maximum insight into the latter and its inherent divergence, it is important to understand that vacuum activity signifies the *uncertainty in field-values when the quantum field is in its vacuum state*. At infinite space-time separations all properties of $|0\rangle$ are manifest and no vacuum fluctuations exist. This does not imply a single field-value as a naive analogy with quantum mechanics would suggest. The vacuum state of the quantum field invariably comprises infinitely many possible field-values at each space-time point. As we discussed in the preceding chapter, at a given moment x^0 each point $\vec{x}$ labels infinitely many amplitudes in $|0_{p_1}\rangle|0_{p_2}\rangle...|0_{p_n}\rangle... \equiv |0\rangle$ which are interpreted as field-values. The physical significance of the vanishing uncertainty at infinite space-time separations is precisely the fact that $|0\rangle$ is a manifest and empirically-concrete reality (absence of actual particles, zero momentum, zero energy) only at infinite space-time separations. It is in this sense that the concomitant ambiguity at finite space-time separations signifies vacuum fluctuations and the emergence of uncertainty in field-values: The quantum field is always in its vacuum state, but the latter is no longer manifest. Lack of information as to $|0\rangle$ over finite space-time separations is equivalent to an apparent state of propagating particles and fluctuating momentum and energy.

It becomes obvious, for that matter, that although void of actual quantum particles, the vacuum state of a free quantum field is not the zero-field state (empty space-time), as a naive analogy with the classical vacuum would suggest. The transition probability inherent in the propagator $\langle 0|\hat{\phi}(x_2)\hat{\phi}(x_1)|0\rangle$ is finite and non-vanishing over finite and non-vanishing space-time separations signifying the emergence and annihilation of transient quantum particles with the implication that the dynamical behavior of the Klein-Gordon vacuum state detracts from the state of no-particles, zero momentum and (renormalized) zero energy in Minkowski space-time. At the coincidence space-time limit

---

[67]Divergences occur exclusively at vanishing space-time separations which for conceptual clarity I consider identical to the intuitive "coincidence space-time limit". I should, nevertheless, alert the discriminating reader to the fact that not only do divergences occur when two space-time points are coincident, they also occur when two space-time points are in null separation - that is, when they can be joined by a null geodesic (refer to the Appendix and to the discussion in the opening section of the fourth chapter).

the fluctuations of the vacuum state become infinitely large yielding the divergence in (4.I). This situation may cause some perplexity, as it appears to imply that the quantum field diverges at each space-time point. Indeed, as each space-time point x labels an entire probability distribution of field-values (Fig.20), should it not be plausible to deduce from (4.I) that arbitrarily high field-values have non-vanishing probabilities for occuring? The answer to this rather frivolous question lies in the coincidence space-time limit $x_2 \to x_1$. Vacuum fluctuations are inextricably linked to space-time separations and it is exclusively at the vanishing space-time separations of the coincidence space-time limit that they become infinitely large. A space-time point x which does not signify the coincidence space-time limit $x_2 \to x_1$ is not associated with propagation over vanishing space-time separations. The arbitrarily high field-values in the probability distribution which that x labels may very well be associated with arbitrarily small probabilities for occuring and indeed they are!

The activity of the vacuum state is contingent upon the space-time separation, the probability which propagates over that separation and the quantum particles which represent that propagating probability. Before we proceed to the examination of such particles' precise nature, it is worth making a few critical remarks at the approach to vacuum fluctuations which abounds in "popular" accounts. At the coincidence space-time limit the vacuum activity becomes infinitely large. At the opposite end of the spectrum, all vacuum activity vanishes precisely because at infinite space-time separations the vacuum state and all its properties are manifest. As already stated, the implication is that vacuum activity progressively increases as the space-time separation progressively decreases. The consolidated approach which aspires to an intuitive explanation of such dynamical behavior is the rendition of the latter in terms of the uncertainty relations $\Delta p \sim \frac{\hbar}{\Delta x}$ and $\Delta E \sim \frac{\hbar}{\Delta t}$. The first criticism which must be levelled at such an approach is that the uncertainty Δp in the momentum of the vacuum state over spatial distance Δx does not have any concrete physical content as, unlike the quantum-mechanical wavefunction of momentum, the propagator $\langle 0|\hat{\phi}(x_2)\hat{\phi}(x_1)|0\rangle$ does not signify observable probabilities for momentum. It is very easy to give $\Delta p \sim \frac{\hbar}{\Delta x}$ the verisimilitude of physical rigor through the claim that over spatial distance Δx the dominant contribution to vacuum fluctuations stems from the plane wave whose wavelength is equal to Δx, a quantity which (by assumption) coincides with the uncertainty for position in Fig.28. Such an approach trivially allows for the interpretation of the uncertainty Δp in the corresponding ground-state wavefunction for momentum as a fluctuation about the fixed momentum p of the dominant plane wave of probability. Moreover, such an approach does appear to yield the expected $\Delta p \to 0$ at $\Delta x \to \infty$ and $\Delta p \to \infty$ at $\Delta x \to 0$. Nevertheless, such an approach dismisses the infinitely many degrees of freedom which constitute the essence of a quantum field and which define the dynamical behavior of the vacuum state at any space-time separation. Worse still, the claim as to such a dominant contribution is physically unacceptable, since over a finite Δx no information as to the "dominant" plane wave exists. This objection is, of course, just as relevant to $\Delta E \sim \frac{\hbar}{\Delta t}$. In fact, this latter uncertainty relation is all the more ambiguous: Although the operator for position $\hat{x}$ is rigorously defined, no operator for time exists... Yet, the most serious drawback of this approach is its conflict with quantum field theory itself. Indeed, particles emerge from and vanish into $|0\rangle$ only as superpositions of all infinitely many momentum states none of which is singled out by the mathematical formalism of quantum field theory. In conclusion, despite the partial intuitive value which the "uncertainty principle" as a vehicle of vacuum fluctuations has, it is more of a problem than what it is worth!...

We have thus renormalized the infinite vacuum energy to zero value only to discover that the infinitely many degrees of freedom in the theory have conspired to generate a new infinity! Worse still, this new divergence cannot be removed by subtracting an infinite constant as we did for the vacuum energy. The reason is that vacuum fluctuations at vanishing space-time separations are infinite with respect to the observable vacuum state $|0\rangle$ at infinite space-time separations where vacuum fluctuations vanish. The momentum and energy at the coincidence space-time limit are infinite with respect to the established zero momentum and zero energy of $|0\rangle$. These zero values which characterize the vacuum state in Minkowski space-time and are, for that matter, irrelevant to space-time separations and associated vacuum fluctuations are very much observable at infinite space-time separations! In the fourth section of the preceding chapter we saw that only energies relative to an observable minimum energy are observable. As, for that matter, the value of $\langle 0|\hat{\phi}^2(x)|0\rangle$ is arbitrarily high with respect to the observable zero uncertainty of the vacuum state (vanishing vacuum fluctuations) at infinite space-time separations, it cannot be removed through the subtraction of an infinite constant. The infinitely high fluctuations of $\hat{\phi}(x)$ at vanishing space-time separations cannot be offset by renormalization.

Implicit to our analysis on vacuum fluctuations is a result which concerns yet another fundamental aspect of the Klein-Gordon vacuum. From the uncertainty in field-values

$$\langle 0|\hat{\phi}(x_2)\hat{\phi}(x_1)|0\rangle - \langle 0|\hat{\phi}(x_2)|0\rangle\langle 0|\hat{\phi}(x_1)|0\rangle$$

which vacuum activity signifies and the fact that the latter is expressed exclusively by the propagator $\langle 0|\hat{\phi}(x_2)\hat{\phi}(x_1)|0\rangle$, it folows that

$$\langle 0|\hat{\phi}(x)|0\rangle = 0 \qquad\qquad (4.IV)$$

This result can be independently confirmed through direct replacement of (1.XI) in this chapter. The quantity $\langle 0|\hat{\phi}(x)|0\rangle$ is, clearly, the expectation value of the Klein-Gordon field in its vacuum state (refer to (11.XV) in the first chapter), or simply *the vacuum expectation value* of the Klein-Gordon field. It actually expresses the average value of the quantum Klein-Gordon field in its vacuum state at each point x in Minkowski space-time. In fact, the vacuum expectation value in (4.IV) is a statement to the effect that *the average value of the quantum Klein-Gordon field in its vacuum state is zero*. At each x, all field-values yield a vanishing average.

How then is the paradox of the observable divergences at vanishing space-time separations resolved? Quantum field theory has been formulated on the assumption of a classical, smooth and continuous space-time. Such an assumption amounts ultimately to an idealization since space-time does not possibly remain classical in the domain of excessively high energies and momenta which quantum field theory accommodates at minute space-time scales[68]. The divergences which emerge at the single point which signifies the coincidence space-time limit are indicative of the limitations which quantum field theory formulated on a classical background space-time has. Quantum field theory rigorously describes the physical reality at space-time scales which lie above a certain minimum. Below that minimum space-time scale - an allusion to the scales of quantum gravity - quantum field theory in its present formulation ceases to be valid and must be superseded by a deeper, as of yet unknown, theory[69]. The formal and centrally important procedure of *renormalization* (the advance into which is a desirable future objective) obviates the divergences which emerge at a single space-time point in order to yield a physically consistent theory whose behavior at large space-time scales is not influenced by its limitations at small space-time scales. Nevertheless, I stress again that infinities such as that in (4.I) signify the limitations of quantum field theory as a *local theory* formulated on a classical background space-time. Any consistent theory which will supersede quantum field theory at space-time scales below the stated minimum, will necessarily be finite (see also (N2) in the preceding chapter).

5. The Quantum Vacuum - II. Actual and Virtual Particles

The exclusive physical reality is that of quantum fields! Particles are only a matter of interpretation, the exclusive consequence of that formal procedure which singles out the excited momentum states and acknowledges them as distinct from the infinitely many void momentum states of the vacuum state. All particles which we have hitherto studied are actual. They have been specified at the outset as a matter of choice and boundary conditions in momentum states satisfying the classical relativistic relation $p^2 = m^2$; ($\hbar = c = 1$). Such a specification is formally expressed through the action of creation and annihilation operators according to (1.IV) and (1.VI) in the present chapter. The classical relation $p^2 = m^2$ renders actual particles observable[70] and their existence as plane waves in the specified energy-momentum eigenstate renders them incompatible with incipient emergence and final annihilation in space-time. On the contrary, when we examine the transition of an actual particle over a given finite space-time separation, we relinquish all information as to that particle's fixed momentum. It is logically inconsistent to consider an actual particle at once over a finite space-time separation and over infinite space-time as a plane wave! Absence of knowledge as to the actual momentum eigenvalue of the actual particle - or, more rigorously, as to the actual energy-momentum eigenvalue of the free quantum field - is equivalent to a superposition of momentum eigenstates of occupation number equal to one and, thereby, to the emergence of a particle from $|0\rangle$ at space-time point x_1 and its subsequent annihilation into $|0\rangle$ at space-time point x_2.

---

[68]This is open to several possibilities. The dominant one is summed up in Joseph Polchinski's *String Theory*: "In general relativity, the geometry of spacetime is dynamical. The three spatial dimensions we see are expanding and were once highly curved. It is a logical possibility that there are additional dimensions that remain small" (vol. 1, p. 231).

[69]Despite shortcomings in its formulation, string theory remains the leading candidate.

[70]Again, the reason that in the theory of free quantum fields actual particles are observable despite the absence of interactions, has been stated in the eighth section of the preceding chapter.

Particles emerge from $|0\rangle$ at space-time point x_1 and vanish into $|0\rangle$ at space-time point x_2 respectively in the one-particle state which the superpositions of momentum states in (2.VI) and (2.VII) signify. As we discussed in the eighth section of the preceding chapter and rigorously established through (1.XII) in the present chapter, the emergence and annihilation of a quantum particle respectively involve the entire three-dimensional space at x_1^0 and x_2^0. The specification of two points $\vec{x}_1$ and $\vec{x}_2$ respectively at these two moments results in the formal transition amplitude from $x_1 = (x_1^0, \vec{x}_1)$ to $x_2 = (x_2^0, \vec{x}_2)$ described by the Klein-Gordon propagator in (2.XVI).

Such statement as we have just made should not be perceived as implying that each actual particle in the energy-momentum eigenstate of the quantum field "turns to" a propagating actual particle over finite space-time separations. We reiterate that over finite space-time separations no information as to the actual energy-momentum eigenvalue and, thereby, as to each actual particle exists. This does not signify merely a limitation of our knowledge over finite space-time separations, but an objective fact! Metaphorically, over finite space-time separations the free quantum field itself "does not know" which energy-momentum eigenvalue it has in the entirety of Minkowski space-time! Outstanding example of this is the vacuum state. We have established that $|0\rangle$ is manifest as the no-particle state only in the entirety of Minkowski space-time and that lack of information as to $|0\rangle$ over finite space-time separations is equivalent to an apparent state of propagating particles and fluctuating momentum and energy. A finite space-time separation $x_2 - x_1$ "smears" the vacuum state to a superposition of infinitely many one-particle energy-momentum eigenstates respectively at x_1 and x_2. Of course, it likewise "smears" any energy-momentum eigenstate of the free quantum field. However, the vacuum state stands out! As we have stressed from the outset, it is of particular importance because all other energy-momentum eigenstates are excitations of it. Pursuant to (1.XII) and to the analysis associated with it, for that matter, the probability which *formally* (by dint of interpretation) emerges at x_1 and vanishes at x_2 receives the interpretation of a quantum particle propagating in the vacuum state, a particle which emerges from $|0\rangle$ at x_1 and respectively vanishes in $|0\rangle$ at x_2 in the one-particle state which the superpositions in (2.VI) and (2.VII) signify.

Thus, the premise upon which we established particle propagation and the propagator itself, although fundamentally correct, is not altogether accurate since over finite space-time separations not only does no information as to the actual momentum eigenvalue of a single particle exist, no information as to the actual energy-momentum eigenvalue of the free quantum field itself at all exists. The result is particle propagation in the vacuum state which, for the Klein-Gordon field, is expressed by (2.XVI). In "popular" literature more so than in quantum field theory itself, such particles are known as "virtual". Their defining property is that, unlike the actual particles, virtual particles do not satisfy the classical energy-momentum relation $p^2 = m^2$ and are thus unobservable. In order to gain as much insight into their essence, we note that the mathematical statement (2.VII) in the third chapter has the direct implication:

$$\langle 0|\hat{\phi}(x_2)\hat{\phi}^\dagger(x_1)|0\rangle = \int d^3x \langle 0|\hat{\phi}(x_2)\hat{\phi}^\dagger(x)|0\rangle \langle 0|\hat{\phi}(x)\hat{\phi}^\dagger(x_1)|0\rangle \qquad (5.I)$$

for the first term in (2.XVI) with a corresponding statement for the second term. The relation in (5.I) is a statement to the effect that the propagation of a π^+, for example, from x_1 to x_2 is equivalent to its emergence from $|0\rangle$ at x_1, its subsequent annihilation into $|0\rangle$ at $x \equiv x^\mu = (x^0, x^1, x^2, x^3)$; $x_1^0 < x^0 < x_2^0$, its re-emergence from $|0\rangle$ at x and its subsequent annihilation into $|0\rangle$ at x_2 on the understanding that annihilation and emergence at intermediate moment x^0 concern space in its entirety and that, for that matter, the transition amplitudes to and from each point $\vec{x} = (x^1, x^2, x^3)$ in space at x^0 independently contribute to the transition amplitude from x_1 to x_2.

Thus, the essence of particle propagation can be summarized in the following four premises:

(i) Propagating (virtual) particles are transient entities. They occur exclusively over finite space-time separations.

(ii) Their transient existence is a continuous process of annihilation into and re-emergence from the vacuum state at each space-time point according to (5.I).

(iii) Pursuant to the theory in the preceding third section of the present chapter, (3.I) ensures that they are away from classicality (they do not satisfy $p^2 = m^2$).

(iv) As a consequence of (iii), they are unobservable. Or are they?

In the first section of this chapter we demonstrated in an intuitive context that the measurement of field-strength as a probability distribution independently at two space-time points x_1 and x_2 allows for that measurement's interpretation as observation of the end-points of propagation. In this sense both, actual and virtual particles, are observable since observation of propagation end-points is equivalent to observation of propagation itself. Nevertheless, the observation of virtual particles is formal, invariably the interpretation of the measurement of the actual dynamical variable, of the field-strength. On the contrary, the observation of actual particles as idealized entities in the entirety of Minkowski space-time is explicit. Such a distinction may appear to be arbitrary in view of the fact that actual particles in Minkowski space-time are also a formal interpretation, that of excited momentum states. Note, however, that in no respect does that interpretation involve the position. In the preceding chapter we interpreted an excited momentum state of four-momentum $p^\mu = (\frac{E}{c}, \vec{p})$ and occupation number n as n particles the fixed momentum $p \equiv |\vec{p}|$ and energy E of each of which relate to each other through $E^2 = (pc)^2 + (mc^2)^2$. It is precisely the explicit measurement of these two fixed attributes, p and E, which amounts to the explicit observation of idealized actual particles. In "popular" literature the distinction between actual and virtual particles is frequently rendered as the difference between the observable actual particles and the unobservable virtual particles. Such a statement is not altogether consistent. The consistent statement is that actual particles, being excited momentum states by dint of interpretation, are explicitly observable in the entirety of Minkowski space-time, whereas virtual particles are merely a formal interpretation of observable field-strength. This reflects the essential difference to the effect that actual particles have indefinite existence with no beginning and no end in space-time, whereas virtual particles only have transient existence (an example is provided in (N1)). We reiterate, nevertheless, that however essential the difference between actual and virtual particles may be, it is of minor importance in comparison to their central reality: Particles, actual or virtual, are only expressions of underlying quantum fields. No procedure can elevate them to fundamental physical entities!

Up to this point we have established that lack of information as to the actual energy-momentum eigenvalue of the free quantum field over finite space-time separations is formally equivalent to propagation. In order to advance this fundamental premise to a concrete physical statement we must examine the conditions within which propagation occurs and the implications which those conditions have to propagation. Recall in this respect that the homogeneity of space implies conservation of momentum and time-translational invariance implies conservation of energy. Since the emergence and annihilation of propagating particles are the exclusive consequence of lack of information over finite space-time separations, the homogeneity of Minkowski space-time implies the conservation of the zero energy-momentum of the vacuum state.

Energy-momentum conservation in the vacuum state imposes the condition that p in $\tilde{\Delta}_F(p)$ cancel against a p. That is, each π^+ which propagates with energy-momentum magnitude p enforces the simultaneous existence of an additional particle which propagates with the same energy-momentum magnitude in such a manner as to satisfy the condition $p - p = 0$. At first sight, this may appear to be paradoxical in view of our interpretation to the effect that $\tilde{\Delta}_F(p)$ expresses the probability amplitude that a single particle of rest mass m propagate with energy-momentum magnitude p. We have seen, however, that this interpretation does neither specify whether propagation occurs forward or backward in time, nor whether it involves positive or negative energies. Indeed, we refuted this ambiguity with the boundary condition that positive-energy states propagate exclusively forward in time and negative-energy states propagate exclusively backward in time. For that matter, $\tilde{\Delta}_F(p)$ may just as well be interpreted as the probability amplitude that a single particle of rest mass m propagate with energy-momentum magnitude p as a positive-energy state forward in time and as a negative-energy state backward in time, as though at some moment to the future of its emergence that positive-energy particle bounces back in time and returns to its emergence as a negative-energy state. It is precisely this interpretation which upholds the conservation of energy-momentum in the vacuum state. For the moment at which the positive-energy π^+ "bounces back" is the moment at which that π^+ vanishes with respect to that arrow of time which points from past infinity to future infinity and the moment at which the negative-energy π^+ emerges with respect to that arrow of time which points from future infinity to past infinity with the implication that at that moment it is $p - p = 0$. This is also the case at the moment the positive-energy π^+ emerges in forward-time propagation and the negative-energy π^+ vanishes in backward-time propagation (recall that no reference to space-time points is possible in momentum space). Equivalently now, by dint of CPT invariance, the momentum-space propagator $\tilde{\Delta}_F(p)$ describes two positive-energy virtual particles in forward-time propagation, the π^+ and the π^-, in perfect conformity with our conclusion in the third section to the effect that the Feynman propagator describes the propagation of a particle and its antiparticle at once. In conclusion, energy-momentum conservation translates in momentum space to a single-particle loop, the single particle (π^+ or π^-) propagating with energy-momentum magnitude $p = \sqrt{(p^0)^2 - |\vec{p}|^2}$ forward in time and with energy-momentum magnitude $p = \sqrt{(-p^0)^2 - |-\vec{p}|^2}$ backward in time. This is equivalent to a $\pi^+ - \pi^-$ pair in which the particle and its antiparticle, π^+ and π^-, propagate with four-momentum magnitude $p = \sqrt{(p^0)^2 - |\vec{p}|^2}$ forward in time (Fig.29(a)).

This calls into question the relation between $\Delta_F(x_2 - x_1)$ and $\tilde{\Delta}_F(p)$ in the context of energy-momentum conservation. It is always (3.I) of course, despite the fact that $\tilde{\Delta}_F(p)$ now expresses two virtual particles in forward-time propagation for each value of p. The simple reason is that, whatever the case, the mathematical expression of $\tilde{\Delta}_F(p)$ is always that in (3.II). Thus, $\Delta_F(x_2 - x_1)$, *the mathematical expression of vacuum activity over space-time separation* $x_2 - x_1$, *also describes the propagation of a virtual particle-antiparticle pair, now in a single inertial frame.* The virtual π^+ (accordingly, π^-), whose emergence at x_1, propagation and annihilation at x_2 $\langle 0|\hat{\phi}(x_2)\hat{\phi}^\dagger(x_1)|0\rangle$ (accordingly $\langle 0|\hat{\phi}^\dagger(x_2)\hat{\phi}(x_1)|0\rangle$) formally describes, enforces the emergence at x_1, propagation and annihilation at x_2 of a virtual π^- (accordingly, π^+). Energy-momentum conservation implies that lack of information as to the zero energy-momentum of the vacuum state over finite space-time separations is formally equivalent to propagation of a virtual particle-antiparticle pair (Fig.29(b)). We reiterate that since no information as to the actual energy-momentum eigenvalue of the quantum field and, thereby as to each actual particle exists over finite space-time separations, the zero value $0 = p - p$ is always the energy-momentum magnitude of the vacuum state. Likewise, the total zero charge at x_1 and x_2 is the zero charge of the vacuum state.

The lesson here is that the propagator must be interpreted with caution. Although a single particle in forward-time propagation between two space-time points is formally described by the propagator in Minkowski space-time, the propagator in Minkowski space-time describes a single particle in forward-time propagation between two space-time points only in principle. When energy-momentum conservation in free fields is a consideration, the propagator necessarily receives the interpretation of a particle-antiparticle pair in forward-time propagation.

The expression of a particle-antiparticle pair by a single propagator reflects the fact that the vacuum state of a free quantum field is exact. It emerges from the quantization of the exact solution to the classical field equations. As that solution is exact, so is the propagator[71]. As the exact vacuum state is known, so is the mathematical expression of propagation over finite space-time separations regardless of whether that propagation is interpreted as one particle, or as a pair of particles. This is not the case in the presence of interactions. The exact vacuum state of an interacting system of fields is unknown precisely because the exact solution to the highly non-linear field equations is practically unattainable. All we can do is approach that unknown vacuum state - and thereby, the unknown exact propagator - through successive perturbations of the known vacuum state of each free field which enters the interacting system.

Let us now descend from abstract quantum field theory to the "real world" in which plane waves represent extended but finite sinusoidal waveforms. Since contact with intuition is the objective now, we shall grant ourselves license to use the unqualified "uncertainty principle". Formally, propagation and the concomitant particle-pairs occur

---

[71]We shall see later in this chapter that the propagator is a *Green function* to the classical field equations. Knowledge of the Green function is equivalent to knowledge of the solution itself.

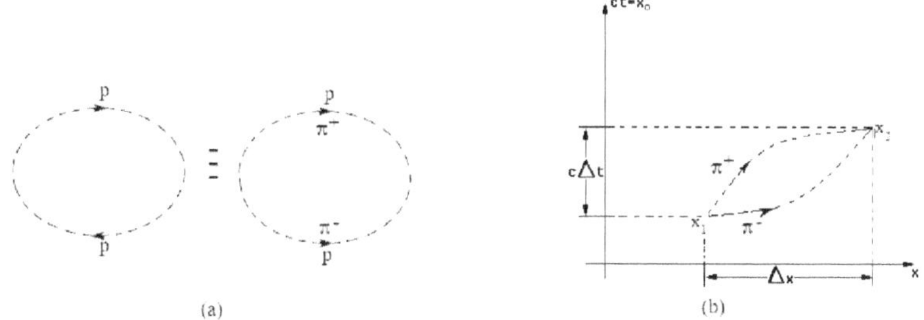

FIGURE 29. Complex Klein-Gordon propagation in momentum space (a) and in Minkowski space-time (b) (real Klein-Gordon propagation replaces π^+ and π^- by π^0). Energy-momentum conservation $p - p = 0$ for any value of p applies as the zero value of the vacuum state gives rise to a single particle propagating forward and backward in time, an effect whose irrelevance to any momentum state signifies the disconnected character of free propagation. CPT invariance renders the single-particle loop for either π^+, or π^-, equivalent to a positive-energy $\pi^+ - \pi^-$ pair propagating forward in time.

at any finite space-time separation simply because finite space-time separations signify the demise of plane waves of probability. However, in a realistic context it is physically legitimate to inquire as to the space-time scale at which the idealized representation of actual extended but finite sinusoidal waveforms as plane waves of probability ceases to be valid. Looking into the example of the Klein-Gordon field, we advance the following reasoning. The rest mass of propagating particles, being an attribute of the field itself, is independent of space-time separations and the fluctuations within them. As, in general, the Klein-Gordon field has non-zero rest mass m, the fluctuation ΔE in the energy over space-time separation $x_2 - x_1$; $x_1 = (ct_1, \vec{x}_1)$, $x_2 = (ct_2, \vec{x}_2)$, must be at least equal to $2mc^2$ for the particle-antiparticle pair in Fig.29 to be produced. From $\Delta E \sim \frac{\hbar}{\Delta t}$ and the condition $\Delta E \geq 2mc^2$ then it follows that $\frac{\hbar}{\left(\frac{\Delta x}{c}\right)} \geq 2mc^2$ and thereby that $\Delta x \leq \frac{\hbar}{2mc}$. Thus, the spatial distance

$$\lambda = \frac{\hbar}{mc} \qquad (5.II)$$

signals the space-time scale above which there are effectively no $\pi^+ - \pi^-$ pairs and below which there is an increasing number of such pairs at progressively decreasing space-time separations. It signals the space-time scale above which the extended waveform qualifies for a plane wave in "infinite" space-time and below which "finite" space-time separations signify lack of information. The spatial distance λ is known as the *Compton wavelength* of the $\pi^\pm$-particle[72]. From (5.II) it follows that particle-antiparticle pairs occur at any space-time scale if the rest mass is equal to zero (for example, in the photon field).

Implicit throughout our entire analysis on quantum fields is the premise that quantum mechanics qualifies at best as an approximate description of Nature in the event that infinitely many degrees of freedom are disregarded. We have seen, in fact, that a free quantum field resembles a quantum-mechanical system at the non-relativistic limit of very low energies and that, for that matter, quantum mechanics emerges as an approximate description if we *arbitrarily* disregard the ever-present infinitely many void momentum states at that limit. Yet, what are the limitations of such an approximation? To what extent does it remain valid over finite and progressively decreasing space-time separations? Our analysis reveals the answer: Formally, the full force of the infinitely many degrees of freedom of quantum fields manifests itself at any finite space-time separation. Realistically, the space-time scale which the Compton wavelength signifies is the effective scale below which that full force manifests itself. For that matter, *finite space-time separations formally and space-time separations below that of the Compton wavelength effectively, preclude any semblance of quantum-mechanical behavior.* In quantum mechanics particles are as immutable as their classical counterparts. They are neither created, nor annihilated and their number remains fixed. This is not the case in quantum field theory where particles are only a matter of interpretation, never a matter of fact. As we have seen pursuant to (5.I), propagation is a continuous process of annihilation into and re-emergence from the vacuum state at each space-time point. Be it at finite space-time separations formally, or at space-time separations below the scale of the relevant Compton wavelength effectively, lack of information is equivalent to a multiplicity of particle-antiparticle pairs. All of them are represented by the diagram in Fig.29(b) and, as we have stressed, are observable only formally, only by dint of interpretation. "Popular" accounts of physics abound with the statement to the effect that the virtual particle-antiparticle pairs which occur below the scale of the Compton wavelength preclude the observation of an actual particle's position and that, for that matter, the Compton wavelengh signifies a "fundamental limitation on the measurement of position", as though somehow the position of the actual particle exists but virtual particle-antiparticle pairs render it unobservable. This erroneous standpoint fails to appreciate the difference between particles and fields: If even for a "quantum-mechanical" particle at the non-relativistic limit of very low energies the position is only a label, how much more so below the scale of the Compton wavelength!

The activity of the vacuum state has observable consequences in conformity either with its interpretation as virtual particles, or with its significance as a detraction from the zero energy of the vacuum state. For example, the *Schwinger effect*, the physical process in the

---

[72]It is physically equivalent to the wavelength of that momentum state of the photon field whose energy $p^0 = |\vec{p}|$; $(c = 1)$ is that of the rest mass m of either π^+ or π^-.

context of which strong electric fields induce the spontaneous production of positively-charged and negatively-charged actual particle-antiparticle pairs, is consistent with the interpretation of vacuum activity in terms of virtual pairs elevated to actual by the external electric field (Fig.30). On the contrary, the *Casimir effect*, the attractive force which two parallel plates exert on each other in vacuum, is interpreted as the exclusive consequence of the fact that the plates disturb the void plane waves of probability, restricting the possible wavelengths between them, thereby reducing the vacuum activity between them (Fig.31) (N2).

Up to this point we have used the Klein-Gordon field as a conceptual device for the analysis on the most fundamental aspects of quantum fields. In what follows, we shall examine certain fundamental aspects of the semi-classical limit of the free quantum boson fields using, as always, the Klein-Gordon field as a prototype. The particular semi-classical limit of the free quantum Klein-Gordon field will be examined by comparison to the semi-classical limit of the photon field.

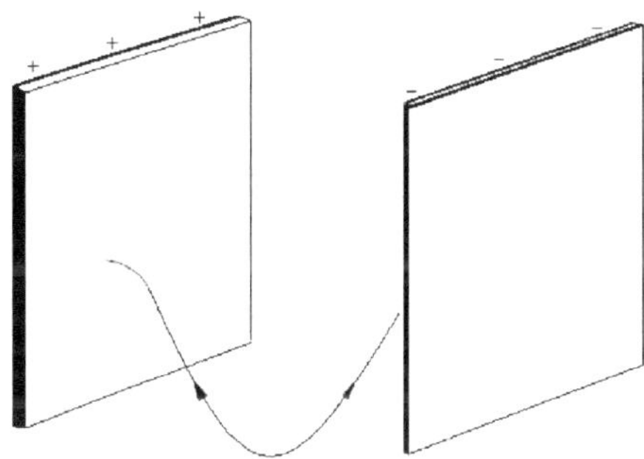

FIGURE 30. The Schwinger effect: A strong electrostatic field renders $|0\rangle$ unstable by providing the energy which elevates the virtual pair in Fig.29(b) to actual. In practice, the energy required for the spontaneous production of a $\pi^+ - \pi^-$ pair is prohibitively high. A more realistic example is obtained through the replacement of the Klein-Gordon field by the electron field (the smooth curved line signifies the much lighter actual pair of an electron e^- and a positron e^+).

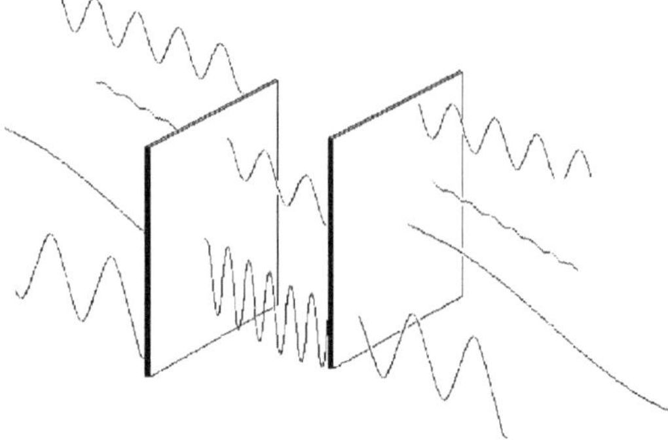

FIGURE 31. The Casimir effect : The presence of two thin parallel plates at initial distance d from each other disrupts vacuum activity. Whereas out of the plates all void plane waves exist, in between the plates only waveforms of an integer number of wavelengths are allowed such that the length of the waveforms is equal to d. Thus, the inward pressure exerted by $|0\rangle$ on each plate, being higher than the outward pressure, drives the two plates toward each other.

6. The Semi-classical Limit

Our analysis has repeatedly stressed that the semi-classical limit $\hbar \to 0$ only has formal significance. The Planck constant $\hbar$ cannot be treated as a variable, let alone as one which tends to zero. The limit $\hbar \to 0$ is, in fact, a formal expression of an actual semi-classical situation characterized by the absence of $\hbar$. It is a fairly straightforward matter to see what that actual situation is in boson fields. We have seen in the seventh section of the preceding chapter that the quantum states of a boson field are never full. Actual particles such as π^0, $\pi^\pm$ can occupy each momentum state of the Klein-Gordon field in arbitrary numbers. If the occupation number n is small, the quantum field is manifest. In a given energy-momentum eigenstate of the quantum field, a momentum state is a plane wave of probability specified by fixed momentum $\vec{p}$ and associated energy E. The amplitude of that plane wave is exclusively determined by n and is irrelevant to $\vec{p}$ and E. As such, a momentum state manifests itself in any of the infinitely many possible energy-momentum eigenstates of the abstract simple quantum harmonic oscillator which represents it, an eigenstate of energy $E_n = nE$ and momentum $\vec{p}_n = n\vec{p}$ (Fig.22). Any change $\Delta n < n$ by a few quanta yields $0 < \frac{\Delta n}{n} < 1$ and, for that matter, also $0 < \frac{\Delta E}{nE} < 1$.

The fixed values of these two ratios express the fact that the creation or annihilation of a quantum in the momentum state of occupation number n manifestly modifies the amplitude of that momentum state.

Let us now postulate a momentum state of an excessively high occupation number $n >> 1$ of the order of, to say the least, trillions of actual particles ($n \sim 10^{12}$). Then, with Δn being always the difference generated by the creation or annihilation of a few quanta in that momentum state, it is $\frac{\Delta n}{n} \to 0$. In turn, it also is $\frac{\Delta E}{nE} \to 0$. These vanishing results for the relative difference $\frac{\Delta n}{n}$ in the occupation-number spectrum and for the relative difference $\frac{\Delta E}{nE}$ in the energy spectrum of that momentum state stand in sharp contrast to the corresponding fixed values which emerge in the context of small occupation numbers. In order to uncover their physical significance we first observe that the relative difference $\frac{\Delta E}{nE} = \frac{\hbar \Delta n \omega}{\hbar n \omega} = \frac{\Delta n}{n}$ is independent of $\hbar$. Thus, the limit $\frac{\Delta E}{nE} = \frac{\Delta n}{n} \to 0$ is conceptually identical to the semi-classical contexts which we studied in the first and the third chapter. Being independent of $\hbar$, it constitutes the physical content of the formal limit $\hbar \to 0$ for boson fields. Of immediate relevance in this respect is the analysis in the first chapter in view of the abstract representation which plane waves of probability receive in quantum field theory. The semi-classical limit $\frac{\Delta E_n}{E_n} \to 0$ which signifies the absence of relative-energy discontinuities for the simple quantum harmonic oscillator at arbitrarily high energy levels ($n >> 1$) constitutes the abstract representation of $\frac{\Delta n}{n} \to 0$ and $\frac{\Delta E}{nE} \to 0$ for the relative difference in occupation number and energy of the boson momentum state respectively. The continuous spectra of occupation number and energy which effectively emerge for $n >> 1$ receive an abstract representation as the continuous semi-classical energy spectrum of the corresponding simple quantum harmonic oscillator at arbitrarily high energy levels. In line with the premise stated in the first chapter, the vanishing quantum fluctuations $\frac{\Delta n}{n} \to 0$ which characterize the occupation number $n >> 1$ and corresponding energy nE signify the emergence of semi-classical behavior for boson quantum fields. The relative differences in the occupation number and in the energy of the momentum state respectively vanish, yielding effectively continuous occupation-number and energy spectra which remain unaffected by the creation or annihilation of an insignificant number of quanta. This is precisely in accordance with expectations of a smooth classical wave of energy.

Let us confirm such expectations. It is not difficult to see that at the limit of vanishing quantum fluctuations $\frac{\Delta n}{n} \to 0$ and spectral continuity, the incipient plane wave of probability reduces indeed to a smooth classical wave of energy. Qualitatively, as the occupation number tends to infinity ($n \to \infty$) and becomes effectively a continuous variable, there are as many bosons in the momentum state as there are points in infinite space! These infinitely many degrees of freedom propagate infinite energy ($nE \to \infty$) in space-time. At once, such an "infinitely-occupied" momentum state stands "infinitely above" the infinitely many void momentum states in the given energy-momentum eigenstate of the quantum field which, for that matter, effectively become physically irrelevant. Thus, the momentum state reduces to a classical wave field! Which wave field? In order to answer this question, it is necessary to elucidate the difference between the unobservable potential of a field and the observable field-strength, both of which concepts describe a field and which concepts we have hitherto indifferently described as

274

field-values. In this respect, it suffices to state that the derivative - a concept which we introduced in the third chapter - is the essence of field-strength $\mathbf{E}(x)$. The strength $\mathbf{E}(x)$ of a field is a vector which essentially expresses the rate of change of the potential $\phi(x)$ in space-time. In field theory, both classical and quantum, the unobservable potential is necessarily a primary concept and the observable field-strength, being the potential's rate of change, a derivative concept. The infinitely many degrees of freedom which signify a field are, in Minkowski space-time, primarily expressed by the potential $\phi(x)$ before $\phi(x)$ itself is observed in terms of its strength. It follows, for that matter, that the classical wave field which emerges at $n \to \infty$ is an unobservable plane wave of oscillatory potential which signifies an observable semi-classical plane wave of oscillatory field-strength. The latter is, of course, the observable plane wave of energy which emerges at $n \to \infty$. The remaining task is to establish that this semi-classical plane wave is, indeed, characterized by the energy density which the classical theory predicts.

We first note that the mass-dimensionality of the Klein-Gordon field $\phi(x)$ is equal to one, an attribute which we formally state as:

$$[\phi] = [m]$$

It is a straightforward matter to establish the validity of this statement through (3.I) (present chapter) since d^4p clearly features four mass-dimensions, whereas $p^2 - m^2$ in the denominator features two mass-dimensions ($\vec{p} = m\vec{v}$ and $E = mc^2$ are clearly of mass-dimensionality equal to one) with the direct implication that $\Delta_F(x_2 - x_1)$ in (2.XVI) is of mass-dimensionality equal to two, whence $[\phi] = [m]$. We note further that since the first derivative of the potential $\phi(x)$ involves an additional length-dimension in the denominator, the field-strength is of mass-dimensionality equal to two[73]:

$$[\mathbf{E}] = [m]^2$$

The energy density $\frac{dE}{dV}$ now, being the first derivative (the rate of change) of the energy with respect to volume, is clearly of mass-dimensionality equal to four (since $dV = dx^1 dx^2 dx^3$) and, thus, of the same mass-dimensionality as that of the square of the strength's magnitude ($[\mathbf{E}^2] = [m]^4$). It follows, for that matter, that the square of the strength's maximum amplitude is essentially identical to the energy density of the observable semi-classical plane wave. The semi-classical limit $n \to \infty$ has eventuated in a plane wave of oscillating field-strength $\mathbf{E}(x)$, the maximum amplitude $\mathbf{E}$ of which determines an energy density $\mathbf{E}^2$. Some reflection on Fig.22(a) readily reveals that the sharp curves which emerge at $n \to \infty$ imply an arbitrarily high rate of change for $\phi(x)$ which is independent of the wavelength λ. Equivalently, the arbitrarily high $\mathbf{E}$ is independent of the frequency w and energy E of the incipient momentum state. This is the signature of a classical plane wave of energy: Unlike the energy $nE = \hbar nw$ of a momentum state, its energy density is independent of its frequency w. Instead, it is determined exclusively by its maximum amplitude $\mathbf{E}$, a conclusion to which we first arrived in the context of simple quantum mechanics in the first chapter (N1). The "classic" example of

---

[73]The mass-dimensionality can be trivially shown to be equal to inverse length-dimensionality. For example, in $e^{i\vec{p}.\vec{x}}$, $[x] \sim [m]^{-1}$ ensures that the "angle" $\vec{p}.\vec{x}$ has no mass (or length) dimensions. Another example can be seen in (5.2) of the preceding section.

a classical plane wave of oscillating field-strength is that in the classical Maxwell theory. The energy density U of an electromagnetic plane wave, featuring electric strength $\mathbf{E}(x)$ and magnetic strength $\mathbf{B}(x)$, is

$$U = \frac{1}{8\pi}(\mathbf{E}^2 + \mathbf{B}^2) \qquad (6.I)$$

and can be seen to have all the above-stated attributes of the semi-classical limit.

The preceding analysis applies to all boson fields. The Bose-Einstein statistics with its unrestricted occupation numbers eventuates in smooth classical potentials and classical wave fields at the semi-classical limit $n \to \infty$. When we later discuss the quantum fields in the realistic presence of interactions we shall see how, as a consequence of such a semi-classical behavior, vacuum activity and virtual particles yield the illusion of smooth classical fields produced by sources. For the specific case of the Klein-Gordon field which is the object of our examination in this chapter, the following statement must be made. Although idealized classical waves of energy do emerge at the semi-classical limit of the idealized free quantum Klein-Gordon field, classical π-meson waves do not exist. The underlying reason is that the π^+, π^- and π^0 are consistent with the Klein-Gordon field only as idealized entities themselves. The actual π^+, π^- and π^0 are bound states of a quark and an anti-quark, with the strong interaction between them providing the relatively high rest mass of those π-mesons. As a result, unlike the photon, they are - like all mesons - unstable, with lifetimes of the order 10^{-8} sec for the $\pi^\pm$ and 10^{-17} sec for the π^0. The reason we observe classical electromagnetic waves but not classical π-meson waves is the instability of the latter. Even if high-energy collisions (in the Earth's atmosphere from impeding cosmic rays, or in particle accelerators) produced the high occupation numbers required for the semi-classical limit, the transient existence of π^+, π^- and π^0 would preclude the emergence of classical π-meson waves. It is interesting to note in this respect that the composite structure of the actual π-mesons which precludes the existence of classical π-meson waves is also responsible for the short-range of the nuclear interaction which is realized by virtual π-mesons. For, the composite structure which causes the π-mesons to rapidly disintegrate to other particles is also the underlying reason for the π-mesons' non-zero rest mass. Whereas at its semi-classical limit the photon field emerges as electromagnetic waves produced by actual photons as well as electric and magnetic fields produced by long-range virtual photons (refer to (5.II) in the preceding section), the π-meson field admits no semi-classical limit.

7. Quantum Field Theory versus Quantum Mechanics

Why are quantum fields the exclusive reality in Nature and particles merely an expression of them? Why is the dichotomy between particles and fields in contradiction with the theory of Relativity? Why is a theory of relativistic quantum mechanics physically unacceptable? These are all equivalent expressions of the issue which motivated the formulation of quantum field theory and which we are now ready to address. By now it is clear that the dichotomy between particles and fields is incompatible with quantum field theory. Particle-antiparticle pairs and variable particle numbers in the presence of interactions directly contradict the fixed number of particles of quantum mechanics. There are compelling conceptual arguments which point to the incompatibility of quantum mechanics with special relativity. We have seen, for example, that the probability for transition over space-like separations is necessarily non-vanishing and that, for that matter, causality implies the existence of antiparticles. A relativistic single-particle quantum theory such as "relativistic quantum mechanics", has no inherent capacity for antiparticles. It is as impervious to antiparticles as the non-relativistic quantum mechanics. This conceptual argument alone strongly advocates the necessity for the field outlook. Yet, the rigorous proof that quantum mechanics is physically inconsistent and that, for that matter, the formal dichotomy between particles and fields is resolved only within the framework of quantum field theory, lies in the mathematical formalism of "relativistic quantum mechanics". The experience which we have hitherto gained from quantum field theory allows us to understand the contradictions inherent in that mathematical formalism and appreciate their rigorous resolution within the framework of quantum field theory. In what follows, we shall approach this issue in - what I deem - the most succinct and conceptual manner. In the process, we shall arrive at the enunciated Schrödinger equation and the Klein-Gordon equation. This new conceptual advance requires a commensurate advance in mathematical content. The mathematical formalism which permeates the analysis in this section may, at first sight, impress upon the reader as arduous. Yet, despite appearances, there is nothing intimidating about it. It consists mostly in the formal and rigorous rendition of concepts already introduced. The few new mathematical concepts which are necessary for the ensuing analysis, will, as always, be rendered transparent.

In non-relativistic quantum mechanics, the probability that the position of a quantum particle be point $\vec{x}$ within volume element d^3x at moment t is $|\psi(t,\vec{x})|^2 d^3x$. The reasons, as we have seen, are that (i) it is positive definite, as all probabilities are, and (ii) it yields a conserved total probability throughout space, a total probability which remains constant in the course of time. We express this as

$$\frac{\partial}{\partial t}\int d^3x |\psi(t,\vec{x})|^2 = 0 \qquad (7.I)$$

with the integral expressing the total probability throughout three-dimensional space and with the *first partial derivative* $\frac{\partial}{\partial t}$ expressing the rate of change of that total probability with respect to time. Some elucidation in this respect, is in order.

As stated in the third chapter, the partial derivative $\frac{\partial f}{\partial x_i}$ signifies differentiation of a multi-variable function $f(x_1, x_2, ..., x_n)$ with respect to variable x_i (i being any integer between 1 and n), on the understanding that in the operation $\frac{\partial}{\partial x_i} f(x_1, x_2, ..., x_n)$ all variables other than x_i are kept constant. For that matter, the first partial derivative of function $f(t, \vec{x}) \equiv f(t, x_1, x_2, x_3)$ with respect to t at $t = t_0$ is

$$\frac{\partial}{\partial t_{|t_0}} f(t,\vec{x}) = lim_{t \to t_0} \frac{f(t,\vec{x}) - f(t_0,\vec{x})}{t - t_0} \qquad (7.II)$$

if this limit exists.

Function $\frac{\partial}{\partial t} f(t, \vec{x})$ is the first partial derivative of $f(t, \vec{x})$ at t_0, with t_0 now being any value of the independent variable t in a subset of function's $f(x_1, x_2, ..., x_n)$ domain of definition. The first partial derivative of $f(t, \vec{x})$ with respect to x_1, or x_2, or x_3, is accordingly defined.

This elucidates the statement in (7.I) to the effect that $\int d^3x |\psi(t, \vec{x})|^2$ has zero rate of change at each moment and is, for that matter, conserved in the course of time. Note in this respect, that all space dependence in that total probability has been summed out, leaving the latter exclusively a function of time t. Nevertheless, the use of the partial derivative $\frac{\partial}{\partial t}$, as opposed to the use of the derivative $\frac{d}{dt}$, is consistent as it serves a purpose which will become apparent when we introduce the Schrödinger equation.

We shall also need the *second partial derivative* of $f(t, \vec{x})$, the definition of which trivially follows now from the first partial derivative. The second partial derivative of $f(t, \vec{x})$ is merely the first partial derivative of the first partial derivative:

$$\frac{\partial^2}{\partial t^2} f(t, \vec{x}) = \frac{\partial}{\partial t}\left(\frac{\partial}{\partial t} f(t, \vec{x})\right) \qquad (7.III)$$

Higher-order partial derivatives are accordingly defined.

After this brief digression, we return to the physical content of (7.I). The validity of the latter rests exclusively with the Schrödinger equation the solution to which is the wavefunction $\psi(t, \vec{x})$. In line with a conceptual approach of minimum mathematical content, we have, thus far, refrained from stating that equation. We have seen in the first chapter, however, that it is consistent with the non-relativistic relation $E = \frac{p^2}{2m}$ between the momentum magnitude $p = |\vec{p}|$ of a free particle and that particle's energy E. As we shall see now, the single power of E in this relation implies that the Schrödinger equation is a linear differential equation which involves the first derivative of the wavefunction with respect to time as a single term and is, thus, of the form $\frac{\partial}{\partial t}\psi(t, \vec{x}) = ...$ It is precisely for this reason that the first derivative in (7.I) allows for the use of the Schrödinger equation.

The fact that the single power of E implies that the Schrödinger equation is of the form $\frac{\partial}{\partial t}\psi(t, \vec{x}) = ...$ is a direct consequence of the physical demand that free quantum-mechanical particles be represented by plane waves of probability the fixed momentum p and fixed energy E of which satisfy the classical non-relativistic relation $E = \frac{p^2}{2m}$. For that matter, plane waves of probability satisfy the prospective quantum-mechanical equation and we can use them in order to formulate the latter through the demand that it reproduce $E = \frac{p^2}{2m}$. Indeed, the single power of t in $\psi_p(t, \vec{x}) = Ce^{\frac{i}{\hbar}(\vec{p}\cdot\vec{x}-E_p t)}$ implies that the Schrödinger equation must necessarily involve the first derivative $\frac{\partial}{\partial t}$ and no higher derivatives with respect to t if the Schrödinger equation is to reproduce the single power of $E \equiv E_p$ in $E = \frac{p^2}{2m}$. Readers familiar with differential calculus recognize in this statement the basic rule of differentiation

$$\frac{d}{dx} e^{cx} = ce^{cx}$$

If you do not fall in that category, you may safely accept the heuristic rule to the effect that the derivative of a sinusoidal function is the sinusoidal function itself multiplied by whatever multiplies the independent variable. For a plane wave of probability $\psi_p(t, \vec{x}) = Ce^{\frac{i}{\hbar}(\vec{p}\cdot\vec{x}-E_p t)}$ then, this implies

$$\frac{\partial}{\partial t}\psi_p(t, \vec{x}) = \frac{\partial}{\partial t}Ce^{\frac{i}{\hbar}(\vec{p}\cdot\vec{x}-E_p t)} = (-\frac{i}{\hbar}E_p)\psi_p(t, \vec{x})$$

This first partial derivative with respect to time is the result of immediate relevance to the comparison between quantum field theory and quantum mechanics. However, since this comparison amounts essentially to a comparison between the Klein-Gordon equation and the Schrödinger equation, it is necessary that we arrive at the explicit statement of both equations in order to grasp the deep physical implications inherent in their different mathematical structure. We proceed, for that matter, to the right side (...) of the Schrödinger equation $\frac{\partial}{\partial t}\psi_p(t, \vec{x}) = ...$

On the same lines as above, the right side of the Schrödinger equation must yield $\sim \frac{p^2}{2m}\psi_p(t, \vec{x})$, so that the Schrödinger equation will, indeed, reproduce the classical relation $E = \frac{p^2}{2m}$. As a direct consequence of the second power of p on the right side of this relation, this is only possible if the right side of the Schrödinger equation involves the second derivative with respect to space - that is, if the right side is of the form $\frac{\partial^2}{\partial x^{1^2}}\psi_p(t, \vec{x}) + \frac{\partial^2}{\partial x^{2^2}}\psi_p(t, \vec{x}) + \frac{\partial^2}{\partial x^{3^2}}\psi_p(t, \vec{x})$. We have, thus, arrived at the Schrödinger equation for a free quantum-mechanical particle:

$$i\hbar\frac{\partial}{\partial t}\psi(t, \vec{x}) = -\frac{\hbar^2}{2m}\nabla^2\psi(t, \vec{x}) \qquad (7.IVa)$$

with the multiplicative factors on either side stemming always from the demand that (7.IVa) imply $E\psi_p(t, \vec{x}) = \frac{p^2}{2m}\psi_p(t, \vec{x})$, as you yourselves can effortlessly confirm and with the *Laplace operator* ∇^2 on the right side being merely a short-hand notation for the above-mentioned second derivative with respect to space:

$$\nabla^2 \equiv \frac{\partial^2}{\partial x^{1^2}} + \frac{\partial^2}{\partial x^{2^2}} + \frac{\partial^2}{\partial x^{3^2}}$$

The form of (7.IVa) suggests directly that the latter generalizes to

$$i\hbar\frac{\partial}{\partial t}\psi(t, \vec{x}) = -\frac{\hbar^2}{2m}\nabla^2\psi(t, \vec{x}) + V(\vec{x})\psi(t, \vec{x}) \qquad (7.IVb)$$

for a quantum-mechanical particle evolving against a static classical potential $V(\vec{x})$. The free Schrödinger equation (7.IVa) emerges as a special case of the Schrödinger equation (7.IVb) in the event that $V(\vec{x}) = 0$. Although the former does not determine any $\psi(t, \vec{x})$ other than the plane waves of probability, it must be stressed that any $\psi(t, \vec{x})$ satisfies the former since any $\psi(t, \vec{x})$ is a superposition of plane waves of probability. It should be clear, of course, that this is a physical demand, not a consequence. We demand at the outset that the Schrödinger equation be a linear differential equation so that, in analogy with the classical wave equation, the superposition of two solutions be also a solution. For that matter, it must be stressed that although we have arrived at the Schrödinger equation through plausibility arguments such as the stated physical demand for linearity and the physical demand that it reproduce the classical relation

$E = \frac{p^2}{2m}$, we have in no way derived the Schrödinger equation. Like all fundamental "physical laws", the Schrödinger equation can neither be derived, nor proved (think of Newton's laws, for example). It is merely the most plausible choice made on the basis of empirically concrete facts and theoretical expectations. The ultimate judge is, of course, the experiment. Speaking of theoretical expectations, note the striking feature of the Schrödinger equation: Space and time enter in a manifestly asymmetrical manner. The derivative with respect to time is first order, whereas the derivative with respect to space is second order. This clearly underscores the non-relativistic character of the Schrödinger equation.

Before we prove (7.I), it is necessary that we briefly examine the operators which the Schrödinger equation has introduced. The above-stated

$$\frac{\partial}{\partial t}\psi_p(t, \vec{x}) = -\frac{i}{\hbar}E_p\psi_p(t, \vec{x})$$

for a plane wave of probability $\psi_p(t, \vec{x})$, implies the energy operator

$$\hat{E} = i\hbar\frac{\partial}{\partial t} \tag{7.V}$$

the eigenfunctions of which are the plane waves $\psi_p(t, \vec{x})$ and the eigenvalues of which are the corresponding energies E_p:

$$\hat{E}\psi_p(t, \vec{x}) = E_p\psi_p(t, \vec{x}) \tag{7.VI}$$

Accordingly, the momentum operator for each dimension of space is

$$\hat{p}^i = -i\hbar\frac{\partial}{\partial x^i} \quad ; \quad i = 1, 2, 3 \tag{7.VIIa}$$

or, concisely

$$\hat{\vec{p}} = -i\hbar\nabla \tag{7.VIIb}$$

with

$$\nabla = \vec{x}_0^1\frac{\partial}{\partial x^1} + \vec{x}_0^2\frac{\partial}{\partial x^2} + \vec{x}_0^3\frac{\partial}{\partial x^3} \qquad (\vec{x}_0^i \; ; \; i = 1, 2, 3 : \textit{unit vectors})$$

and satisfies the, familiar from the first chapter, eigenvalue equation for momentum

$$\hat{\vec{p}}\psi_p(t, \vec{x}) = \vec{p}\psi_p(t, \vec{x}) \tag{7.VIII}$$

The operator on the right side of (7.IVb) is

$$\hat{H} = -\frac{\hbar^2}{2m}\nabla^2 + V(\vec{x}) \equiv \frac{\hat{\vec{p}}^2}{2m} + V(\vec{x}) \tag{7.IX}$$

and we immediately recognize it as the Hamiltonian operator for a quantum-mechanical particle, the very same operator which in the first chapter we had loosely labelled as the "energy operator". For $V(\vec{x}) = 0$, the Hamiltonian operator for the Schrödinger equation reduces to the Hamiltonian operator for the free Schrödinger equation. The appellation "energy operator" for $\hat{H}$ is actually not quite consistent. The reason can be seen in the precise distinction between the energy operator $\hat{E}$ and the Hamiltonian operator $\hat{H}$. Both operators are associated with eigenvalues of energy. Mathematically, however, they are manifestly different. Thus, only one of the two qualifies for the energy operator. Physically, $\hat{E}$ is the energy operator because it is defined by the temporal derivative $\frac{\partial}{\partial t}$. We have repeatedly stressed the intimate connection between energy and time. Energy conservation implies time-translational invariance. This intimate connection is manifest in the temporal derivative. The Hamiltonian operator $\hat{H}$, on the contrary, is not the sum of a "kinetic operator" and a "potential operator". It is an "energy operator" only because of (7.IVb):

$$\hat{E}\psi(t,\vec{x}) = \hat{H}\psi(t,\vec{x})$$

This is inevitable since, as we have seen and stressed (especially in the second chapter), the total energy of a quantum state can neither be separated in a kinetic term $\frac{p^2}{2m}$ and a potential term $V(\vec{x})$ at each point $\vec{x}$, nor can it even be considered to consist in a "kinetic term" and a "potential term".

The validity of (7.I) follows now directly from (7.IVb). Since the integral is over space but the derivative is over time, the latter can be taken "inside" the integral to act directly on $\psi^*(t,\vec{x})\psi(t,\vec{x})$. The result is determined by the rule for differentiation concerning products of functions to the effect that the derivative of a product of two functions is the derivative of the first function multiplied by the second function plus the derivative of the second function multiplied by the first function. For that matter, $\frac{\partial}{\partial t}|\psi(t,\vec{x})|^2 = \psi(t,\vec{x})\frac{\partial}{\partial t}\psi^*(t,\vec{x}) + \psi^*(t,\vec{x})\frac{\partial}{\partial t}\psi(t,\vec{x})$. Using $\frac{\partial}{\partial t}\psi(t,\vec{x}) = \frac{1}{i\hbar}\hat{H}\psi(t,\vec{x})$ from (7.IVb) as well as $\frac{\partial}{\partial t}\psi^*(t,\vec{x}) = -\frac{1}{i\hbar}(\hat{H}\psi(t,\vec{x}))^*$ from complex-conjugating both sides of (7.IVb), the right side of (7.I) is:

$$\frac{\partial}{\partial t}\int d^3x |\psi(t,\vec{x})|^2 = -\frac{1}{i\hbar}\int d^3x(\hat{H}\psi(t,\vec{x}))^*\psi(t,\vec{x}) + \frac{1}{i\hbar}\int d^3x\psi^*(t,\vec{x})(\hat{H}\psi(t,\vec{x}))$$

which is equal to zero since $\hat{H}$ is a self-adjoint operator, as all operators which express physical quantities are, and pursuant to (5.VIII) in the first chapter, for that matter, it is

$$\int d^3x\psi^*(t,\vec{x})(\hat{H}\psi(t,\vec{x})) = \int d^3x(\hat{H}\psi(t,\vec{x}))^*\psi(t,\vec{x})$$

Let us now try to formulate a quantum-mechanical theory consistent with the relativistic relation $E^2 = (pc)^2 + (mc^2)^2$. The second power of E in this relation accordingly implies that a relativistic plane wave of probability $\psi_p(ct,\vec{x}) = Ce^{\frac{i}{\hbar}(\vec{p}\cdot\vec{x}-p^0x^0)} = Ce^{\frac{i}{\hbar}(\vec{p}\cdot\vec{x}-E_pt)}$ satisfies a linear differential equation of the form $\frac{\partial^2}{\partial t^2}\psi(ct,\vec{x}) = \ldots$. It is clear that such a relativistic differential equation is in conflict with (7.I) which involves only the first derivative with respect to time. Consequently, $|\psi(ct,\vec{x})|^2$ no longer qualifies for a positive-definite probability density for position which yields a conserved total probability throughout three-dimensional space. A necessary condition for a mathematical expression $P(ct,\vec{x})$ to qualify in this respect is that $P(ct,\vec{x})$ involve itself the first derivatives $\frac{\partial}{\partial t}\psi(ct,\vec{x})$ and $\frac{\partial}{\partial t}\psi^*(ct,\vec{x})$. For, only then is it possible that $\frac{\partial}{\partial t}$ provide the necessary $\frac{\partial^2}{\partial t^2}\psi(ct,\vec{x})$ and $\frac{\partial^2}{\partial t^2}\psi^*(ct,\vec{x})$ in order to yield zero when operating on the total probability throughout three-dimensional space:

$$\frac{\partial}{\partial t}\int d^3x\, P(ct,\vec{x}) = 0 \qquad (7.X)$$

For that matter, before we determine $P(ct,\vec{x})$, we derive the linear differential equation of relativistic quantum mechanics upon which the validity of (7.X) rests. On exactly the same lines as those on which the non-relativistic Schrödinger equation follows from $E = \frac{p^2}{2m}$, that relativistic linear differential equation follows from the demand that it reproduce $E_p^2 = (pc)^2 + (mc^2)^2$:

$$-\hbar^2\frac{\partial^2}{\partial t^2}\psi(ct,\vec{x}) = (-\hbar^2\nabla^2 c^2 + m^2 c^4)\psi(ct,\vec{x}) \qquad (7.XI)$$

or, equivalently

$$\left[\Box + \left(\frac{mc}{\hbar}\right)^2\right]\psi(ct,\vec{x}) = 0 \qquad (7.XII)$$

with the *d' Alembert operator* being

$$\Box \equiv \frac{\partial^2}{\partial x^{0\,2}} - \frac{\partial^2}{\partial x^{1\,2}} - \frac{\partial^2}{\partial x^{2\,2}} - \frac{\partial^2}{\partial x^{3\,2}} \equiv \frac{\partial^2}{\partial x^{0\,2}} - \nabla^2$$

(7.XII) is the *Klein-Gordon equation* the solution to which describes the classical relativistic scalar wave field in (1.I) of the present chapter. In this section's treatment it is, of course, tentatively interpreted as the equation for a relativistic quantum-mechanical particle. Like the non-relativistic Schrödinger equation, the Klein-Gordon equation can neither be derived, nor proved. We have merely arrived at it through plausibility arguments. Note that, as an immediate consequence of $\frac{\partial^2}{\partial t^2}$, the complex conjugate $\psi_p^*(ct,\vec{x}) = C^* e^{-\frac{i}{\hbar}(\vec{p}.\vec{x}-E_p t)}$ is also a solution to the Klein-Gordon equation. In the context of the latter's tentative interpretation as the equation for a relativistic quantum-mechanical particle, $\psi_p^*(ct,\vec{x})$ describes a plane wave of probability propagating with energy $-E_p$.

The task now is to determine a positive-definite function $P(ct,\vec{x})$ such that it satisfies (7.X) on the basis of the Klein-Gordon equation. The calculation in this direction formally advances by multiplying (7.XII) by $\psi^*(ct,\vec{x})$, the complex conjugate equation to (7.XII) by $\psi(ct,\vec{x})$ and subtracting the two results. However, that calculation and the implications inherent in it are predicated on the *continuity equation* and the *conserved current* which that equation implies. We would like to avoid extending the already conceptually laden analysis in that direction (the mathematically-literate reader can find that formal derivation in [25], [26] and the first volume of [30]). We shall advance, instead, a rather heuristic derivation of the possible $P(ct, \vec{x})$.

Given that the prospective probability density $P(ct, \vec{x})$ must involve $\frac{\partial}{\partial t}\psi(ct, \vec{x})$ and $\frac{\partial}{\partial t}\psi^*(ct, \vec{x})$, it is not difficult to conjecture its form. Since the non-relativistic quantum-mechanical probability density for position is $|\psi(t, \vec{x})|^2 = \psi^*(t, \vec{x})\psi(t, \vec{x})$, it is reasonable to expect that (up to the inessential $\frac{\hbar}{2mc^2}$) the function is

$$P(ct, \vec{x}) = i\frac{\hbar}{2mc^2}\left[\psi^*(t, \vec{x})\frac{\partial}{\partial t}\psi(t, \vec{x}) - \psi(t, \vec{x})\frac{\partial}{\partial t}\psi^*(t, \vec{x})\right] \qquad (7.XIII)$$

The formal calculation which proves that $P(ct, \vec{x})$ in (7.XIII) does indeed satisfy (7.X) is simple and parallels the calculation which proves[74] (7.I). Qualitatively, however, our expectation is justified on the following grounds:

(i) Since both $\psi(ct, \vec{x})$ and $\psi^*(ct, \vec{x})$ are solutions to the Klein-Gordon equation, $P(ct, \vec{x})$ must involve $\psi(ct, \vec{x})$ and $\psi^*(ct, \vec{x})$ on equal footing. Thus, the required temporal derivative $\frac{\partial}{\partial t}$ must operate symmetrically on $\psi(t, \vec{x})$ and $\psi^*(t, \vec{x})$. This attribute is, in addition, supported by the fact that at the non-relativistic limit $v << c$ of speeds insignificant in comparison to that of light, $P(ct, \vec{x})$ must reduce to the non-relativistic probability density which manifestly features $\psi(t, \vec{x})$ and $\psi^*(t, \vec{x})$ on equal footing.

(ii) We do not need the simple calculation which actually proves that $P(ct, \vec{x})$ in (7.XIII) reduces indeed to $|\psi(t, \vec{x})|^2$ at $v << c$. Let it suffice to state that any higher powers of either $\psi(t, \vec{x})$ or $\psi^*(t, \vec{x})$ in (7.XIII) would clearly be incompatible with $\psi^*(t, \vec{x})\psi(t, \vec{x})$.

(iii) The two terms in (7.XIII) are of opposite sign. This is consistent with the demand that the two integrals in (7.X) cancel against each other.

(iv) (7.XIII) is consistent with the demand for relativistic covariance. By way of elucidating this, we first briefly look at the relativistic length contraction (see Appendix) in the following context: A rod moving relative to an inertial frame K in a direction perpendicular to its length will suffer no contraction. The same rod moving with constant speed relative to K in the direction of its length will be observed with smaller length l than its rest-length l_0. We merely state the length contraction as $l = \frac{1}{\gamma}l_0$; $\gamma > 1$. It follows that the volume V of a moving object as observed in K is related to that body's rest-volume V_0 by $V = \frac{1}{\gamma}V_0$. Now, if Pd^3x is to qualify for the probability that the position be a point within volume d^3x (a quantity which characterizes the quantum system and is, for that matter, frame-independent), P must itself, under a Lorentz transformation, transform in such a manner as to offset the contraction of the volume element $d^3x \rightarrow \frac{1}{\gamma}d^3x$ - that is, it must transform as $P \rightarrow \gamma P$. That this is indeed the case follows from the fact that $\psi(ct, \vec{x})$ is invariant under Lorentz transformations ($\psi'(x') = \psi(x)$ is the solution to a differential equation which, being consistent with $E^2 = (pc)^2 + (mc^2)^2$, can hardly help being Lorentz invariant itself!) and the fact

---

[74]For the mathematically-inclined reader: The temporal derivative $\frac{\partial}{\partial t}$ in (7.X) and that in (7.XIII) yield $\frac{\partial^2}{\partial t^2}$ in (7.XI). The subsequent application of Gauss' *divergence theorem* and the fact that $\psi(ct, \vec{x})$ vanishes "at infinity", completes the proof.

that $\frac{\partial}{\partial t} \to \gamma \frac{\partial}{\partial t}$ (which effortlessly follows from $d\tau = \sqrt{1 - \frac{v^2}{c^2}}\,dt$, revealing at once that $\gamma = \frac{1}{\sqrt{1 - \frac{v^2}{c^2}}}$).

(v) The multiplicative factor $\frac{\hbar}{2mc^2}$ in (7.XIII) is, for all intents and purposes, inessential, but the i-factor is crucial. Without it, $P(ct, \vec{x})$ is not even a real number at each space-time point. The reason is always the fact that the solution to the Klein-Gordon equation is $\psi(t, \vec{x}) = Ce^{-\frac{i}{\hbar}E_p t + \frac{i}{\hbar}\vec{p}.\vec{x}}$, as well as the complex conjugate $\psi^*(t, \vec{x}) = C^* e^{\frac{i}{\hbar}E_p t - \frac{i}{\hbar}\vec{p}.\vec{x}}$. For that matter, $\frac{\partial}{\partial t}$ in (7.XIII) will respectively result in a multiplicative $\pm i$-factor. This is the factor which the i-factor in (7.XIII) aims at offsetting.

Thus, the conjectured $P(ct, \vec{x})$ in (7.XIII) satisfies (7.X) and qualifies for a conserved quantity. In addition, it satisfies almost all physical requirements which a prospective relativistic probability density is expected to satisfy. All except one, that is. We would interpret $P(ct, \vec{x})$ as the probability density for a relativistic quantum-mechanical particle, were it not for the fact that $P(ct, \vec{x})$ is not positive-definite! Mere inspection of (7.XIII) reveals that $P(ct, \vec{x})$ may receive negative values. Without meeting the crucial requirement of positivity, a quantity does not qualify for probability density. Thus, it is impossible to formulate a relativistic quantum-mechanical theory which yields a physically acceptable probability for position. The demand for relativistic covariance inevitably leads to unphysical negative probabilities. The particle-wavefunction $\psi(ct, \vec{x})$ is incompatible with Lorentz covariance! Quantum mechanics is incompatible with the theory of relativity!

If we had carefully reflected on the term "relativistic quantum mechanics", we would have sensed this deadlock from the outset! For, as our analysis has repeatedly stressed, the relativistic relation $E^2 = (pc)^2 + (mc^2)^2$ entails both, positive-energy and negative-energy values. The negative-energy values were bound to result in negative-energy eigenvalues. The reason that negative energies are physically unacceptable was stressed in the third section of the present chapter. If they were available to a physical system evolving in the observable course of time from past infinity to future infinity, that system would collapse to a state of infinite negative energy. In the final analysis, it is those unphysical negative-energy eigenvalues which underlie the emergence of the unphysical negative probabilities. Recall, in this respect, a relevant point. As we discussed in the first chapter, the total probability for position is the quantum analogue to total classical wave energy. Both of them are conserved in the course of time.

We would hardly overstate the matter if we stressed again that the conflict between quantum mechanics and the theory of relativity, the reason the term "relativistic quantum mechanics" is a contradiction of terms, lies in the second temporal derivative $\frac{\partial^2}{\partial t^2}$ of the Klein-Gordon equation and, for that matter, in the underlying $E^2 = (pc)^2 + (mc^2)^2$. Why not then follow the example of the classical theory and merely discard the unphysical negative energies? Why not exclusively consider the physically acceptable positive energies $E = +\sqrt{(pc)^2 + (mc^2)^2}$? The single power of E in this simple relation would again imply a linear differential equation which, as in non-relativistic quantum mechanics, involves the first derivative $\frac{\partial}{\partial t}\psi(t, \vec{x})$. A logically legitimate argument indeed! That was, in fact, the first attempt at the formulation of a consistent theory of relativistic quantum mechanics. It takes, however, a closer look at that differential equation in order to see that such an argument collapses. For, although the demand that the negative square root be discarded does - through the replacement $p \to -i\hbar\nabla$ (see (7.VIIb)) - yield the differential equation

$$i\hbar\frac{\partial}{\partial t}\psi(ct,\vec{x}) = \sqrt{-\hbar^2 c^2 \nabla^2 + m^2 c^4}\,\psi(ct,\vec{x}) \qquad (7.XIV)$$

which involves only the first temporal derivative on its left side, the physical legitimacy of this equation rests exclusively on its right side. Although at classical level the square root of the number $(pc)^2 + (mc^2)^2$ is rigorously defined, at quantum level, where all observable attributes are represented by operators, the corresponding square root on the right side of (7.XIV) is, to say the least, ambiguous. It is not clear how the second derivatives in ∇^2 act on $\psi(ct,\vec{x})$ when $-\hbar^2 c^2 \nabla^2 + m^2 c^4$ is under the square root. The direct interpretation of that square root as an expansion in powers of $-\hbar^2 c^2 \nabla^2$ - an elementary identity in differential calculus[75] - is a disaster! We only list the consequences of such an interpretation: (i) It yields a differential equation which involves only the first derivative with respect to time, but derivatives of all orders (first, second, third...) with respect to space, thus treating space and time in an utterly asymmetrical manner and in direct contradiction to the spirit of relativity. (ii) Mathematically, such a differential equation is extremely intractable. (iii) Such a differential equation results in a non-local quantum theory. This third consequence is the most devastating! For, it allows for violation of causality. The problem is not in the prediction that quantum particles propagate at higher-than-light speeds (they do!), but in the prediction that such a propagation has observable consequences. We have seen how quantum field theory safeguards causality through its prediction of antiparticles. As stated, this is a prediction for which "relativistic quantum mechanics" has no inherent capacity.

This then is the contradiction to which we have arrived: Wavefunctions describing relativistic quantum-mechanical particles necessarily satisfy the Klein-Gordon equation, yet relativistic quantum mechanics is physically inconsistent on account of the second temporal derivative $\frac{\partial^2}{\partial t^2}$ in the Klein-Gordon equation. On the evidence, there are two possible explanations. Either our interpretation of the solution $\psi(ct,\vec{x})$ to the Klein-Gordon equation as a particle-wavefunction is incorrect, or, after all, relativistic particle-wavefunctions do satisfy a differential equation which - like the Schrödinger equation - involves only the first temporal derivative $\frac{\partial}{\partial t}\psi(ct,\vec{x})$, yet without the problems which plague (7.XIV). Clearly, if such a differential equation exists, it must be consistent with the relativistic relation $E^2 = (pc)^2 + (mc^2)^2$: The operation of "squaring" both sides of that differential equation $(\frac{\partial^2}{\partial t^2}\psi(ct,\vec{x}) = ...$; see (7.III)), must yield the Klein-Gordon equation. Although the resolution of the apparent contradiction is inherent in the first possibility, by a quirk of history it was the pursuit of the second possibility which eventuated in that resolution. It is instructive, for that matter, to make at least a passing reference to the course of history which led to quantum field theory.

Ironically, the first to arrive at the Klein-Gordon equation for a relativistic quantum-mechanical particle was Erwin Schrödinger in 1926. Confronted by the above-stated

---

[75] $\sqrt{1+x} = 1 + \frac{1}{2}x - \frac{1}{8}x^2 + \frac{1}{16}x^3 - ...$; if $|x| < 1$, explicitly stated in the third section.

inconsistencies, however, he abandoned it and formulated, instead, his equation which is predicated on the non-relativistic $E = \frac{p^2}{2m} > 0$ and does yield a positive-definite probability density for position. In the same year, Walter Gordon and Oscar Klein independently arrived at the same equation which eventually came to bear their names. The inherent contradictions and inconsistencies stemming from the persistent refusal of physicists to interpret the solution to the Klein-Gordon equation as anything other than a wavefunction for a relativistic quantum-mechanical particle led eventually to the demise of the Klein-Gordon equation. In 1928, in what was arguably one of the most significant developments in all Physics, Paul Dirac managed to arrive at a linear differential equation of the form $\frac{\partial}{\partial t}\psi(ct,\vec{x}) = \dots$ which, for that matter, did allow for a positive-definite probability density for position. Being consistent with $E^2 = (pc)^2 + (mc^2)^2$, the *Dirac equation* yielded the Klein-Gordon equation: In a very concrete sense which we shall examine in the next chapter, the solution $\psi(ct,\vec{x})$ to the Dirac equation was also the solution to the Klein-Gordon equation. Yet, the "square root" of the Klein-Gordon equation yields (7.XIV) and its counterpart with a minus sign on the right side. Thus, although the Dirac equation constituted a highly non-trivial interpretation of (7.XIV), it was still physically inconsistent. It had dispensed with the negative probabilities, but not with the negative energies.

In a stroke of intuition and ingenuity, Dirac confronted the unphysical negative energies with the Pauli exclusion principle. As we have seen, the latter is a statement to the effect that no two identical fermions may simultaneously occupy the same quantum state. On this basis, the *hole theory*, put forward by Dirac in 1930, advanced two premises to the effect that the predicted negative-energy eigenstates are (i) unobservable and (ii) occupied exclusively by one particle ($n_p = 1$). In effect, the reason electrons do not collapse to infinite negative energies is the fact that the unobservable *Dirac sea* of negative-energy eigenstates is already fully occupied and cannot, for that matter, accommodate more particles. In addition, upon absorbing adequate energy unobservable electrons in the Dirac sea make a transition to observable energy eigenstates leaving behind, as it were, a hole of observable positive energy. Not only had the Dirac equation led to the prediction of infinitely many unobservable electrons, it had led to the prediction of the positron. It was a matter of time, for that matter, before the profound implication underlying the Dirac equation was realized[76]. The infinitely many occupied but unobservable negative-energy eigenstates were indistinguishable from a set of infinitely many degrees of freedom. The Dirac sea was indistinguishable from the vacuum state of a quantum electron field. The dichotomy between particles and fields, very much an article of faith since the sixteenth century, is spurious. Quantum fields are the exclusive reality in Nature. Particles are merely expressions of underlying quantum fields.

Thus, the unqualified prediction of negative energies had, all along, been a consequence of the arbitrary interpretation of $\psi(ct,\vec{x})$ as a single-particle wavefunction. For that matter, the radical reinterpretation of the solution $\psi(ct,\vec{x})$ to the Dirac equation as a quantum electron field, unequivocally replaced the unphysical negative energies by the positive-definite energies in (1.XI) of the present chapter (refer also to the discussion associated with Fig.25). Of course, (1.XI) is the quantized solution to the Klein-Gordon

---

[76]In retrospect, Dirac facetiously remarked: *My equation proved to be more intelligent than I am.*

equation interpreted as a quantum field. The argument, however, remains intact in view of the fact that, like all fields, the Dirac feld is a superposition of plane waves, as we shall explicitly see in the next chapter. Let it suffice to reiterate a premise which has permeated the entire analysis in this chapter: *As long as the arrow of time points from past infinity to future infinity, no negative-energy states are admissible!*

Having dispensed with the negative energies through the reinterpretation of the solution to the Dirac equation as a quantum field, it was a matter of time before the Klein-Gordon equation itself was restored to physical legitimacy through its reinterpretation as a field equation. We have seen, of course, that the Klein-Gordon field is a boson field, a fact which precludes the interpretation of its stability as a consequence of the Pauli exclusion principle. Nevertheless, by the time the Klein-Gordon equation was rediscovered, the issue of the negative energies had vanished. The Klein-Gordon field admits no negative-energy states merely because the arrow of time points from past infinity to future infinity.

The interpretation of the solution as a quantum field - instead of, as a particle-wavefunction - was, at once, the solution to the issue of probabilities. No longer were the negative probabilities inherent in (7.XIII) an issue. The probabilities for position simply do not exist! The Klein-Gordon equation expresses a field, a physical system of infinitely many degrees of freedom in which the position is merely a label. Nor, for that matter, was the positive-definite probability density of the Dirac equation relevant any more. The dynamical variable of the Dirac field is - as in all fields - the field-strength. The position is merely a label.

It is in light of such conclusions that we must revisit the relation which quantum mechanics has to quantum field theory. Recall from the analysis on the Fock space of the theory that in the interpretation of an energy-momentum eigenstate of the quantum field as a particle state, in that formal procedure which singles out the excited momentum states and acknowledges them as distinct from the infinitely many void momentum states, the semblance of quantum mechanics emerges only at the non-relativistic limit of very low energies. For, although that formal procedure invariably singles out a finite number of degrees of freedom, the excited momentum states which those degrees of freedom signify satisfy the Schrödinger equation with its inherent probabilistic interpretation of the wavefunction only at that limit. Considering the one-particle state in (6.I) for example, the semblance of quantum mechanics in the dynamical behavior of the quantum field, emerges exclusively from those terms which feature a non-relativistic excited momentum state. By extension, it is only when in each such term we dismiss all infinitely many void momentum states - an arbitrary and unjustified operation - that quantum mechanics emerges *at best* as an approximate description of Nature. Away from that non-relativistic limit, quantum mechanics emerges neither as a semblance, nor as an approximation. The full significance of this fact transcends the Schrödinger equation: The excited momentum states which remain upon dismissing the infinitely many void momentum states are relativistic plane waves of probability characterized by the unphysical negative energies and negative probabilities associated with (7.XIII). *The semblance of probability density and of quantum mechanics emerges at the non-relativistic limit of very low energies precisely because it is at that limit that the Klein-Gordon equation reduces to the Schrödinger equation.* Such a semblance emerges exclusively from the excited non-relativistic momentum states. Quantum mechanics itself emerges at the non-relativistic limit when we arbitrarily take the next step and dismiss the void momentum states altogether! However, as that arbitrary step is not logical, as it contradicts the essence of quantum fields, quantum mechanics emerges, at best, as a low-energy approximation to quantum field theory - never as the "low-energy limit" of quantum field theory! In retrospect, for that matter, is it not plausible to reflect on the reason that of all infinitely many possible coefficients in (6.I), we chose only those which conform with (6.II)?

We would like to elevate such rigorous statement as we have just made to complete proof. It takes, in fact, a brief calculation to show that at the non-relativistic limit of very low energies, the excited momentum states appear "quantum-mechanical". First, we put the relativistic differential equation in which our analysis has eventuated, into its proper physical perspective. (7.XII), now properly interpreted as a differential equation for a classical relativistic wave field $\phi(ct, \vec{x})$:

$$[\Box + (\frac{mc}{\hbar})^2]\phi(x) = 0 \qquad (7.XV)$$

is the Klein-Gordon equation. Its solution (1.I) expresses the classical Klein-Gordon field the canonical quantization of which yields the free quantum Klein-Godon field, a scalar (spin-0) field expressed by the operator

$$\hat{\phi}(ct, \vec{x}) = \int d^3p \frac{1}{\sqrt{(2\pi)^3 E_p}}[\hat{a}(p)e^{\frac{i}{\hbar}\vec{p}.\vec{x} - \frac{i}{\hbar}E_p t} + \hat{a}^\dagger(p)e^{-\frac{i}{\hbar}\vec{p}.\vec{x} + \frac{i}{\hbar}E_p t}] \qquad (7.XVI)$$

which is, of course, none other than (1.XI).

There is nothing intimidating about the ensuing brief calculation which proves that at the non-relativistic limit of very low energies, an excited momentum state of the free quantum Klein-Gordon field satisfies the Schrödinger equation. I present it because it is a highly important calculation laden with information about the dynamical behavior of free quantum fields at the non-relativistic limit while, at once, accessible to everyone who was able to follow the analysis thus far in this section. However, for the benefit of the less confident readers, I first advance a heuristic argument which conveys the essence of that calculation.

The non-relativistic limit is, by definition, that physical context which involves speeds negligible in comparison to the speed of light: $v << c$. With

$$\gamma = \frac{1}{\sqrt{1 - \frac{v^2}{c^2}}}$$

being the characteristic factor of special relativity (stated primarily in the Appendix, also in the preceding analysis), it is for that matter:

$$E = \sqrt{(pc)^2 + (mc^2)^2} = \sqrt{(\gamma mvc)^2 + (mc^2)^2} =$$

$$mc^2\sqrt{1+(\frac{\gamma mvc}{mc^2})^2} = mc^2\sqrt{1+\frac{1}{1-\frac{v^2}{c^2}}\frac{v^2}{c^2}}$$

Since $v^2 << c^2$, we can exploit the elementary *expansion* in differential calculus:

$$\frac{1}{1-x} = 1+x+x^2+\ldots$$

which does converge to a real number for $-1 < x < 1$, in order to state:

$$E \sim mc^2\sqrt{1+(1+\frac{v^2}{c^2})\frac{v^2}{c^2}} \sim mc^2\sqrt{1+\frac{v^2}{c^2}}$$

We can further exploit the expansion:

$$\sqrt{1+x} = 1+\frac{1}{2}x-\frac{1}{8}x^2+\frac{1}{16}x^3-\ldots\ ;\ -1 < x < 1$$

in order to state:

$$E \sim mc^2(1+\frac{1}{2}\frac{v^2}{c^2}) = mc^2+\frac{1}{2}mv^2 \equiv mc^2+\frac{p^2}{2m}$$

Of course, in this relation, it is $\frac{p^2}{2m} << mc^2$. As the rest-mass energy mc^2 is a purely relativistic quantity, it must be subtracted at the non-relativistic limit $v << c$ in order to bring the latter in line with Newtonian physics which acknowledges no relation between mass and energy. Thus, at the non-relativistic limit of very low energies, the relativistic energy $E = \sqrt{(pc)^2+(mc^2)^2}$ reduces to the non-relativistic Newtonian kinetic energy $\frac{p^2}{2m}$:

$$E \sim \frac{p^2}{2m}$$

Since the Klein-Gordon equation and the Schrödinger equation have been respectively founded upon these two relations, it follows that at the non-relativistic limit of very low energies the former reduces to the latter with the implication that the free quantum Klein-Gordon field appears at that limit as a quantum-mechanical system.

After this "warm-up exercise", we proceed to the actual calculation which rigorously establishes the above conclusion and reveals the precise manner in which at the non-relativistic limit the free quantum field appears as a quantum-mechanical system. Without recourse to (1.I), let us state the solution to (7.XV) merely as

$$\phi(ct,\vec{x}) = \tilde{\phi}(ct,\vec{x})e^{-\frac{i}{\hbar}mc^2t} \qquad (7.XVII)$$

which manifestly features the rest-mass energy in the oscillatory exponential and use the relation $E \sim mc^2+\frac{p^2}{2m}$ which emerged in the preceding elementary calculation in order to define the kinetic energy:

$$E' = E - mc^2\ ;\ E' << mc^2$$

Since in (7.XVII) the rest-mass energy has been factored, the kinetic energy E' is necessarily inherent in:

$$\tilde{\phi}(ct,\vec{x}) = \tilde{\phi}(\vec{x})e^{\pm\frac{i}{\hbar}E't}$$

with $\tilde{\phi}(\vec{x})$ carrying all dependence of $\tilde{\phi}(ct,\vec{x})$ on space. It follows that:

$$\frac{\partial}{\partial t}\tilde{\phi}(ct,\vec{x}) \sim \pm\frac{i}{\hbar}E'\tilde{\phi}(ct,\vec{x}) \Rightarrow$$

$$|i\hbar\frac{\partial}{\partial t}\tilde{\phi}(ct,\vec{x})| \sim E'\tilde{\phi}(ct,\vec{x}) << mc^2\tilde{\phi}(ct,\vec{x}) \qquad (7.XVIII)$$

Now, (7.XVII) yields:

$$\frac{\partial}{\partial t}\phi(ct,\vec{x}) = -\frac{i}{\hbar}mc^2\tilde{\phi}(ct,\vec{x})e^{-\frac{i}{\hbar}mc^2t} + \frac{\partial}{\partial t}\tilde{\phi}(ct,\vec{x})e^{-\frac{i}{\hbar}mc^2t}$$

which, on account of (7.XVIII), becomes:

$$\frac{\partial}{\partial t}\phi(ct,\vec{x}) = -\frac{i}{\hbar}mc^2\tilde{\phi}(ct,\vec{x})e^{-\frac{i}{\hbar}mc^2t} \pm \frac{i}{\hbar}E'\tilde{\phi}(ct,\vec{x})e^{-\frac{i}{\hbar}mc^2t}$$

Accordingly, the second temporal derivative is:

$$\frac{\partial^2}{\partial t^2}\phi(ct,\vec{x}) = -(\frac{mc^2}{\hbar})^2\tilde{\phi}(ct,\vec{x})e^{-\frac{i}{\hbar}mc^2t} \pm \frac{mc^2}{\hbar^2}E'\tilde{\phi}(ct,\vec{x})e^{-\frac{i}{\hbar}mc^2t}$$

$$\pm\frac{mc^2}{\hbar^2}E'\tilde{\phi}(ct,\vec{x})e^{-\frac{i}{\hbar}mc^2t} \pm \frac{1}{\hbar^2}E'^2\tilde{\phi}(ct,\vec{x})e^{-\frac{i}{\hbar}mc^2t}$$

with the non-relativistic limit:

$$\frac{\partial^2}{\partial t^2}\phi(ct,\vec{x}) \sim -(\frac{mc^2}{\hbar})^2\tilde{\phi}(ct,\vec{x})e^{-\frac{i}{\hbar}mc^2t} \pm \frac{2mc^2}{\hbar^2}E'\tilde{\phi}(ct,\vec{x})e^{-\frac{i}{\hbar}mc^2t} \qquad (7.XIX)$$

since $E'^2 << 2E'mc^2$ renders the last term negligible.

Replacing now (7.XIX) in the Klein-Gordon equation (7.XV) stated as (see also (7.XI)):

$$\frac{1}{c^2}\frac{\partial^2}{\partial t^2}\phi(ct,\vec{x}) = \nabla^2\phi(ct,\vec{x}) - (\frac{mc}{\hbar})^2\phi(ct,\vec{x})$$

yields:

$$-\frac{1}{c^2}[(\frac{mc^2}{\hbar})^2\tilde{\phi}(ct,\vec{x}) \pm \frac{2mc^2}{\hbar^2}E'\tilde{\phi}(ct,\vec{x})]e^{-\frac{i}{\hbar}mc^2t} \sim [\nabla^2 - (\frac{mc}{\hbar})^2]\tilde{\phi}(ct,\vec{x})e^{-\frac{i}{\hbar}mc^2t}$$

$$= [\nabla^2 - \frac{1}{c^2}(\frac{mc^2}{\hbar})^2]\tilde{\phi}(ct,\vec{x})e^{-\frac{i}{\hbar}mc^2t}$$

which, upon using (7.XVIII) on the second term of the left side, finally results in:

$$ i\hbar \frac{\partial}{\partial t} \tilde{\phi}(ct, \vec{x}) = -\frac{\hbar^2}{2m} \nabla^2 \tilde{\phi}(ct, \vec{x}) \qquad (7.XX) $$

We recognize (7.XX) as the Schrödinger equation, albeit for a classical scalar wave field. Of course, we expected that the equation which emerges at the non-relativistic limit will be an aspect of the Klein-Gordon equation. The physical content of this calculation is that *the non-relativistic limit of the Klein-Gordon equation is the Schrödinger equation.* In view of the fact that each plane wave in the superposition (1.XI) independently satisfies the Klein-Gordon equation on account of the linear nature of the latter, this directly implies that at the non-relativistic limit, each such momentum state satisfies the Schrödinger equation. Consequently, each momentum state considered in isolation is at that limit a quantum-mechanical plane wave of probability. Yet, in addition to whichever excited momentum states, infinitely many void momentum states also satisfy the Schrödinger equation at that limit. It is precisely these infinitely many non-relativistic degrees of freedom which most emphatically underscore the conclusion which we have already reached: At the non-relativistic limit of very low energies, the relativistic free quantum Klein-Gordon field reduces to a non-relativistic free quantum Klein-Gordon field. Although the latter bears the semblance of quantum mechanics, no quantum-mechanical system emerges at the non-relativistic limit!

Although the semblance of quantum mechanics at the non-relativistic limit of very low energies leaves no margin for misinterpretation, it is inconsistently interpreted by certain authors who take the view that quantum mechanics is, in fact, the low-energy limit of quantum field theory. We have thoroughly demonstrated the physical irrelevance of such a standpoint. Unlike the quantum-mechanical momentum states, the momentum states of quantum fields are degrees of freedom. An excited momentum state is but one degree of freedom in a product of infinitely many. Degrees of freedom do not evaporate! The view which maintains that quantum mechanics is the low-energy limit of quantum field theory dismally fails to explain how a physical system of infinitely many degrees of freedom "at high energies" reduces to a physical system of a finite number of degrees of freedom "at low energies". We reiterate the standpoint which we have established. We really have no right to disregard the infinitely many void momentum states and declare the existence of a quantum-mechanical wavefunction. It is only when at the non-relativistic limit we dismiss the infinitely many void momentum states that the semblance of quantum mechanics gives way to quantum mechanics itself. Such a formal operation lacks physical justification. The essence of the matter is inherent in the conclusions in which our analysis has eventuated:

(i) Relativistic quantum mechanics signifies a contradiction of terms. A relativistically covariant quantum theory is necessarily a quantum field theory.

(ii) Non-relativistic quantum mechanics is conceptually consistent, but physically irrelevant. Not a single system in Nature comprises a finite number of degrees of freedom!

8. The Action Functional and the Path Integral in Field Theory

We have already studied the concept of path integral in the third chapter. There, we remarked that to each trajectory the dynamical behavior of the quantum-mechanical system assigns an action, a number which expresses that path's contribution to the dynamical behavior, with the minimum (classical) action corresponding to the classical trajectory (Fig.9). It is natural to expect, for that matter, that the extension of this abstract mathematical concept to quantum fields is accomplished if the "path" is replaced by the "field configuration". We shall now examine what the action functional in quantum field theory actually is and does! As always, our advance into the relevant mathematical formalism will be sufficiently rigorous.

Formally, in order to construct the action functional for the Klein-Gordon field we would have to repeat that procedure in the first section of the third chapter which took us from (1.V) to (1.VIII) for n degrees of freedom $q_i(t)$ and finally arrive at the continuous limit at which $i = 1, 2, ..., n, ...$ gives way to $\vec{x}$ and the field $\phi(ct, \vec{x})$ emerges as the dynamical variable. Nevertheless, our qualitative treatment allows for the following shortcut. Contrasting the equation of motion (1.V) for an oscillating particle against the classical field equation (7.XV) for the Klein-Gordon field, we note that both equations involve a second derivative, the former with respect to time, the latter with respect to space-time. This allows us to make this erudite conjecture: If the action functional which yields (1.V) is (1.VIII), the action functional which yields (7.XV) must be of the form

$$S[\phi] = \int_{x_1^0}^{x_2^0} dx^0 \int d^3x \frac{1}{2} \left[\left(\frac{\partial\phi(x)}{\partial x^0}\right)^2 - \left(\frac{\partial\phi(x)}{\partial x^1}\right)^2 - \left(\frac{\partial\phi(x)}{\partial x^2}\right)^2 - \left(\frac{\partial\phi(x)}{\partial x^3}\right)^2 - V(\phi(x)) \right]$$

on the understanding that:

(i) As all time dependence in (1.III) in general and in (1.VIII) in particular has been integrated out, all space-time dependence must be integrated out in $S[\phi]$. Note in this respect that since the integral over all space yields the Lagrange function - henceforth abbreviated to merely *the Lagrangian* - the integrated expression (that in the brackets multiplied by $\frac{1}{2}$) is necessarily a *Lagrangian density*.

(ii) The pattern $+, -, -, -$ in the first four of the five terms in the bracket is justified by the fact that $S[\phi]$ is Lorentz invariant (the same for all inertial observers) and, for that matter, the sum-total of these four terms must yield the square of the magnitude of a vector in Minkowski space-time. This being the case, we can refine the mathematical notation on the lines of $ds^2 = \eta_{\mu\nu} dx^\mu dx^\nu$ (refer to the Appendix):

The existence of a contravariant metric tensor $\eta^{\mu\nu}$ such that

$$\eta^{\mu\nu'} \eta_{\nu'\nu} = \delta^\mu_\nu$$

implies that

$$\left(\frac{\partial\phi(x)}{\partial x^0}\right)^2 - \left(\frac{\partial\phi(x)}{\partial x^1}\right)^2 - \left(\frac{\partial\phi(x)}{\partial x^2}\right)^2 - \left(\frac{\partial\phi(x)}{\partial x^3}\right)^2 = \eta^{\mu\nu}\partial_\mu\phi(x)\partial_\nu\phi(x) = \partial^\nu\phi(x)\partial_\nu\phi(x)$$

with the *contravariant derivative* being the vector

$$\eta^{\mu\nu}\partial_\mu\phi(x) = \partial^\nu\phi(x) \equiv \frac{\partial\phi(x)}{\partial x_\nu} = \left(\frac{\partial\phi(x)}{\partial x_0}, \frac{\partial\phi(x)}{\partial x_1}, \frac{\partial\phi(x)}{\partial x_2}, \frac{\partial\phi(x)}{\partial x_3}\right)$$

and the *covariant derivative* being the vector

$$\partial_\nu \phi(x) \equiv \frac{\partial \phi(x)}{\partial x^\nu} = \left(\frac{\partial \phi(x)}{\partial x^0}, -\frac{\partial \phi(x)}{\partial x^1}, -\frac{\partial \phi(x)}{\partial x^2}, -\frac{\partial \phi(x)}{\partial x^3}\right)$$

This leaves us with the task of determining the potential term $V(\phi(x))$. This is somewhat more subtle. The action functional necessarily has dimensions of action! It is impervious to mass-dimensionality and to length-dimensionality. Thus, the four length-dimensions which stem from d^4x must be offset by the length-dimensionality of $V(\phi(x))$ which must, for that matter, be equal to -4. Recall in this respect that - pursuant to the preceding sixth section - $\phi(x)$ is of mass-dimensionality equal to one. Since mass-dimensions are equivalent to inverse length-dimensions[77], $\phi(x)$ is equivalently of length-dimensionality equal to -1. This leaves $V(\phi(x))$ open to several possibilities. One such possibility, for example, is the famous self-interaction term $\phi^4(x)$, since the four length-dimensions introduced by d^4x cancel against the four inverse length-dimensions introduced by $\phi^4(x)$. Other possibilities involve combinations of the mass m (of mass-dimensionality equal to one!), of the space-time derivatives (clearly of inverse length-dimensionality equal to one) and relevant powers of $\phi(x)$. None of them but one qualifies for $V(\phi(x))$. The reason is that the requirement for zero length- and mass-dimensionality of $S[\phi]$ is only a necessary condition that a mathematical expression qualify for a term of $S[\phi]$. A candidate expression must also be consistent with the requirement of Lorentz invariance and with all symmetries inherent in the physical context which $S[\phi]$ aspires to describe. Terms such as $m\phi(x)\partial^\mu\phi(x)$, or $m^2\partial^\mu\phi(x)$, although dimensionally consistent, do not qualify because they are not Lorentz invariant (notice the "free" index μ: a Lorentz transformation would take the initial vector $\frac{\partial \phi(x)}{\partial x_\mu}$ to the distinct vector $\frac{\partial \phi(x')}{\partial x'_\mu}$). The stated $\phi^4(x)$ qualifies in all respects but one: It signifies a non-trivial self-coupling, a scalar field in interaction with itself. As such, it is incompatible with the action functional of the free Klein-Gordon field.

Up to the term $V(\phi(x))$, we have constructed the action functional $S[\phi]$ for the Klein-Gordon field by analogy to the action functional $S[x]$ for the simple harmonic oscillator (1.VIII). We could possibly extend this analogy to $V(\phi(x))$. Since a free field is a collection of abstract simple harmonic oscillators, it is reasonable to expect that $V(\phi(x))$ must likewise be quadratic in $\phi(x)$, accordingly expressing a parabolic potential, a potential which yields the same graph for $S[\phi]$ versus ϕ as that for $S[x]$ versus x in Fig.9. For that matter, we extend our initial conjecture to ($\hbar = c = 1$):

$$V(\phi(x)) = m^2\phi^2(x)$$

since it is this term which satisfies all requirements imposed upon the action functional for a free scalar field. In this expression, the parameter m (of mass-dimensionality equal to one, but otherwise arbitrary) is interpreted as the mass of the Klein-Gordon field. This $V(\phi(x))$ still signifies a self-interaction for the Klein-Gordon field. This self-coupling, however, is trivial. It is the self-coupling which generates the mass of the free Klein-Gordon field.

---

[77]Refer to the footnote in the sixth section.

Putting all pieces together now, we state that the *action functional* for the free Klein-Gordon field is

$$S[\phi] = \int d^4x \frac{1}{2}\left[\partial^\mu\phi(x)\partial_\mu\phi(x) - m^2\phi^2(x)\right] \qquad (8.I)$$

where the integration extends over the entire Minkowski space-time $(d^4x = dx^0 dx^1 dx^2 dx^3)$ since it is over the entire Minkowski space-time that $\phi(x)$ is defined.

For all its simplicity and plausibility, the statement in (8.I) remains an erudite conjecture as it has yet to be subjected to the ultimate test of an action functional's veracity: the classical equations of motion! In order to realize this objective, we first derive the general Euler-Lagrange field equations[78].

The general action functional for a scalar, spinor, or vector field $\phi(x)$ is

$$S[\phi] = \int d^4x \mathcal{L}(\phi, \partial_\mu\phi)$$

In fact, $\phi(x)$ in this general $S[\phi]$ may even represent a collection of such fields. The function

$$\mathcal{L}(\phi, \partial_\mu\phi)$$

is, as stated, the Lagrangian density of that physical system.

A one-parameter family of field configurations

$$\phi(x, \epsilon) = \phi_0(x) + \epsilon\eta(x) \implies \partial_\mu\phi(x, \epsilon) = \partial_\mu\phi_0(x) + \epsilon\partial_\mu\eta(x)$$

defines the variation

$$\delta\phi(x) \equiv \epsilon\eta(x) \implies \delta(\partial_\mu\phi(x)) \equiv \epsilon\partial_\mu\eta(x) = \partial_\mu(\epsilon\eta(x)) = \partial_\mu\delta\phi(x)$$

of $\phi(x)$ about the classical field configuration $\phi_0(x)$. Formally, the condition of a vanishing such variation:

$$\eta(x_1) = 0 \iff \delta\phi(x_1) = 0 \;;\; \eta(x_2) = 0 \iff \delta\phi(x_2) = 0$$

may be imposed at any two space-time points x_1 and x_2. For the purpose of illustration, we directly consider that this variation vanishes over infinite space-time separations.

The calculation which eventuates in the Euler-Lagrange field equations, advances on the same lines as those which eventuated in (E - L) in the third chapter: Since

$$d[\phi_0(x) + \epsilon\eta(x)]_{|\epsilon=0} = \frac{d\phi(x, \epsilon)}{d\epsilon}\bigg|_{\epsilon=0}\epsilon = [\phi_0(x) + \epsilon\eta(x)] - \phi_0(x) = \epsilon\eta(x) = \delta\phi(x) \implies$$

$$d[\partial_\mu\phi_0(x) + \epsilon\partial_\mu\eta(x)]_{|\epsilon=0} = \frac{d[\partial_\mu\phi(x, \epsilon)]}{d\epsilon}\bigg|_{\epsilon=0}\epsilon =$$

$$[\partial_\mu\phi_0(x) + \epsilon\partial_\mu\eta(x)] - \partial_\mu\phi_0(x) = \epsilon\partial_\mu\eta(x) = \delta(\partial_\mu\phi(x)) = \partial_\mu\delta\phi(x)$$

---

[78]Again, although the importance of this derivation cannot be overstated, readers who find it difficult to follow may disregard it at first reading.

Hamilton's principle

$$\frac{dS}{d\epsilon}\Big|_{\epsilon=0} = 0 \iff \frac{\delta S}{\delta \phi}\Big|_{\phi=\phi_0} = 0 \iff \delta S = 0$$

implies

$$\int d^4x \Big[\frac{\partial \mathcal{L}}{\partial \phi}\delta\phi + \frac{\partial \mathcal{L}}{\partial[\partial_\mu\phi]}\delta(\partial_\mu\phi)\Big]_{\phi=\phi_0} = 0 \iff \int d^4x \Big[\frac{\partial \mathcal{L}}{\partial \phi}\delta\phi + \frac{\partial \mathcal{L}}{\partial[\partial_\mu\phi]}\partial_\mu(\delta\phi)\Big]_{\phi=\phi_0} = 0 \iff$$

$$\int d^4x \Big[\frac{\partial \mathcal{L}}{\partial \phi}\delta\phi + \partial_\mu(\frac{\partial \mathcal{L}}{\partial[\partial_\mu\phi]}\delta\phi) - \partial_\mu\frac{\partial \mathcal{L}}{\partial[\partial_\mu\phi]}\delta\phi\Big]_{\phi=\phi_0} = 0 \iff$$

$$\int d^4x \Big[\frac{\partial \mathcal{L}}{\partial \phi} - \partial_\mu\frac{\partial \mathcal{L}}{\partial[\partial_\mu\phi]}\Big]_{\phi=\phi_0}\delta\phi + \oint_\sigma d\sigma_\mu \frac{\partial \mathcal{L}}{\partial[\partial_\mu\phi]}\Big|_{\phi=\phi_0}\delta\phi = 0$$

In advancing from the second to the third line we have used the *divergence theorem*:

$$\int d^4x \partial_\mu(\frac{\partial \mathcal{L}}{\partial[\partial_\mu\phi]}\delta\phi) = \oint_\sigma d\sigma_\mu \frac{\partial \mathcal{L}}{\partial[\partial_\mu\phi]}\delta\phi$$

which we shall soon prove. In this statement, integration on the right side extends over a three-dimensional spherical surface σ of infinite radius with $d\sigma^\mu$ being the vector which represents the outward-oriented surface element.

Note now that the *surface integral* over σ vanishes as a direct consequence of the condition

$$\delta\phi = 0$$

over infinite space-time separations. The direct implication is the Euler-Lagrange equations

$$\frac{\partial \mathcal{L}}{\partial \phi_0} - \partial_\mu\frac{\partial \mathcal{L}}{\partial[\partial_\mu\phi_0]} = 0 \qquad\qquad (E-L)$$

for the physical system described by $S[\phi]$.

For the particular Lagrangian density

$$\mathcal{L} = \frac{1}{2}[\partial^\mu\phi(x)\partial_\mu\phi(x) - m^2\phi^2(x)]$$

in (8.I) now, it is:

$$\frac{\partial \mathcal{L}}{\partial \phi_0} = \frac{\partial}{\partial \phi}\Big|_{\phi=\phi_0}[-\frac{1}{2}m^2\phi^2(x)] = -m^2\phi_0(x)$$

and

$$\partial_\mu\frac{\partial \mathcal{L}}{\partial[\partial_\mu\phi_0]} = \partial_\mu\frac{\partial}{\partial[\partial_\mu\phi]}\Big|_{\phi=\phi_0}[\frac{1}{2}\partial^\mu\phi(x)\partial_\mu\phi(x)] = \partial_\mu\frac{\partial}{\partial[\partial_\mu\phi]}\Big|_{\phi=\phi_0}[\frac{1}{2}\eta^{\mu'\nu'}\partial_{\mu'}\phi(x)\partial_{\nu'}\phi(x)] =$$

$$\partial_\mu[\frac{1}{2}\eta^{\mu'\nu'}\delta_{\mu'\mu}\partial_{\nu'}\phi_0(x) + \frac{1}{2}\eta^{\mu'\nu'}\partial_{\mu'}\phi_0(x)\delta_{\nu'\mu}] = \partial_\mu[\frac{1}{2}\eta^{\mu\nu'}\partial_{\nu'}\phi_0(x) + \frac{1}{2}\eta^{\mu'\mu}\partial_{\mu'}\phi_0(x)] =$$

$$\partial_\mu[\frac{1}{2}\partial^\mu\phi_0(x) + \frac{1}{2}\partial^\mu\phi_0(x)] = \partial_\mu\partial^\mu\phi_0(x) \equiv \Box\phi_0(x)$$

so that (E - L) implies

$$-m^2\phi_0(x) - \Box\phi_0(x) = 0 \iff [\Box + m^2]\phi_0(x) = 0$$

that is, the classical field equation (7.XV) for the Klein-Gordon field the solution to which is the field configuration $\phi_0(x)$, that field configuration to which the minimum action $S_0[\phi_0]$ corresponds. This confirms that $S[\phi]$ in (8.I) is, indeed, the action functional of the free Klein-Gordon field.

We now digress to the proof of the divergence theorem. The procedure which follows is understandably somewhat involved and necessitates some caution. For that matter, subsequent to this formal derivation, I also advance a heuristic argument for the validity of the divergence theorem in the procedure leading to (8.III). Readers who find it difficult to follow the ensuing formal procedure, may disregard the latter at first reading and reflect, instead, on that heuristic argument.

In Minkowski space-time, we consider a Cartesian coordinate system and a closed three-dimensional surface σ such that a line parallel to the coordinate axes intercepts that surface in at most two points. Specifically, if the line is parallel to the temporal x^0-axis, one of the two points lies in the lower portion S_1 of σ and the other in the upper portion S_2 (Fig.32). The equation of S_1 is of the form $x^0 = f_1(x^1, x^2, x^3)$ and the equation of S_2 is of the form $x^0 = f_2(x^1, x^2, x^3)$. We further consider a contravariant vector field (i.e. a vector function) $A^\mu(x) \equiv (A^0(x), A^1(x), A^2(x), A^3(x))$ (of which $\frac{\partial \mathcal{L}}{\partial[\partial_\mu\phi]}$ is an example). *The divergence* of this vector field is:

$$\partial_\mu A^\mu(x) = \partial_0 A^0(x) - \partial_1 A^1(x) - \partial_2 A^2(x) - \partial_3 A^3(x) \equiv \frac{\partial A^0(x)}{\partial x^0} - \frac{\partial A^1(x)}{\partial x^1} - \frac{\partial A^2(x)}{\partial x^2} - \frac{\partial A^3(x)}{\partial x^3}$$

We wish to evaluate the integral of this divergence over the volume V bounded by the closed three-dimensional surface σ. Commencing with $\partial_0 A^0(x)$, we have

$$\int_V d^4x \frac{\partial A^0}{\partial x^0} = \int_{\mathcal{R}} d^3x \int_{f_1(x,y,z)}^{f_2(x,y,z)} dx^0 \frac{\partial A^0}{\partial x^0} =$$

$$\int dx^1 dx^2 dx^3 [A^0(f_2, x^1, x^2, x^3) - A^0(f_1, x^1, x^2, x^3)] =$$

$$\int_{S_2} dx^1 dx^2 dx^3 A^0(f_2, x^1, x^2, x^3) - \int_{S_1} dx^1 dx^2 dx^3 A^0(f_1, x^1, x^2, x^3) \qquad (DT1)$$

where (i) the projection of σ on the three-dimensional space (the set of points (x^1, x^2, x^3) at a certain moment) is $\mathcal{R}$ and (ii) in advancing from the first to the second equality, use has been made of the Fundamental Theorem of Calculus stated in the first section of the third chapter.

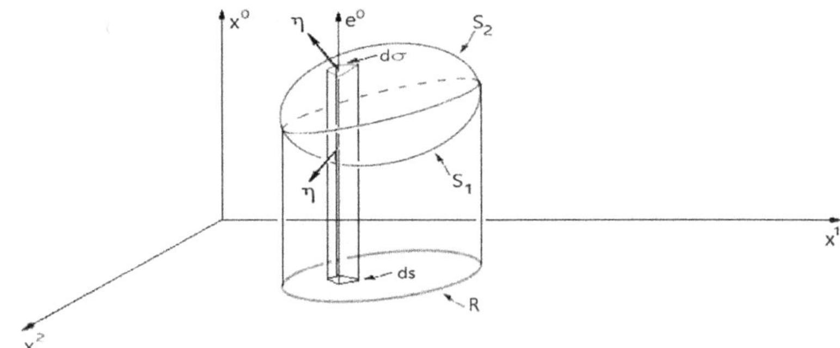

FIGURE 32. The Divergence Theorem in Minkowski space-time (one dimension has been supressed). The angle between the unit vector η^μ and the future-oriented unit vector e^0 has a positive cosine over segment S_2, but a negative cosine over segment S_1.

Simple reflection on the geometry of the situation in question now, reveals that if η^μ is the outward-oriented unit vector at each point of the closed three-dimensional surface and e^0 is the future-oriented unit vector then the relation between the surface element $dse^0 \equiv (dx^1 dx^2 dx^3)e^0$ of region $\mathcal{R}$ and the surface element $d\sigma\eta^\mu$ of the closed three-dimensional surface is

$$ds \equiv dx^1 dx^2 dx^3 = d\sigma\eta^\mu.e^0 \equiv d\sigma\eta^0 \tag{DT2}$$

It should be clear that since on S_2 the angle θ between η^μ and e^0 is smaller than ninety degrees, ds is positive on $\mathcal{R}$ $(cos\theta > 0)$. Likewise, since on S_1 the angle θ between η^μ and e^0 is bigger than ninety degrees, ds is negative on $\mathcal{R}$ $(cos\theta < 0)$. For that matter, (DT2) implies:

$$\int_{S_2} dx^1 dx^2 dx^3 A^0(f_2, x^1, x^2, x^3) = \int_{S_2} d\sigma\eta^0 A^0(f_2, x^1, x^2, x^3) \tag{DT3a}$$

and

$$\int_{S_1} dx^1 dx^2 dx^3 A^0(f_1, x^1, x^2, x^3) = -\int_{S_1} d\sigma\eta^0 A^0(f_1, x^1, x^2, x^3) \tag{DT3b}$$

on the understanding that e^0 on the right side of (DT3b) is now the past-oriented unit vector e^0 (so that $cos\theta > 0$).

Replacing (DT3) in (DT1) yields

$$\int_V d^4x \frac{\partial A^0}{\partial x^0} = \int_{S_2} d\sigma \eta^0 A^0(f_2, x^1, x^2, x^3) + \int_{S_1} d\sigma \eta^0 A^0(f_1, x^1, x^2, x^3) =$$

$$\int_\sigma d\sigma \eta^0 A^0(x^1, x^2, x^3) \qquad (DT4a)$$

By likewise projecting the three-dimensional closed σ on the remaining three-dimensional coordinate planes, we obtain

$$\int_V d^4x \frac{\partial A^1}{\partial x^1} = \int_\sigma d\sigma \eta^1 A^1(x^0, x^2, x^3) \qquad (DT4b)$$

$$\int_V d^4x \frac{\partial A^2}{\partial x^2} = \int_\sigma d\sigma \eta^2 A^2(x^0, x^1, x^3) \qquad (DT4c)$$

$$\int_V d^4x \frac{\partial A^3}{\partial x^3} = \int_\sigma d\sigma \eta^3 A^3(x^0, x^1, x^2) \qquad (DT4d)$$

Adding (DT4b), (DT4c), (DT4d) and subtracting the result from (DT4a) results in

$$\int_V d^4x [\partial_0 A^0(x) - \partial_1 A^1(x) - \partial_2 A^2(x) - \partial_3 A^3(x)] =$$

$$\int_\sigma d\sigma [\eta^0 A^0(x) - \eta^1 A^1(x) - \eta^2 A^2(x) - \eta^3 A^3(x)] \iff$$

$$\int_V d^4x \partial_\nu A^\nu = \int_\sigma d\sigma \eta_{\nu'} A^{\nu'}$$

That is

$$\int_V d^4x \partial\nu A^\nu = \int_\sigma d\sigma_{\nu'} A^{\nu'} \qquad (DT5)$$

We note now that this result has rendered the initial inertial Cartesian frame irrelevant as both sides of (DT5) are manifestly Lorentz invariant (note that neither side features free indices). This concludes our digression to the proof of the divergence theorem in Minkowski space-time.

In addition to realizing the objective of constructing the action functional for the Klein-Gordon field, the procedure which eventuated in (8.I) avails this basic lesson: The action functional for a field is constructed in conformity with the demand for zero mass- and length-dimensionality, in conformity with the demand for Lorentz invariance and in conformity with whichever other symmetries are inherent in the physical context. The ultimate test is invariably the classical field equations, although in practice a dimensionally consistent Lorentz-invariant action functional in conformity with all symmetries present is known in advance to yield the expected classical field equations at its stationary points. In this respect, *the kinetic term of $S[\phi]$ (the integral over $\partial^\mu \phi(x) \partial_\mu \phi(x)$) and* *the mass term of $S[\phi]$* (the integral over $m^2\phi^2(x)$) could have been constructed not by analogy to the action functional of the simple harmonic oscillator, but purely on dimensional arguments. You can confirm this yourselves merely by inspection and proceed to answer the simple question as to the reason that terms such as $m\Box\phi \equiv m\partial^\mu\partial_\mu\phi \equiv m\partial^2\phi$ are, despite their dimensional consistency, unacceptable in $S[\phi]$.

We may now take a closer look at the physical content of the action functional. The implication inherent in the preceding analysis is that the premise in theoretical mechanics to the effect that to each trajectory in configuration space corresponds an action, extends directly to quantum field theory: To each field configuration $\phi(x)$ corresponds the action $S[\phi]$, that number which expresses the contribution of $\phi(x)$ to the dynamical behavior of the free Klein-Gordon field. In turn, the action functional $S[\phi]$ expresses entirely the dynamical behavior of the free Klein-Gordon field. Up to this point we have studied the dynamical behavior of free quantum fields from the standpoint of canonical quantization - in particular, as a consequence of the quantization of the classical field equations. Since the latter emerge from the action functional at the action functional's stationary points, it follows that the action functional expresses the dynamical behavior of quantum fields at a level which precedes the interpretation of that dynamical behavior in terms of particles - at primary level, as it were. We reiterate, in this respect, Julian Schwinger's profound premise which has provided the perspective of our analysis on quantum fields from the outset: *When you begin with field equations, you operate on a level where the particles are not there from the start. It is when you solve the field equations that you see the emergence of particles.* In fact, we do not even begin with field equations. We begin, instead, with the action functional from which the field equations are derived. It is in dealing with the action functional $S[\phi]$ that we operate on a level where no interpretation in terms of particles is possible (because momentum states are not defined) and the field configurations $\phi(x)$ - the field itself as a fundamental and primary physical entity - are the exclusive reality. We reiterate that this profound physical essence of the action functional, the action functional as the fundamental expression of a quantum system, was not known before the development of quantum theory. When it was introduced in 1827, it was considered merely an elegant way of arriving at already known physics, of rederiving Newton's equations of motion without the concept of force.

The basic reason we introduced the action functional is the latter's central importance to the procedure of renormalization. However, its significance to quantum field theory extends far beyond renormalization. As the language of field configurations is equivalent to the language of particles, as both languages independently express the dynamical behavior of quantum fields, the action functional necessarily gives rise to a quantization procedure equivalent to that of canonical quantization. It is the procedure of path-integral quantization, of course! This comes as no surprise. As we discussed in the third section of the fourth chapter, the set of infinitely many possible evolution patterns for a given initial field configuration $\phi_1(x_1^0 = ct_1, \vec{x})$ which is obtained through the demand that these evolution patterns also eventuate in a given final field configuration $\phi_2(x_2^0 = ct_2, \vec{x})$, parallels the set of infinitely many possible paths for a quantum particle ((2.XII), third chapter) which intercept two given space-time points: As the unique classical trajectory is replaced by infinitely many possible paths for the quantum-mechanical particle, so the unique evolving classical field configuration is replaced by infinitely many possible evolving field configurations of the quantum field. Do note, nevertheless, that this analogy is deceptively simple! Physically, the procedure of path-integral quantization for fields is far from mere extension of the procedure of path-integral quantization for particles. The infinitely many degrees of freedom render the transition amplitude

$$\langle \phi_2, x_2^0 | \phi_1, x_1^0 \rangle = \int \mathcal{D}[\phi] e^{\frac{i}{\hbar} S[\phi]} \tag{8.II}$$

inherently distinct from - and infinitely more complicated than - the quantum-mechanical transition amplitude $\langle x_2^0, \vec{x}_2 | x_1^0, \vec{x}_1 \rangle$. In order to appreciate the difficulties associated with the rigorous definition of the path integral in (8.II), we turn to (2.X) in the third chapter. Already in non-relativistic quantum mechanics, the rigorous definition of the measure $\int \mathcal{D}[x]$ is a formidable challenge. In quantum field theory, the path integral over particle-trajectories does not even exist for the transition amplitude $\langle x_2^0, \vec{x}_2 | x_1^0, \vec{x}_1 \rangle$. We reiterate (refer to the last section of the preceding chapter) that as the position is only a label, the path in the concrete sense of trajectory is altogether meaningless. It is no wonder then that in abandoning the language of particles in favor of the equivalent language of field configurations, we abandon all hope for a rigorous definition of the measure $\mathcal{D}[\phi]$ in (8.II)!

Yet, notwithstanding such insurmountable difficulties, the use of the path integral in practical calculations reduces to a few relatively simple rules in quantum mechanics and to certain - not so simple, but tractable - rules in quantum field theory. In what follows, we shall examine the concept which underlies these rules in quantum field theory using as guidelines the corresponding concept and techniques which we developed in the third chapter. The approach which I shall present is the one which I consider the most succinct and conceptually accessible. The more technical and rigorous approach can be found in [28].

We begin with the simple mathematical statement:

$$\partial^\mu (\phi(x)\partial_\mu \phi(x)) = \partial^\mu \phi(x)\partial_\mu \phi(x) + \phi(x)\partial^\mu \partial_\mu \phi(x) \equiv \partial^\mu \phi(x)\partial_\mu \phi(x) + \phi(x)\partial^2 \phi(x)$$

where we have denoted the d' Alembert operator $\Box$, also introduced in the preceding section, by ∂^2 ($\Box \equiv \partial^2$) in order to stress its character as the square of the magnitude of a vector in Minkowski space-time. The implication for (8.I) is

$$S[\phi] = \frac{1}{2} \int d^4x [\partial^\mu (\phi(x)\partial_\mu \phi(x)) - \phi(x)\partial^2 \phi(x) - m^2 \phi^2(x)]$$

Note now that the first term in the brackets is a *total derivative* in the sense that the derivative operator ∂^μ acts on function $\phi(x)\partial_\mu \phi(x)$ through all four of the latter's variables x^μ ; $\mu = 0, 1, 2, 3$. Thus, the integral effectively "offsets" it! This heuristic statement conveys the essence of the divergence theorem in advanced differential calculus which we formally proved above. We can grasp that essence if we consider that in the one-dimensional case it simply is

$$\int_a^b dx\, f(x) = F(b) - F(a)$$

with function $F(x)$ being such that

$$\frac{dF(x)}{dx} = f(x)$$

From the Fundamental Theorem of Calculus, for that matter, it directly follows that

$$\int_a^b dx\, \frac{df(x)}{dx} = f(b) - f(a)$$

- a statement already made in the first section of the third chapter. If we now consider that (i) the operation of $\partial^\mu \equiv \frac{\partial}{\partial x_\mu}$ on function

$$\phi(x)\partial_\mu\phi(x) \equiv \phi(x)\frac{\partial}{\partial x^\mu}\phi(x)$$

is essentially a four-dimensional extension of the simple derivative $\frac{d}{dx}f(x)$ and (ii) a three-dimensional sphere in Minkowski space-time is a boundary to the enclosed four-dimensional region as points a and b are boundaries to the one-dimensional segment between them then it becomes obvious[79] that the extension of the above statement in Minkowski space-time is:

$$\int_V d^4x\, \partial^\mu(\phi(x)\partial_\mu\phi(x)) = \oint_{sph} d\sigma n^\mu \phi(x)\partial_\mu\phi(x) \equiv \oint_{sph} d\sigma^\mu \phi(x)\partial_\mu\phi(x)$$

where the integral on the left side is considered over a four-dimensional volume V bounded by a three-dimensional sphere and on the right side we have projected the covariant vector $\phi(x)\partial_\mu\phi(x)$ onto the contravariant unit vector n^μ normal (perpendicular) to the three-dimensional spherical boundary in order to render the right side the Lorentz invariant (scalar) quantity which the left side is. Minkowski space-time in its entirety may now be recovered by letting the radius of this sphere become infinitely large. At this infinite limit, we impose the physical demand that the rate of change $\partial_\mu\phi(x)$ along each dimension vanish. The result is

$$S[\phi] = \frac{1}{2}\int d^4x[-\phi(x)\partial^2\phi(x)-m^2\phi^2(x)] = -\frac{1}{2}\int d^4x[\phi(x)(\partial^2+m^2)\phi(x)] \qquad (8.III)$$

Let us now examine the transition amplitude for the Klein-Gordon field. Replacing (8.I) in (8.II) yields

$$\langle \phi_2, x_2^0 | \phi_1, x_1^0 \rangle = \int \mathcal{D}[\phi]e^{\frac{i}{\hbar}\frac{1}{2}\int_{x_1^0}^{x_2^0} dx^0 \int d^3x[\partial^\mu\phi(x)\partial_\mu\phi(x)-m^2\phi^2(x)]} \qquad (8.IV)$$

since now the relevant dynamical evolution is that between the two specified field configurations at moments x_1^0 and x_2^0. Recall that we have specified (most notably in the fifth section of the fourth chapter), the precise relation between the language of field

---

[79]I reiterate that this heuristic argument is by no means a proof of the divergence theorem.

configurations and the equivalent language of particles (energy-momentum eigenstates). At a given moment x^0, each point $\vec{x}$ labels infinitely many amplitudes in the infinitely many plane waves of probability which compose the energy-momentum eigenstate of the quantum field. The infinitely many evolving field configurations emerge through the interpretation of these amplitudes as field-values labeled by $\vec{x}$. The choice of an amplitude at each $\vec{x}$ at moment x^0 yields a field configuration throughout space. The totality of the infinitely many field configurations at x^0 is the quantum field $\phi(x^0, \vec{x})$ in the specific energy-momentum eigenstate at that moment. The representation of each field configuration as an abstract ket $|\phi, x^0\rangle$ reflects the fact that $\vec{x}$ labels, at once with all possible field-values, the corresponding probability distribution for these values. Inherent in these considerations is the implication for the transition between states of the quantum field rather than between field configurations. This situation is an extension of our discussion in the second section of the third chapter, to quantum fields. The two kets $|\phi_1, x_1^0\rangle$ and $|\phi_2, x_2^0\rangle$ respectively involve all infinitely many field configurations at the two moments. The specification of two configurations respectively, yields the transition amplitude in (8.IV). In particular, if the free quantum field is in its vacuum state then (8.IV) implies a simple expression for the *vacuum-to-vacuum transition amplitude* $\langle 0|0 \rangle$ - the amplitude that the vacuum state make the transition from past infinity to future infinity - as a path integral. We already know, of course, that $\langle 0|0 \rangle = 1$. In direct analogy to (3.VI) (third chapter) for the quantum-mechanical amplitude $\langle \Omega_{+\infty}|\Omega_{-\infty}\rangle$, we define:

$$\langle 0|0 \rangle = N \int \mathcal{D}[\phi] e^{\frac{i}{\hbar}\frac{1}{2} \int d^4x [\partial^\mu \phi(x)\partial_\mu \phi(x) - m^2\phi^2(x)]} = 1 \qquad (8.Va)$$

on the understanding, of course, that the reason at past and future infinity $\langle 0|0 \rangle$ in (8.Va) and $\langle \phi_2, x_2^0 \to +\infty | \phi_1, x_1^0 \to -\infty \rangle$ in (8.IV) receive the same expression as a path integral is the same as the reason $\langle \Omega_{+\infty}|\Omega_{-\infty}\rangle$ in (3.VI) (third chapter) and $\langle x'_{+\infty}|x_{-\infty}\rangle$ in (3.IV) also do.

Relation (8.Va) expresses the obvious invariance of the free vacuum state ($\langle 0|0 \rangle = 1$) as a path integral! Nevertheless, this "trivial" path integral is the incipient point of a procedure which yields the entire dynamical behavior of quantum fields - in fact, not only the dynamical behavior of the free quantum fields, also the enormously more complicated dynamical behavior of the interacting quantum fields. For we reserve the right to act upon a free quantum field! The immediate consequence of such action is the refutation of the stated invariance: If we act upon the free quantum field, it will no longer be certain that the latter will remain in the energy-momentum eigenstate specified at the outset. In particular, if we drive the vacuum state from past infinity to future infinity in any manner of our choice, all excited energy-momentum eigenstates - the particle states by dint of interpretation - will eventually emerge. Thus, in the presence of a classical source $J(x)$ (a classical field signifying whichever manner in which we act upon the vacuum state), the vacuum-to-vacuum transition amplitude is $\langle 0|0 \rangle_J \neq 1$. In anticipation of this calculation, for that matter, we use (8.III) in order to recast (8.Va) in the form

$$\langle 0|0 \rangle = N \int \mathcal{D}[\phi] e^{-\frac{i}{\hbar}\frac{1}{2} \int d^4x [\phi(x)(\partial^2 + m^2)\phi(x)]} = 1 \qquad (8.Vb)$$

and merely note that the presence of the "bilinear" expression $\phi(x)(\partial^2 + m^2)\phi(x)$ in the action functional points to the propagator $\langle 0|\phi(x_2)\phi(x_1)|0\rangle$, suggesting a non-trivial outcome in the presence of $J(x)$.

We now extend the definition (8.Va) to the definition of the *vacuum-to-vacuum transition amplitude in the presence of a driving source* $\langle 0|0\rangle_J$ by adding in the action functional the term $J(x)\phi(x)$ which signifies the operation of $J(x)$ on the free Klein-Gordon field and impose the physical condition that $J(x)$ vanish at past infinity and future infinity where no operation on the vacuum state is physically possible. As we are at liberty to operate on the Klein-Gordon field at will, $J(x)$ is interpreted as a classical field of mass-dimensionality equal to three (so that $J(x)\phi(x)$ is of mass-dimensionality equal to four), but arbitrary in all other respects:

$$\langle 0|0\rangle_J = N \int \mathcal{D}[\phi] e^{-\frac{i}{\hbar}\int d^4x[\frac{1}{2}\phi(x)(\partial^2+m^2)\phi(x)-J(x)\phi(x)]} \tag{8.VI}$$

Although not difficult, the evaluation of $\langle 0|0\rangle_J$ requires additional information on the mathematical foundations of quantum field theory which may be distracting to our conceptual endeavor. We opt instead for an appoach of minimal mathematical content which, nevertheless, emphasizes the underlying physical concepts involved (again, the rigorous evaluation of $\langle 0|0\rangle_J$ can be found in [28]).

As it stands, the expression (8.VI) for $\langle 0|0\rangle_J$ is not transparent as to the effect of driving source $J(x)$ on the quantum field $\phi(x)$. We are seeking, instead, an expression which dissociates the dependence of $\phi(x)$ on $J(x)$ from $\langle 0|0\rangle_{J(x)=0} \equiv \langle 0|0\rangle$. Let us first consider what we require of such an expression as a consequence of the preceding analysis: (i) For $J(x) = 0$, it must reduce to $\langle 0|0\rangle$. (ii) Since the source $J(x)$ excites the vacuum state at space-time point x, $\langle 0|0\rangle_J$ must involve the Feynman propagator $\Delta_F(x-x')$. (iii) $\Delta_F(x-x')$ must be determined by operator (∂^2+m^2), since it is this operator which essentially determines $\langle 0|0\rangle_J$. From the outset, we claim that the only expression which satisfies these three requirements is:

$$\langle 0|0\rangle_J = e^{-\frac{i}{\hbar}\frac{1}{2}\int d^4x\int d^4x' J(x)(\partial^2+m^2)^{-1}_{x,x'}J(x')}\langle 0|0\rangle = e^{-\frac{i}{\hbar}\frac{1}{2}\int\int d^4x d^4x' J(x)(\partial^2+m^2)^{-1}_{x,x'}J(x')} \tag{8.VII}$$

In order to establish this claim, we need to elucidate the peculiar $(\partial^2 + m^2)^{-1}_{x,x'}$, an operator which deserves attention in its own merit indeed!

In general, each operator $\hat{L}$ has an inverse operator $\hat{L}^{-1}$ such that

$$\hat{L}\hat{L}^{-1} = \hat{L}^{-1}\hat{L} = \hat{I} \tag{8.VIII}$$

with $\hat{I}$ being the *unit operator*. This elementary identity makes an obvious statement to the effect that the successive action of $\hat{L}$ and $\hat{L}^{-1}$ on a function yields the function itself (the action of the unit operator $\hat{I}$ on a function yields the number one multiplied by the function itself: $\hat{I}f(x) = 1.f(x) = f(x)$). Operator $(\partial^2 + m^2)^{-1}$, for that matter, is the inverse of operator $(\partial^2 + m^2)$:

$$(\partial^2+m^2)(\partial^2+m^2)^{-1} = (\partial^2+m^2)^{-1}(\partial^2+m^2) = \hat{I} \tag{8.IX}$$

The operator $(\partial^2 + m^2)^{-1}_{x,x'}$ which appears in (8.VII) is essentially a representation of inverse operator $(\partial^2 + m^2)^{-1}$ as a function. Again, that function must be $\Delta_F(x - x')$ if (8.VII) is to be the correct expression for $\langle 0|0\rangle_J$. In order to prove that this is indeed the case, we consider the differential equation

$$\hat{L}_x f(x) = g(x) \tag{8.X}$$

with $\hat{L}_x$ being an operator which differentiates with respect to single variable x (not to be confused with space-time variable x) and $g(x)$ being a given single-variable function. Clearly, the solution to this differential equation is

$$f(x) = \hat{L}_x^{-1} g(x) \tag{8.XI}$$

(since $\hat{L}_x f(x) = \hat{L}_x \hat{L}_x^{-1} g(x) = 1.g(x) = g(x)$). The concrete content of this abstract statement is:

$$f(x) = \int dx' g(x') G(x, x') \tag{8.XII}$$

with function $G(x, x')$ satisfying the *Green equation*:

$$L_x G(x, x') = \delta(x - x') \tag{8.XIII}$$

(since $L_x f(x) = \int dx' g(x') L_x G(x, x') = \int dx' g(x') \delta(x - x') = g(x)$). The two-variable function $G(x, x')$ which clearly represents the inverse operator $\hat{L}_x^{-1}$ is the *Green function to operator* $\hat{L}_x$ and is of fundamental importance to the theory of differential equations. As (8.XII) directly reveals, knowledge of the Green function to operator $\hat{L}_x$ is equivalent to knowledge of the solution $f(x)$ itself.

The implication of these elementary considerations is that if $\hat{L}_x = (\partial_x^2 + m^2)$ (∂_x differentiates with respect to space-time variable x) then the concrete content of (8.IX) is (8.XIII), the discrepancy between the δ-function in the latter and the unit operator in the former merely accounting for the fact that operator $\hat{L}_x^{-1} = (\partial_x^2 + m^2)^{-1}$ is now represented by Green function $G(x, x')$. That is:

$$(\partial^2 + m^2)^{-1}_{x,x'} = G(x, x') \tag{8.XIV}$$

with

$$(\partial_x^2 + m^2) G(x, x') = -\delta(x - x') \tag{8.XV}$$

where we have been careful to take into consideration the minus sign which appeared in the action functional $S[\phi]$ in (8.III) and has been with us all along! The operator which determines $\langle 0|0\rangle_J$ for the Klein-Gordon field is really $\hat{L}_x = -(\partial_x^2 + m^2)$ and accounts for the minus sign on the right side of (8.XV). Note in this respect the central importance of this operator to the Klein-Gordon field: It is also the differential operator of the Klein-Gordon equation (7.XV) ($c = \hbar = 1$).

In order to determine the Green function $G(x, x')$ in (8.XV) now, we introduce the *integral representation of the δ-function*:

$$\delta(x-x') = \int \frac{d^4p}{(2\pi)^4} \, e^{-ip.(x-x')} \qquad (8.XVI)$$

in which each of the four components of p^μ extends from $-\infty$ to $+\infty$. Since the right side of (8.XV) is an integral, we seek an integral representation of $G(x, x')$ on the left side. For that matter, we state (8.XV) as:

$$(\partial_x^2 + m^2) \int \frac{d^4p}{(2\pi)^4} \, \tilde{G}(p)e^{-ip.(x-x')} = -\int \frac{d^4p}{(2\pi)^4} \, e^{-ip.(x-x')} \qquad (8.XVII)$$

In this relation, we are at liberty to swap the order between the differential operator (which differentiates with respect to x) and the integral (which continuously sums over p):

$$\int d^4p\, \tilde{G}(p)(\partial_x^2 + m^2)e^{-ip.(x-x')} = -\int d^4p \, e^{-ip.(x-x')} \qquad (8.XVIII)$$

The remaining task is to evaluate $(\partial_x^2 + m^2)e^{-ip.(x-x')}$. To this end, we need the elementary rule from simple calculus to the effect that $\frac{d}{dx}e^{ax} = ae^{ax}$. In view of this and the fact that $p.(x - x') = p^0(x^0 - x^{0'}) - p^1(x^1 - x^{1'}) - p^2(x^2 - x^{2'}) - p^3(x^3 - x^{3'})$, you can trivially show yourselves that

$$\partial_x^2 e^{-ip.(x-x')} = -p^2 e^{-ip.(x-x')}$$

and that, for that matter, (8.XVIII) directly implies that

$$\tilde{G}(p) = \frac{1}{p^2 - m^2}$$

which we immediately recognize as the momentum space Feynman propagator $\tilde{\Delta}_F(p)$ for the Klein-Gordon field in (3.II).

Thus, as announced at the outset of this calculation, the physical content of $(\partial^2 + m^2)^{-1}_{x,x'} = G(x, x')$, is the Feynman propagator in Minkowski space-time:

$$(\partial^2 + m^2)^{-1}_{x,x'} = \Delta_F(x - x') = \int \frac{d^4p}{(2\pi)^4} \frac{e^{-\frac{i}{\hbar}p.(x-x')}}{p^2 - m^2} \qquad (8.XIX)$$

Equivalently, (8.XIX) makes a statement to the effect that *the Feynman propagator $\Delta_F(x - x')$ is the Green function to operator $(\partial_x^2 + m^2)$*. The procedure which eventuated in (8.XIX) constitues an independent and equally rigorous derivation of the relation between the Feynman propagator in Minkowski space-time $\Delta_F(x - x')$ and the Feynman propagator in momentum space $\tilde{\Delta}_F(p)$. Of course, as it stands, the expression in (8.XIX) is not defined! It falls short of being the Feynman propagator by the "$i\epsilon$-prescription". This time, however, we shall not introduce this prescription through contour integration in the complex p^0-plane as we did in the preceding third section. We adhere, instead, to the fundamental premise that $\Delta_F(x - x')$ is - as all dynamical characteristics of the quantum field - inherent in the action functional and return, for that matter, to the path integral in (8.VI).

Recall from the third chapter that the oscillatory character of $e^{-\frac{i}{\hbar}S[\phi]}$ raises an issue of convergence. As the path integral is ambiguous, we must specify what exactly we mean by it. It is, in fact, this ambiguity which "percolates" in the propagator! On the same lines as those in the third chapter, we resolve this issue by adding an ϵ-dependent *convergence factor* $e^{-\frac{1}{2}\epsilon \int d^4x \phi^2(x)}$, $(\epsilon > 0)$ in the action functional which asymptotically damps the oscillations and yields a finite result (refer to (Fig.13)):

$$\langle 0|0 \rangle_J = N \int \mathcal{D}\phi \, e^{-\frac{i}{\hbar}\int d^4x[\frac{1}{2}\phi(x)(\partial^2+m^2-i\epsilon)\phi(x)-J(x)\phi(x)]} \tag{8.XX}$$

on the understanding that once that finite result has been attained, that factor must be "switched off" by taking the limit $\epsilon \to 0$.

The immediately obvious consequence of this operation, is that the parameter ϵ eventually "percolates" in (8.XIX) (we follow exactly the same steps, this time with m^2 replaced by $m^2 - i\epsilon$). Thus, we correct, the latter to the familiar (3.I):

$$(\partial^2 + m^2)^{-1}_{x,x'} = \Delta_F(x - x') = \int \frac{d^4p}{(2\pi)^4} \frac{e^{-\frac{i}{\hbar}p.(x-x')}}{p^2 - m^2 + i\epsilon} \tag{8.XXI}$$

in which the $i\epsilon$-term in the denominator now determines the manner in which we avoid the two poles by deforming the path of integration Rep^0 around them, precisely as we discussed in the preceding third section.

Thus, we have established our claim! The vacuum-to-vacuum transition amplitude in the presence of a driving source:

$$\langle 0|0 \rangle_J = e^{-\frac{i}{\hbar}\frac{1}{2}\int d^4x \int d^4x' J(x)\Delta_F(x-x')J(x')} \tag{8.XXIIa}$$

is manifestly equal to $\langle 0|0 \rangle = 1$ for $J(x) = 0$ and centrally involves the Feynman propagator $\Delta_F(x - x')$ which, as (8.XXI) explicitly states, is the Green function to operator $(\partial^2 + m^2)$ in (8.VI)[80]. Henceforth, we shall refer to (8.XXIIa) in terms of the simplified notation $\langle 0|0 \rangle_J \equiv W[J]$ as:

$$W[J] = W[0] \, e^{-\frac{i}{\hbar}\frac{1}{2}\int\int d^4x d^4x' J(x)\Delta_F(x-x')J(x')} \tag{8.XXIIb}$$

The statement which (8.XXII) makes underscores the essence of path-integral quantization for fields: (i) The Lorentz invariance of the theory remains invariably manifest! $S[\phi]$, $\phi(x)\partial^2\phi(x)$, $J(x)\Delta_F(x - x')J(x')$ and the measure $\mathcal{D}[\phi]$ are manifestly Lorentz invariant (the measure, because $\phi(x)$ itself is) and at no stage of a calculation is the manifest character of that invariance lost. Recall in this respect that manifest Lorentz invariance is not the case in canonical quantization (although the quantized field is always Lorentz invariant) on account of the three-dimensional equal-time surface which is necessarily tied to a specific inertial frame. (ii) The physical significance of $\Delta_F(x - x')$ is

---

[80]In establishing (8.XXIIa) we have at once established (3.IX) in the third chapter.

transparent. In driving the vacuum state of the free Klein-Gordon field from infinite past to infinite future, the classical source $J(x)$ generates particles over space-time separation $x - x'$. Propagating "virtual" particles, described by propagators obtained by taking successive *functional derivatives* of $W[J]$ with respect to $J(x)$ at $J(x) = 0$, emerge over finite space-time separations. This approach transcends propagation! The significant advantage of (8.XXII) is that it extends directly to the description of the dynamical behavior of interacting quantum fields. The basic idea, as we shall see, is to arrive in the interacting context by treating the interaction itself as a small *perturbation* to the *idealized* free quantum fields. In the context of this procedure, the idealized propagator is treated as the zero order, as that order which signifies the absence of interactions. The actual transition amplitude itself emerges as the result of successive additions of infinitely many such small corrections. This is not limited to the propagator. It applies with equal necessity to transition amplitudes which involve more than one particles. In quantum electrodynamics, for example, an incoming photon may give rise to an electron-positron pair. Conversely, an electron-positron pair may give rise to a photon. Such processses are also inherent in that extension of (8.XXII) which includes the interactions between quantum fields with the corresponding n-particle Green functions also treated as infinite series of successive approximations order by order in *perturbation theory*. Knowledge of all transition amplitudes to all orders in perturbation theory is equivalent to knowledge of the interacting system of quantum fields and amounts, for that matter, to the exact solution to that system.

In order to further elucidate these remarks, we consider a specific context of interacting quantum fields. The success of perturbation theory is crucially contingent upon an adequately small coupling. In *quantum chromodynamics*, the theory of interacting quark and gluon fields, neither the quarks, nor the gluons exist as *asymptotic states* (as physical states, as states which can survive in isolation) because the interaction which binds them becomes infinitely strong over infinite space-time separations. Perturbation theory does not apply to the transition amplitudes of quantum chromodynamics over space-time separations in which the coupling exceeds a certain limit. Yet, the answer to the question as to whether the coupling grows stronger with the distance or, on the contrary, weaker (as is the case in quantum electrodynamics) is inherent in the very solution which we have set out to find! Thus, it is imperative that we develop an approach to the solution of the interacting system which does not rely on "a priory" knowledge of the coupling strength and the physical states of the interacting system. This is precisely the merit of (8.XXII) in the presence of interactions. For whatever the behavior of the coupling between the quantum fields, if we drive the unknown vacuum state of the interacting system from past infinity to future infinity with an arbitrary driving source, we shall necessarily generate all transition amplitudes with all the inherent information as to the behavior of the coupling-strength over increasing space-time separations.

This concludes our examination of the Klein-Gordon field which served as a prototype for the analysis on the most general aspects of free quantum fields on account of its comparative simplicity. It is certainly desirable to proceed to the examination of the spin-$\frac{1}{2}$ Dirac field and of the spin-1 photon field before advancing the analysis to the theory of interacting quantum fields and the centrally important procedure of renormalization. It is my intension to address these issues.

NOTES

Chapter 1.

N1. Insight as to the fundamental difference between the plane waves of probability in quantum physics and the plane waves of energy in classical physics can be elicited through physical effects such as the photoelectric. The photons are not the little sinusoidal packets of energy which abound in "popular accounts" and popular imagination. We shall see that from the rigorous perspective of quantum field theory, they are energy-momentum eigenstates of the photon field characterized by specific occupation numbers. At this early stage of our analysis, however, it suffices to merely treat them qualitatively as plane waves of probability in the framework of simple quantum mechanics. This implies that the probability that the position of a photon be in the vicinity of any given point in infinite space is finite and constant throughout space and throughout time.

According to the classical electromagnetic theory, the radiation incident upon a metal surface transfers energy on the electrons in a continuous manner. An increase in the luminosity of incident light implies an increase in the electric and magnetic strength and, thereby, an increase in the force exerted by light on the electrons of the metal. Thus, although the classical theory correctly predicts that an increase in incident luminosity yields an increase in the intensity of the photoelectric current, it incorrectly predicts that the increase in incident luminosity yields an increase in the speed of the electrons emerging from the metal. The fact of the matter is that the speed of the emerging electrons is independent of the incident luminosity. It increases only as the result of an increase in the frequency of the incident light w on account of the fundamental relations $p = \hbar k$ and $E = \hbar w$. An incident photon of higher frequency w signifies a plane wave of probability of higher energy E. It is exclusively that E which is - from the classical perspective - instantaneously transferred on an electron. Part of that energy corresponds to the energy required for the extraction of the electron from the metal lattice and constitutes the minimum energy below which there is no photoelectric current. The remaining part corresponds to the kinetic energy of the free electron. The higher intensity of the photoelectric current which emerges as a result of higher incident luminosity is the direct and exclusive consequence of the higher number of incident photons.

The photoelectric effect is but a special case of an entire range of physical contexts in which light interacts with matter. The nature of light as a collection of photons signifying plane waves of probability - instead of, as a classical plane wave of energy - can also be elicited from common experience. Skin tissue, for example, develops a tan when exposed to ultraviolet radiation, a fact which implies that the relevant chemical reactions are activated only when the frequency of incident light exceeds a minimum value. The reason is identical to that of the photoelectric effect. A chemical reaction is possible only when the energy of the interacting molecules exceeds a minimum value. If light and all electromagnetic waves were actually continuous plane waves continuously transferring energy, as the classical electromagnetic theory predicts, then the frequency of the incident radiation would be an irrelevant parameter. People would develop a tan merely by standing next to a radio antenna!

The precise significance of this qualitative analysis will become transparent in the fifth chapter where we shall see that the waves of the classical electromagnetic theory do indeed emerge as the semi-classical limit of the photon field.

N2. If this is difficult to comprehend, consider the following heuristic argument. Direct a particle toward a given static attractive potential. If the particle is classical, you can determine in advance the total energy which it will have once captured by the classical field. It will be equal to the initial kinetic energy and will remain constant in the course of time as a consequence of time-translational invariance. If, on the contrary, you direct a quantum particle then you have no way of determining in advance which energy eigenvalue within the potential that particle will settle in. All you can determine is the probability which each energy eigenvalue has for being eventually the actual energy value of the particle. The set of all such probabilities determines, in turn, the wavefunction $\psi(t, x)$. As different values of initial kinetic energy determine different $\psi(t, x)$, there are infinitely many possible quantum states for the electron in the static attractive potential.

N3. Does all this not sound somewhat... pompous? "However strong a potential is, it shall never manage!" quantum mechanics assures us. Atoms, alright then! Yet, what about the strongest potential which theoretical physics has ever revealed? What about black holes? General relativity predicts that everything in a black hole must end up right at its gravitational singularity. The statement which general relativity makes is categorical: Time in a black hole shall not advance unless everything in it advances toward the singularity! No electron would have the luxury of keeping itself "aloft" in a quantum state above the singularity and still hope to evolve in time! And what about the singularity itself? A point of infinite space-time curvature and infinite density! It is not space and it is not time. Yet, whatever it is, it is point-like, unimaginably smaller than the electron itself! If we accept that anything of a minute size which is confined in a minute volume is necessarily a quantum system, as the quantum theory insists, should then that point-like singularity not be instantly smeared out as a result of the uncertainty principle? On the evidence, there is a profound conflict between the classical theory of general relativity and the quantum theory. Which one prevails over the other then? General relativity, or the quantum theory? And, how?

Chapter 2.

N1. A dialectical relation between two aspects of a situation in opposition to each other involves either aspect as a *Moment* in the development of the other. The dialectical method found its supreme expression in Hegel's *Phenomenology of the Spirit* and in K. Marx's *Capital*, especially in the first of the three volumes which that work comprises.

N2. Schrödinger's Cat is a thought experiment which underscores the macroscopic consequences of the wavefunction. The following excerpt from [4] (p.431) succinctly renders the essence of the matter:

We assume that the result of the previously discussed quantum-mechanical measurement triggers a rifle pointing at a cat. The total "experimental setup" should be placed in a box to be opened later on to observe the cat's state (alive or dead). Making that *observation, the result of the measurement can be deduced. In analogy to the previous discussion, the wave function of the box is χ_+ or χ_-, depending on whether the cat is alive or dead. [...] This, of course, would have the consequence that the state of the cat before opening the box would be neither alive nor dead. If we reject this absurd conclusion, we have to ask when the wave-function reduction takes place: When the particle enters the apparatus? When the cat dies? Or at some other time?*

I shall only add that the question of "when" as posed here is nebulous. Neither do quantum particles "enter", nor do cats - if also considered as quantum objects - die at specific moments. In the dialectical-materialist approach which I advance, at any given moment the unobserved but objective position of the quantum particle may be that required to trigger the rifle, or not. Equivalently, at the same moment, the unobserved but objective momentum may be that required to trigger the rifle, or not.

Chapter 4.

N1. In the context of the classical premise to the effect that fields are produced by sources, a classical field is free in the region of space within which it extends if there are no points in that region at which that field interacts with particles, currents or other fields, including itself. Otherwise, it is an interacting classical field. The electromagnetic field is a free classical field at least in vacuum (the non-trivial issue of whether it interacts with its "sources" is properly addressed by quantum electrodynamics). For example, the electrostatic Coulomb field produced by a point-like charge at rest is a free classical field throughout space except at the point occupied by its source at which the field-strength becomes infinite and the classical description "breaks down".

The outstanding example of an interacting classical field is the gravitational field as envisaged in the general theory of relativity. The gravitational field is identical to the curvature of space-time produced by concentrations of matter and energy. In turn, space-time curvature causes all forms of matter and energy in empty space to freely fall by setting all particles on space-time curves of minimum length (*geodesics*). Clearly, this description stands in direct opposition to the familiar Newtonian concepts of "gravitational force" and "weight". However, the real challenge to intuition which this standpoint poses, relates to the self-interacting character of the gravitational field. According to the *equivalence principle* which lies at the foundation of general relativity, all forms of matter and energy couple to the background space-time with equal strength (somehow, the rigorous expression of the Newtonian statement "all objects fall at the same rate"). Since this necessarily includes the gravitational energy inherent in the curvature of space-time itself, it follows that gravitational energy also falls and that, for that matter, space-time curvature couples to itself! In an intuitive sense, gravity gravitates! The gravitational field is a self-interacting classical field.

N2. The stated subtraction of the infinite constant from the vacuum energy is not tolerated in the realistic presence of gravity. In general relativity the energy density of the quantum vacuum (the energy E within a given volume V divided by that volume) directly enters the Einstein equations (the solution to which determines the space-time curvature) as a cosmological constant $\Lambda = \frac{E}{V}$. This, in turn, renders the energy density of the vacuum directly observable through its effect on the observable background space-time curvature.

In 1998 observations of remote supernovae type Ia established that the expansion of the Universe is accelerating, with 70% of its energy density being in the form of a cosmological constant (dark energy). This was, subsequently, independently confirmed through the observations of the WMAP satellite. The value of the cosmological constant estimated on the basis of these observations is $\Lambda \sim 10^{-12} ev^4$, remarkably close to zero! The irony of this state of affairs is that - following Einstein, who introduced the cosmological constant Λ in his equations in 1917 and repudiated it in 1931 - the cosmological constant has been conjectured to be strictly equal to zero. The vacuum-energy densities of all quantum fields (realistically coupled to the background space-time geometry) should add up to zero - arguably, by dint of *supersymmetry*. As observations have invalidated that conjecture, any value - however high - of Λ should be as physically legitimate as any other. In fact, estimates stemming from theories of quantum gravity are as high as $\Lambda \sim 10^{108} ev^4$! The cosmological constant is "neither here, nor there", as it were. It is neither zero as a result of some underlying symmetry, nor as enormously high as advanced theories predict in the absence of such a symmetry. It is $\Lambda \sim 10^{-12} ev^4$! Why do the vacuum-energy densities of all quantum fields in space-time add up to something so remarkably close to, but not precisely, zero? What cancels their sum-total to such high, yet not perfect, accuracy? This question is as "dark" as the energy which poses it: The answer is unknown!

Chapter 5.

N1. Virtual particles are the source of serious misconceptions in "popular" accounts of Physics. The most notable is in the argument "the uncertainty principle implies that empty space is full of virtual particles". We have already dealt with the limitations of the "uncertainty principle" as a vehicle for vacuum fluctuations. However, even from the heuristic perspective to which this approach aspires, this argument is false! "Empty space is full of virtual particles" only because "empty space" accommodates at least one quantum field in its vacuum state. However, as the vacuum state is not the zero-field state, such an "empty space" is not really empty at all! Worse still, the uncritical reference to the uncertainty principle in this context is, to say the least, misleading! The uncertainty principle also characterizes quantum mechanics. Yet, quantum mechanics is incompatible with virtual particles! Without the infinitely many degrees of freedom of quantum fields there are no virtual particles.

Another misconception lies in the attempt to blur the difference between actual and virtual particles. It is one thing to claim that the difference between actual and virtual particles is not of primary importance in quantum field theory because both are reducible to the stark actual reality of underlying quantum fields and quite another to claim that there is no such difference! Certain authors argue that actual particles also emerge from and vanish into the vacuum in the presence of interactions and that, for that matter, there is no essential difference between actual and virtual particles. They argue that actual particles exist over substantially longer periods of time than those over which virtual particles exist and that, for that matter, the distinction between the two categories of particles is merely a matter of convenience. This is false! Although actual particles may also emerge and vanish in the realistic presence of interactions between quantum fields they are - unlike their virtual counterparts - always reducible to the idealized momentum states of free quantum fields. It is, in fact, that reduction to "asymptotic states" which allows for the rigorous description of interacting quantum fields. That aside, the most essential flaw in this argument lies in the fact that it defines the concept of "particle" not on any evidence of fact as to which features are essential or inessential to the physical content of that concept, but *merely on the formal procedure of combining properties common to a heterogenous assortment of situations and building abstraction out of analogy* [13]. Indeed, even in the actual presence of interactions the essential difference between actual and virtual particles is not in longevity but in observation: Actual particles are explicitly observable. Virtual particles are merely a formal interpretation of observable field strength.

In order to elucidate this point we consider here the physical context within which such a misconception usually occurs: "Two electrons repel each other through the exchange of virtual photons" (their language). This "popular argument" holds that this process is essentially identical in nature to the production of a photon in a distant galaxy and its subsequent observation here, since this latter process is also represented by the same Feynman diagram as the former. As an actual photon was emitted "there and then" presumably also by an electron and is observed "here and now" by exciting an electron in the optic nerve, we cannot but deduce the absence of any essential difference between that actual photon and the "virtual" photon which mediates the electromagnetic interaction between two electrons.

Let us examine this claim in some detail. Although appearances are indeed deceptive, suggesting an underlying identity between both processes, careful consideration of the latter reveals otherwise. The essential question in these two processes is not whether the respective photons involved emerge at a space-time point and vanish at another, but whether they are asymptotic states or not - that is, whether, having emerged at a space-time point, they have the capacity to transit to infinite distance. In this respect, the photon from the distant galaxy is actual precisely because it would have transited to infinite distance from that galaxy had it not been intercepted by the observer's eye. In fact, it is directly observable precisely because it is actual. On the contrary, the photon which mediates the interaction between two electrons is virtual and unobservable precisely because it never advances to an asymptotic momentum state at infinite space-time separations. Diagrammatically, for that matter, whereas a single Feynman diagram describes the interaction between the two electrons, two distinct Feynman diagrams are necessary in order to describe the "disconnected" events of emission and absosrption of the asymptotic state which describes the actual photon (Fig.N1). Electromagnetic interactions are indeed realized either through virtual or through actual photons. Yet, whereas virtual photons exist conditionally (generated by the relative space-time separation between the two electrons) and - pursuant to (1.XII) - as a matter of interpretation at that, actual photons exist unconditionally (regardless of whether

314

an electron will interrupt their transition to infinite distance, or not) and are explicitly observable. Whereas virtual photons account for the electromagnetic force between
electrically charged particles, actual photons account for the electromagnetic waves and
light.

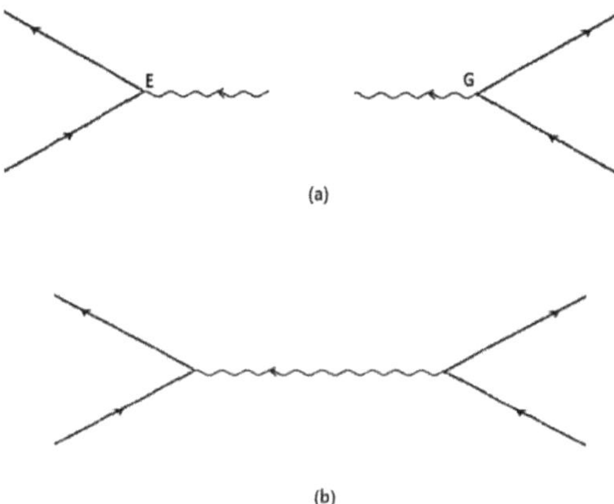

FIGURE N1. A photon emmitted by an accelerated electron in a distant galaxy (G)
ending up absorbed by an electron in the optical nerve (E) is actual and observable,
thereby signifying two independent asymptotic states (infinite space-time separations, indefinite existence) described by two independent Feynman diagrams (a). The
photon which mediates the electromagnetic interaction between two electrons is virtual and unobservable. As such, it exists exclusively over finite space-time separations (b).

N2. It is important to note that the pressure exerted by $|0\rangle$ is negative. If this is not
immediately obvious from the Casimir effect, merely replace one of the two plates by the
fixed vertical wall of a chamber and the other by a piston. The piston is free to slide
toward or away from the fixed vertical wall without friction, thereby determining the
volume of the chamber. If the chamber contains gas then outward pressure is exerted by
the latter on the piston, driving it away from the fixed vertical wall, thereby increasing
the volume of the chamber. In vacuum, on the contrary, the Casimir effect drives the
piston toward the fixed vertical wall, thereby decreasing the volume of the chamber: *The
pressure exerted by the vacuum energy density is negative.*

Note now the implications which this has in the realistic context of gravity. The
outward positive pressure of the gas in the chamber is the pressure of "ordinary matter",
a term commonly used for a collection of actual particles in whichever quantum states.
Ordinary matter, of course, curves the background space-time geometry. In turn, positive
pressure contributes to that curvature, thereby increasing the attractive gravitational
effect of ordinary matter. The positive pressure of ordinary matter tends to reverse the

expansion of the universe and cause space to contract. On the contrary, the negative pressure exerted by the vacuum state has the opposite effect. It causes space to expand. In fact, the negative pressure exerted by the vacuum energy density in the early universe constitutes the basis of the *inflationary cosmological models.*

In concluding the analysis on the quantum vacuum, it is also necessary to stress that formally vacuum fluctuations occur at all finite space-time scales, regardless of whether physical boundaries are present or not. In Minkowski space-time in particular, the fluctuations of $|0\rangle$ have observable consequences exclusively because of the presence of physical boundaries such as the Casimir plates and the electrons in atoms (the actual energy eigenvalues of which differ from those predicted by quantum mechanics on account of the vacuum activity). Without the presence of physical boundaries, the fluctuations of the Klein-Gordon vacuum (or any field's vacuum state) in Minkowski space-time have no observational consequences exclusively because in this idealized physical context the quantum field does not couple to the background space-time geometry. Yet, the vacuum fluctuations exist all the same! In the realistic presence of gravity caused by the coupling of the quantum field to the background space-time geometry the fluctuations of the vacuum state would have an immediate effect on space-time itself (see also (N2) in the fourth chapter).

APPENDIX

BASIC ASPECTS OF SPECIAL RELATIVITY

1. Minkowski Space-time

Einstein's discovery (1905) that space and time are unified in a higher entity finds its rigorous expression in *Minkowski space-time*. Minkowski space-time is a flat space-time (zero curvature, zero gravity). Although all observers are physically equivalent and all observations equally valid, the fundamental character of Minkowski space-time as flat singles out a preferred set of observers in it, the *inertial observers*, in the sense that it is only such observers who register the characteristic flat metric (essentially, the distance between two points in the absence of gravity) of Minkowski space-time[81]. The fundamental postulates of Special Relativity concern inertial observers or, equivalently, *inertial frames*. An inertial frame is a coordinate system which is

(i) based on three mutually orthogonal axes which give coordinates x, y, z in space and an associated set of synchronized clocks at rest in the system which gives a time coordinate t and

(ii) such that when motion is formulated in it, a particle on which no force is exerted remains in uniform motion (zero acceleration).

It follows that an inertial frame is moving with respect to any other without rotation and with constant velocity. The four coordinates t, x, y, z label points in space-time with each space-time point being an *event*.

The fundamental postulates of Special Relativity are:

1. The speed of light c is the same in all inertial frames.

2. The laws of Nature are the same in all inertial frames.

The first postulate is clearly incompatible with Newtonian premises on light propagation. If K and K' are two inertial frames then it implies that

$$c = \frac{dr}{dt} = \frac{dr'}{dt'}$$

where $dr^2 = dx^2 + dy^2 + dz^2$ in K and $dr'^2 = dx'^2 + dy'^2 + dz'^2$ in K'. This relation may be restated as

$$c^2 dt^2 - dx^2 - dy^2 - dz^2 = c^2 dt'^2 - dx'^2 - dy'^2 - dz'^2 = 0$$

and is consistent with the existence of a Lorentz-invariant interval (the same for all inertial observers) ds between neighboring events such that[82]

$$ds^2 = c^2 dt^2 - dx^2 - dy^2 - dz^2 = c^2 dt'^2 - dx'^2 - dy'^2 - dz'^2$$

In view of this relation, it is convenient to introduce new coordinates such that each event in Minkowski space-time is specified as $x^\mu = (x^0, x^1, x^2, x^3)$; $\mu = 0, 1, 2, 3$; $x^0 = ct$, $x^1 = x$, $x^2 = y$, $x^3 = z$. For that matter, $x^\mu = (x^0, x^1, x^2, x^3) = (x^0, \vec{x})$ also signifies the *space-time vector* with which each event is in one-to-one correspondence. In this

---

[81]Although accelerating observers also register the zero curvature of Minkowski space-time, locally they find themselves in a gravitational field according to the *Equivalence Principle*.

[82]Although this particular expression for ds^2 remains invariant only under transformations between inertial frames, ds^2 itself remains invariant under general transformations between frames. The distance between any two neighboring space-time points is an inherent property of space-time itself and does not depend on who observes these points, or with which coordinates he chooses to label them.

coordinate system the light cone of O (the collection of space-time lines representing trajectories of light which pass through the origin O) is symmetrically inclined at 45^0 with respect to the spatial and temporal axis (Fig.A1). This relation implies the existence of a *metric tensor* $\eta_{\mu\nu}$ such that

$$\eta_{00} = 1 \ , \ \eta_{11} = -1 \ , \ \eta_{22} = -1 \ , \ \eta_{33} = -1 \ , \ \eta_{\mu\nu} = 0 \ if \ \mu \neq \nu$$

Following the *Einstein summation convention* to the effect that repeated upper and lower indices are summed over, the above relation is recast into the concise form

$$ds^2 = \eta_{\mu\nu}dx^\mu dx^\nu$$

(first set $\mu = 0$ and successively sum over $\nu = 0, 1, 2, 3$. Repeat for $\mu = 1$, then for $\mu = 2$ and finally for $\mu = 3$. Add up the results).

The stated invariant interval allows for the following three possibilities (Fig.A1)

$$(i) \qquad\qquad\qquad ds^2 = 0$$

if the neighboring events are on the light cone of O,

$$(ii) \qquad\qquad\qquad ds^2 > 0$$

if the neighboring events are within the light cone of O (the set of events any two of which are in causal communication through signals which propagate at speeds smaller than c) and

$$(iii) \qquad\qquad\qquad ds^2 < 0$$

if the neighboring events are out of the light cone of O (the set of events no two of which are in causal relation, as no signal propagates at speeds higher than that of light).

These considerations signify a striking difference between the metric structure of *Euclidean space* (the space of Newtonian physics) and the metric structure of Minkowski space-time. In Euclidean space there are no vectors of imaginary magnitude since the square of a vector's magnitude

$$x^2 + y^2 + z^2$$

is invariably a non-negative quantity. This is not the case in Minkowski space-time. All space-time vectors which lie within the light cone have magnitudes the square of which is a positive-definite quantity (*time-like vectors*), all space-time vectors which lie on the light cone are of zero magnitude (*null vectors*) and all space-time vectors which lie out of the light cone have magnitudes the square of which is a negative quantity (*space-like vectors*). That is,

$$c^2t^2 - x^2 - y^2 - z^2 > 0 \ ; \ for \ time-like \ vectors$$

$$c^2t^2 - x^2 - y^2 - z^2 = 0 \ ; \ for \ null \ vectors$$

$$c^2t^2 - x^2 - y^2 - z^2 < 0 \ ; \ for \ space-like \ vectors$$

or, in concise notation

$$\eta_{\mu\nu}x^{\mu}x^{\nu} = \begin{cases} > 0 \; ; \; for\; time-like\; vectors \\ = 0 \; ; \; for\; null\; vectors \\ < 0 \; ; \; for\; space-like\; vectors \end{cases}$$

Accordingly, *world lines* (particle trajectories in Minkowski space-time) are time-like, null, or space-like, depending on whether their tangent vectors at each point are time-like, null, or space-like. The physical significance of the tangent vectors will be elucidated shortly. It is clear however that the trajectories of massive particles in Minkowski space-time (particles of non-zero rest mass) are time-like world lines and the trajectories of massless particles (particles of zero rest mass such as photons) are null world lines (Fig.A1).

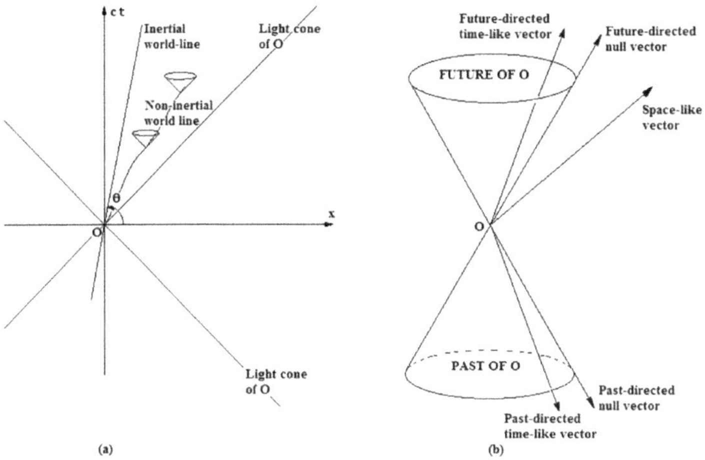

FIGURE A1. A space-time diagram based on a Cartesian inertial frame. Two of the four dimensions of Minkowski space-time have been suppressed (each point on the spatial x-axis represents a plane). The light cone of O consists of all photon paths passing through event O and invariably dissects the diagram at 45^0. World lines of particles in inertial motion are straight lines whose inclination $\theta > 45^0$ determines the particle's constant speed. The world line of a   non-inertial particle which accelerates toward the speed of light is also shown (a). The three-dimensional light cone is a null hypersurface. All vectors on it have zero magnitude (null vectors). All vectors within the light cone have magnitudes whose square is positive (time-like vectors) and all vectors out of the light cone have magnitudes whose square is negative (space-like vectors). The upper half of the light cone is the future of event O. The lower part, is the past of event O. Causal relations exist only for events within the light cone (b).

The effect of time dilation can be directly seen by considering the world line of a (point-like) clock whose coordinate differences relative to some inertial frame K are cdt, dx, dy, dz. The clock has been synchronized at the origin O with the stated synchronized clocks at rest in K. *Time is what the clock shows!* The lapse of time $d\tau$ registered by the clock is the physical, *proper time* in the clock's frame. In its own frame the lapse of the clock's proper time is clearly identical to the invariant (coordinate-independent - observer-independent) length of the time-like world line which the clock delineates (in its own frame the clock moves exclusively in the temporal direction). If the clock moves with speed v as measured by K then that invariant length is

$$ds^2 \equiv c^2 d\tau^2 = c^2 dt^2 - dx^2 - dy^2 - dz^2 \equiv (dx^0)^2 - (dx^1)^2 - (dx^2)^2 - (dx^3)^2 = \eta_{\mu\nu} dx^\mu dx^\nu$$

In view of

$$dx^2 + dy^2 + dz^2 = v^2 dt^2$$

the direct implication is that

$$d\tau = \sqrt{1 - \frac{v^2}{c^2}}\, dt$$

Clearly, the higher the speed v of the clock is, the smaller the lapse $d\tau$ is with respect to the coordinate lapse dt. In fact, $d\tau \to 0$ as $v \to c$! *Moving clocks go slow!* As observed from K the moving clock ticks progressively slower as its speed progressively increases and tends to stop ticking altogether as its speed approaches the unattainable speed of light ($v = c$ is impossible!). Conversely, the smaller v is, the more $d\tau$ tends to the coordinate lapse dt. If the clock is at rest in K then its proper time τ coincides with the *coordinate time t* registered by the stationary synchronized clocks in K.

2. Energy-Momentum Vector

The proper time τ renders a concrete physical content to the tangent vector at each event of a world line. For $v < c$ the proper time τ as defined by its above-stated relation to coordinate time t is real and in one-to-one correspondence with the coordinates which label the space-time points of a world line, that is $x^\mu = x^\mu(\tau)$. In turn, the tangent vector $u^\mu \equiv \frac{dx^\mu}{d\tau}$ defines the *world velocity* of the particle (the reason $v^\mu \equiv \frac{dx^\mu}{dt}$ is not physically relevant will be elucidated shortly). From $c^2 d\tau^2 = \eta_{\mu\nu} dx^\mu dx^\nu$ it follows that $\eta_{\mu\nu} u^\mu u^\nu = c^2$ with the direct implication that the tangent vector u^μ at each event on the world line of a particle of non-zero rest mass is time-like. Accordingly, the tangent vector u^μ at each event on the world line of a photon is null.

The 4-*momentum* p^μ of a particle whose non-zero rest mass is m is now defined in terms of u^μ as

$$p^\mu = mu^\mu$$

Note that for $\mu = i = 1, 2, 3$ it is

$$p^i = m\frac{dx^i}{d\tau} = mv^i\frac{dt}{d\tau} = \frac{m}{\sqrt{1 - \frac{v^2}{c^2}}}v^i$$

a fact which reveals p^i as the 3-momentum of the particle and

$$\frac{m}{\sqrt{1 - \frac{v^2}{c^2}}}$$

as the particle's mass in the inertial frame which registers that momentum. Accordingly

$$p^0 = m\frac{dx^0}{d\tau} = c\frac{m}{\sqrt{1 - \frac{v^2}{c^2}}} = \frac{E}{c}$$

with

$$E = \frac{m}{\sqrt{1 - \frac{v^2}{c^2}}}c^2$$

being the particle's energy in the same frame, so that the 4-momentum of a massive particle is the vector

$$p^\mu = (\frac{E}{c} , \vec{p})$$

which is clearly time-like:

$$\eta_{\mu\nu}p^\mu p^\nu > 0 \iff (p^0)^2 - |\vec{p}|^2 > 0$$

Massless particles are treated as plane waves of energy propagating at speed equal to c (the photon is not defined in the classical theory). Associated with a plane wave of wavelength λ is the 3-vector $\vec{k} = \frac{2\pi}{\lambda}\vec{n}$ of magnitude $|\vec{k}| = \frac{2\pi}{\lambda}$ and orientation along the unit vector (of magnitude equal to one) $\vec{n}$ which signifies the direction of propagation. The four vector

$$k^\mu = (\frac{2\pi}{\lambda} , \vec{k})$$

is null ($\eta_{\mu\nu}k^\mu k^\nu = (\frac{2\pi}{\lambda})^2 - (\frac{2\pi}{\lambda})^2 = 0$) and yields the 4-momentum

$$p^\mu = \hbar k^\mu$$

which is itself a null vector:

$$\eta_{\mu\nu}p^\mu p^\nu = 0 \iff (p^0)^2 - |\vec{p}|^2 = 0$$

Note that this is in identity with the 4-momentum of a plane wave of probability for the photon field in quantum field theory.

Tachyons are particles of speed $v > c$. From $d\tau = \sqrt{1 - \frac{v^2}{c^2}}dt$ it follows that τ is an imaginary quantity and that, for that matter, tachyons are unobservable. Their 4-momentum is

$$p^\mu = mu^\mu$$

with the mass $\dfrac{m}{\sqrt{1-\frac{v^2}{c^2}}}$ now being imaginary. The four-momentum of tachyons is clearly a space-like vector:

$$\eta_{\mu\nu}p^\mu p^\nu < 0 \iff (p^0)^2 - |\vec{p}|^2 < 0$$

On the understaning that tachyons cannot be decelerated to speeds no bigger than c, they are a theoretical possibility.

3. Lorentz Transformations and Space-time diagrams

The second postulate implies the existence of a set of transformations amongst inertial frames, the *Lorentz transformations*. Given a vector in terms of the coordinates of inertial frame K, the Lorentz transformations express that vector in terms of the coordinates of inertial frame K'. Thus, given the observations in K, the Lorentz transformations provide all information as to the observations in K'. In particular, if a Lorentz transformation expresses space-time vector x^μ in K as space-time vector x'^μ in K' then

$$\eta_{\mu\nu}x^\mu x^\nu = \eta_{\mu\nu}x'^\mu x'^\nu \iff$$

$$(x^0)^2 - (x^1)^2 - (x^2)^2 - (x^3)^2 = (x'^0)^2 - (x'^1)^2 - (x'^2)^2 - (x'^3)^2$$

that is, the magnitude $\eta_{\mu\nu}x^\mu x^\nu$ of a space-time vector remains invariant under Lorentz transformations (as the magnitude $x^2 + y^2 + z^2$ of a vector in Euclidean space remains invariant under transformations between coordinate systems). If a Lorentz transformation expresses energy-momentum vector p^μ in K as energy-momentum vector p'^μ in K' then

$$\eta_{\mu\nu}p^\mu p^\nu = \eta_{\mu\nu}p'^\mu p'^\nu \iff$$

$$(p^0)^2 - (p^1)^2 - (p^2)^2 - (p^3)^2 = (p'^0)^2 - (p'^1)^2 - (p'^2)^2 - (p'^3)^2$$

that is, the magnitude $\eta_{\mu\nu}p^\mu p^\nu$ of a 4-momentum remains invariant under Lorentz transformations. This is, in fact, the reason the 4-momentum is defined in terms of the world velocity $u^\mu = \frac{dx^\mu}{d\tau}$ instead of the *coordinate velocity* $v^\mu = \frac{dx^\mu}{dt}$. The latter is clearly not a Lorentz-invariant vector.

The relation between the energy E, the momentum magnitude $p = |\vec{p}|$ and the rest mass m of a relativistic particle as the latter is observed in an inertial frame can be elicited through a simple Lorentz transformation to the particle's rest frame. The energy-momentum vector in the initial inertial frame is $p^\mu = (p^0 = \frac{E}{c}, \vec{p})$ the square of whose magnitude is $\eta_{\mu\nu}p^\mu p^\nu = (p^0)^2 - |\vec{p}|^2$. In the particle's rest frame the energy-momentum vector is $p'^\mu = (\frac{mc^2}{c}, \vec{p} = \vec{0})$ the square of whose magnitude is $\eta_{\mu\nu}p'^\mu p'^\nu = (mc)^2$. Since the magnitude of a vector in Minkowski space-time is Lorentz invariant it follows that

$(\frac{E}{c})^2 - |\vec{p}|^2 = (mc)^2$. Thus, the relativistic kinematic relation between the energy E, the momentum magnitude $p = |\vec{p}|$ and the rest mass m is

$$E^2 = (pc)^2 + m^2c^4$$

The manner in which K and K' are themselves related by Lorentz transformations can be deduced as follows. Since in its own frame the lapse of a massive particle's proper time is identical to the invariant length of its world line, that world line is necessarily the time axis of the orthogonal (Cartesian) inertial frame K' bound to that particle. Thus, as observed from K, the time axis ct' of K' is the straight line inclined at angle θ in Fig.A1(a). The spatial axis x' of K' as observed from K follows as a direct consequence of the first postulate. The light cone is invariant under Lorentz transformations. Thus, as the invariant light cone of the origin O common to K and K' bisects the coordinate system K, so it does the coordinate system K'(Fig.A2).

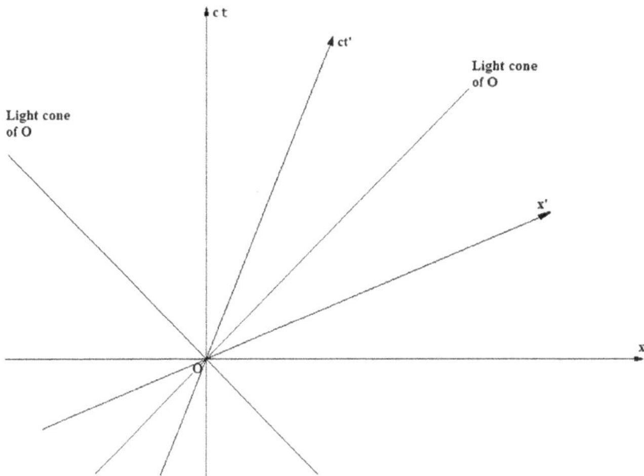

FIGURE A2. Inertial frame $K' \equiv (ct', x')$ relative to inertial frame $K \equiv (ct, x)$.

The relation between inertial frames as established by the Lorentz transformations in Fig.A2 renders the physical content of Special Relativity manifest in space-time diagrams. In the relation

$$c^2t^2 - x^2 - y^2 - z^2 = 1$$

the space-time point $x = y = z = 0$, $ct = 1$ represents the unit of time in inertial frame K. Since this relation remains invariant under Lorentz transformations, this relation determines the unit of time in all inertial systems. Likewise, in the relation

$$c^2t^2 - x^2 - y^2 - z^2 = -1$$

324

each space-time point in the set $t = 0$, $x^2 + y^2 + z^2 = 1$ represents the unit of length in K. Since this relation remains invariant under Lorentz transformations, this relation determines the unit of length in all inertial systems. All implications which these two premises have are inherent in the space-time diagrams of Fig.A3 and Fig.A4. Since space-time diagrams represent Minkowski space-time with two spatial dimensions suppressed, use will be made, in what follows, of the two-dimensional reduction $c^2t^2 - x^2 = \pm 1$ respectively.

In the coordinate system XY which the light cone defines the X-axis is described by $Y = 0$. At once, $x = ct$, or $ct - x = 0$. It follows that the X-axis is equivalently described by $Y = ct - x$. Likewise, the Y-axis is described by $X = ct + x$. As a consequence, $XY = (ct + x)(ct - x) = c^2t^2 - x^2 = 1$. In the XY-system this relation defines two hyperbolae symmetric about O (Fig.A3). As a consequence of the invariant character of $c^2t^2 - x^2 = 1$ under Lorentz transformations, all vectors defined by the pair of space-time points $O \equiv (0,0,0,0)$ and any point on either hyperbola have magnitude equal to 1 - or, in general, to any given positive number. Thus, the two hyperbolae $c^2t^2 - x^2 = 1$ determine the unit of time in all inertial systems. In a similar manner, the relation $c^2t^2 - x^2 = -1$ reduces to $XY = -1$ and determines the unit of length in all inertial systems through the corresponding two hyperbolae about O (Fig.A4).

The two fundamental consequences of the first postulate - time dilation, whose mathematical expression was derived above, and length contraction - can now be studied on the space-time diagrams of Fig.A3 and Fig.A4. The clocks in inertial frames K and K' in Fig.A3 have been synchronized at O. The magnitude of time-like vector OA represents the unit of time in K - by choice, one second. By dint of the above considerations, the magnitude of time-like vector OA' is invariably equal to one second. As events B and A' are simultaneous in K[83] and the magnitude of OB is bigger than the magnitude of OA, frame K observes that the synchronized clock in frame K' has registered one second when the clock in K has ticked in excess of one second. As observed from K, the synchronized clock in K' is retarded and time in K' flows at a slower rate. It is clear from Fig.A3 that this effect of time dilation is more pronounced, the higher the relative speed v between K and K' is. Time-like vector OA' invariably signifies a lapse of one second in K', yet with event A' being simultaneous to a progressively later event B in K as v increases. What this implies at the limit $v \to c$ follows from the fact that the future-directed hyperbola approaches the light cone asymptotically (it continuously approaches the light cone without ever reaching it). At $v \to c$ a one-second lapse in K' necessitates an infinite lapse in K before it is observed from K.

Closely associated with time dilation is the effect of length contraction. Two identical rods of length equal to one meter when at rest are in relative motion in the direction of their length with speed v. Each rod has been supplied with a clock mounted on it. The two clocks are synchronized at event O defined by the coincidence of the two rear ends of the rods (Fig.A4). The magnitude of space-like vector OC represents the unit of length in the rest frame K of either rod - by choice, one meter. The magnitude of

---

[83]Recall that moment ct_B of event B characterizes all points $\vec{x}$ in the sense that all synchronized clocks on that three-dimensional surface have registered the same lapse of time ct_B. The presence of such a synchronized clock at point $x_{A'}$ of event A' offsets the time required for a light signal to cover the distance between the two clocks in K and K' and renders events B and A' simultaneous in K.

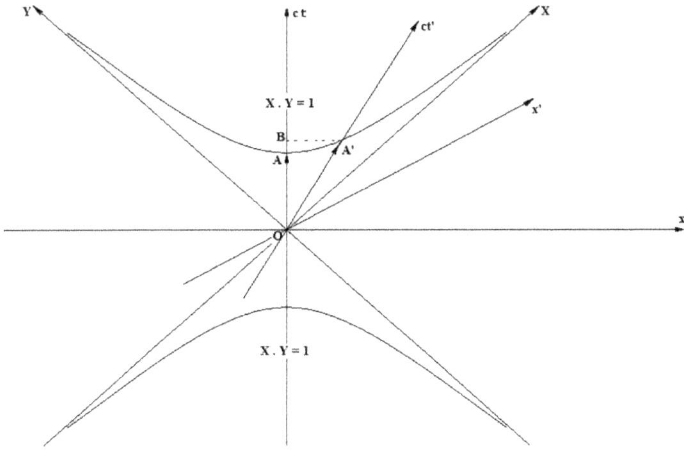

FIGURE A3. Time dilation

space-like vector OC' is invariably equal to one meter. As observed from K, the rod which is at rest in K' occupies space-time positions parallel to the x'-axis defined by the ct'-axis (the world line of the rod's rear end) and the line parallel to the ct'-axis which is tangent to the hyperbola $XY = -1$ at C' (the world line of the rod's front end depicted by the syncopated line which passes through C'). The set of world lines delineated by all points of the rod in K', the set of its successive positions in K', defines a stripe which intercepts the x-axis of the K-system between O and point C''. Again, simultaneity is crucial. All points between O and C'' are simultaneous in K. In view of $OC'' < OC$, it follows that as observed from the rest frame K of either rod the moving rod contracts along the direction of its motion.

It is a simple matter to confirm that the same effect identically occurs when the rest frame K' of what has been the moving rod observes the other rod whose rest frame is K in relative motion to K'. In addition to symmetry considerations, this effect can be independently confirmed in the same space-time diagram of Fig.A4 by observing that the syncopated tangent to $XY = -1$ at C intercepts the x'-axis at a point D such that[84] $OD < OC'$. Note also that, as was the case with time dilation, length contraction is more pronounced, the higher the relative speed between the two frames is.

Finally a few remarks on the relativity of simultaneity. As events which are simultaneous in frame K are represented by a line parallel to the x-axis and events which are simultaneous in frame K' are represented by a line parallel to the x'-axis, it is clear from the space-time diagrams in Fig.A3 and Fig.A4 that events which are simultaneous in

---

[84]By way of conceptualizing this result, imagine two witches on two identical broomsticks. One of them flies past the other as the other is about to set off. Either witch will boast that her broomstick is longer than the broomstick of the other witch. There is no paradox because the comparison has been made with different criteria of simultaneity in each of the two inertial frames.

K - such as B and A' in Fig.A3 - are events which not simultaneous in K'. However, although all frames agree that event Q lies to the future of event P if Q and P are within the light cone of O, there is an ambiguity as to which event precedes the other if either one or both of them lie out of the light cone of O. Fig.A5 illustrates this point. As observed from frame K, event P precedes event Q. On the contrary, as observed from frame K' it is event Q which precedes event P.

The use of space-time diagrams as demonstrated suffices for the study of any situation in Minkowski space-time and for the resolution of apparent contradictions such as the "train-and-platform paradox", or the "twin paradox".

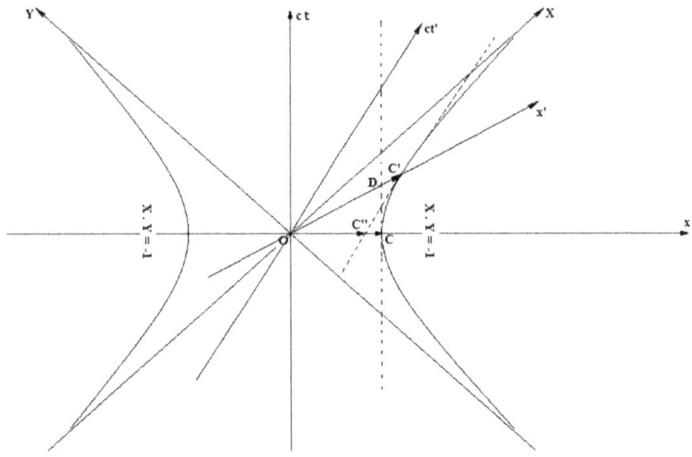

FIGURE A4. Length contraction

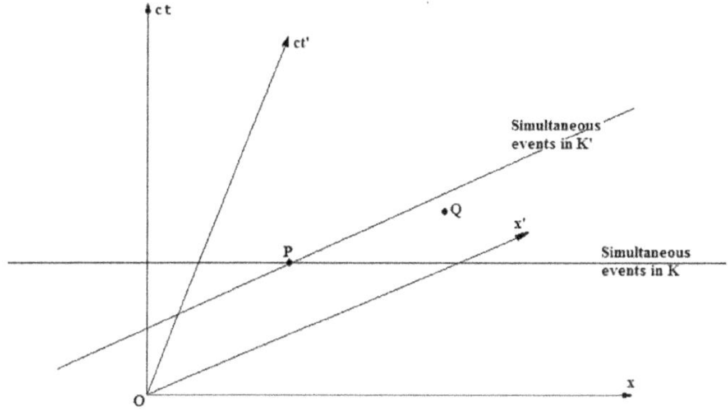

FIGURE A5. Relativity of simultaneity

BIBLIOGRAPHY

The books in the following list, including those citations on the Epistemological Interpretation of which I am critical, are recommended to the reader who wishes to advance the study of the subject matter treated in this book. The level of technical complexity in the textbooks of theoretical physics varies and the mathematically-inclined reader must decide for himself which one best suits his needs.

On Quantum Mechanics:

[1]. *The Feynman Lectures on Physics* R. Feynman, R. B. Leighton and M. Sands, Addison Wesley (1966).
The third volume of this famous three-volume lecture series is an introduction to quantum mechanics. Ideal for undergraduate students of physics and physics enthusiasts alike on account of its simplicity and deep physical insight.

[2]. *Quantum Physics* S. Gasiorowicz, Wiley (1974).
A nice introduction.

[3]. *Quantum Mechanics (Non-relativistic Theory)* L. D. Landau and E. M. Lifshitz, Pergamon Press (1981).
Moderately advanced and most rewarding to the dedicated reader (if only because of its intuitive value).

[4]. *Quantum Mechanics - An Introduction* W. Greiner, Springer-Verlag (1992).
A clear and comprehensive introduction including many solved problems.

[5]. *Quantum Mechanics - Symmetries* W. Greiner and B. Müller, Springer-Verlag (1994).
Advanced approach with a wonderful introduction to group theory.

[6]. *Practical Quantum Mechanics* S. Flügge, Springer-Verlag (1999).
A thorough introduction.

[7]. *Quantum Mechanics - A Modern Development* L. E. Ballentine, World Scientific (2001).
Moderately advanced.

[8]. *Problems in Quantum Mechanics* F. Constantinescu and E. Magyari, Pergamon Press (1985).
A remarkable collection of problems with solutions extending from non-relativistic to relativistic quantum mechanics. Valuable for a serious study of the subject.

[9]. *Problems in quantum mechanics with solutions* G. L. Squires Cambridge University Press (1995).
A wonderful collection of problems with solutions ranging over several

aspects of quantum mechanics.

On the Epistemological Interpretation:

[10]. *The German Ideology* Karl Marx and Friedrich Engels, in *Collected Works*, International Publishers, New York. - Also of interest, the abbreviated edition: *The German Ideology - Part One with selections from Parts Two and Three*, International Publishers, New York (1986).
Establishes the theory of historical materialism and the relation between social practice and consciousness. A landmark work!

[11]. *Atomtheorie und Naturbeschreibung* N. Bohr Springer, Berlin, (1931). Translated as:

[11]. *Atomic Theory and the Description of Nature*, Cambridge University Press, Cambridge (2011).
The definitive statement of the explicitly positivist and implicitly idealist standpoint of the Copenhagen Interpretation.

[12]. *The Nature of Space and Time* Stephen Hawking and Roger Penrose, Princeton University Press (1996).
An exciting and at times amusing book. Basic knowledge of the theory of relativity and of quantum mechanics is required.

[13]. *Materialism and Empirio-criticism - Critical Comments on a Reactionary Philosophy* V. I. Lenin, International Publishers, New York (1970).
Lenin directs his critique against the reactionary "modern positivism" of his time (1908) as expressed primarily by Ernst Mach, but he could very well have targeted the reactionary Copenhagen Interpretation and the philosophical views of N. Bohr and W. Heisenberg which emerged some twenty years later!

[14]. *Early Writings* Karl Marx, Penguin Classics (1992).
Step-by-step inversion of the Hegelian philosophy to historical materialism. Essential for understanding the analytical method which refutes positivist and idealist interpretations of modern theoretical physics.

[15]. *The Development of the Interpretation of the Quantum Theory* Werner Heisenberg, in: *Niels Bohr and the Development of Physics*, Pergamon Press, London (1955).

[16]. *Physics and Philosophy - The Revolution in Modern Science* Werner Heisenberg, Penguin Classics (2000).
A testament of the essentially idealist and reactionary character of the Copenhagen Interpretation.

[17]. *The Poverty of Philosophy* (Chapter II) K. Marx, in *Collected Works*, International Publishers, New York.

[18]. *In Search of Schrödinger's Cat* John Gribbin, Bantam New Age Books (1984).

[19]. *Capital - volume 1, volume 2, volume 3* Karl Marx, Penguin Classics (1990).
A supreme application of the dialectical method in the terrain of political economy
(especially in volume 1). One of the most summital works of all time.
(Also included in *Collected Works* K. Marx and F. Engels, Progress Publishers).

[20]. *Dialectics in Modern Physics* M. E. Omelyanovsky, Progress Publishers (1979).

[21]. *Are There Quantum Jumps?* (Part I) E. Schrödinger,
The British Journal for the Philosophy of Science, vol.3, No.10 (1952) pp. 109-123.

[22]. *Are There Quantum Jumps?* (Part II) E. Schrödinger,
The British Journal for the Philosophy of Science, vol.3, No.11 (1952) pp. 233-242.

[23]. *Complementarity and Scientific Rationality* S. Saunders,
arXive:quant-ph/0412195v1.

[24]. *Die Plancksche Entdeckung und philosophischen Grundfragen der Atomlehre*
W. Heisenberg *Naturwissenschaften*, 45, 10: 230, (1958).

On Quantum Field Theory:

[25]. *Advanced Quantum Mechanics* J. J. Sakurai, Addison-Wesley (reprinted 2008).
An informal introduction to quantum field theory in the language of "relativistic
quantum mechanics". Of high intuitive value.

[26]. *Relativistic Quantum Mechanics* J. D. Bjorken and S. D. Drell,
McGraw-Hill (1964).
Also a "particle-approach" to quantum field theory in general and to the relativistic
propagator in particular stated, in the quantum-mechanical language of "hole
theory". Moderately advanced.

[27]. *Relativistic Quantum Fields* J. D. Bjorken and S. D. Drell, McGraw-Hill (1965).
Replaces the intuitive arguments in [26] by the rigor of formal quantum field theory.
Moderately advanced.

[28]. *Field Theory: A Modern Primer - Second Edition* P. Ramond,
Westview Press (1997).
An outstanding treatment of quantum field theory in terms of path-integral
quantization in imaginary time.

[29]. *An Introduction to Quantum Field Theory* M. E. Peskin and D. V. Schroeder,
Westview Press (1995).
An outstanding introduction indeed!

[30]. *The Quantum Theory of Fields - volume I, volume II, volume III*
S. Weinberg, Cambridge University Press (reprinted 2002).
Outstanding and really advanced.

[31]. *Quantum Field Theory: A Modern Introduction* M. Kaku, Oxford University Press (1993).
A clear introduction.

[32]. *Path Integrals in Field Theory: An Introduction* U. Mosel, Springer-Verlag Berlin Heidelberg (2004).
A wonderful introduction to path-integral formalism, both in quantum mechanics and in quantum field theory.

[33]. *Theory of Spinors: An Introduction* M. Carmeli and S. Malin, World Scientific (2000).
The theory of spinors from quantum mechanics to quantum field theory in flat and curved spacetimes. Really advanced.

[34]. *Quantum Field Theory in a Nutshell* A. Zee, Princeton University Press (2003).
A rather advanced text which stresses the essence of the physical concepts involved.

[35]. *Field Theory: A Path Integral Approach* Ashok Das, World Scientific (1993).
A wonderful introduction to quantum field theory from the perspective of path-integral quantization.

[36]. *Field Quantization* Walter Greiner and Joachim Reinhardt, Springer-Verlag (1996).
Following [4] and [5], this is the third in a five-volume series which essentially covers all quantum Physics, from Quantum Mechanics to Quantum Chromodynamics. An outstanding introduction to quantum field theory treating both, the canonical and the path-integral quantization. Several solved problems complete the comprehensive analysis in each chapter.

[37]. *Quantum Electrodynamics* Walter Greiner and Joachim Reinhardt, Springer-Verlag (1994).

[38]. *Quantum Chromodynamics* Walter Greiner and Andreas Schäfer, Springer-Verlag (1995).

[39]. *Quantum Field Theory* Lewis H. Ryder, Cambridge University Press (1996).
An introduction which renders the basic aspects of quantum field theory readily accessible to readers who understand quantum mechanics and special relativity.

[40]. *Quantum Field Theory* C. Itzykson and J.-B. Zuber, McGraw-Hill Book Company (1980).
A moderately advanced treatment of the most essential aspects of quantum field theory from the perspective of both, the canonical and the path-integral field quantization.

[41]. *Quantum Field Theory* David Tong,
Lecture notes available at http://www.damtp.cam.ac.uk/user/tong/qft.html
A brief but concise and comprehensive introduction.

The following two citations refer to the sporadic references to string theory in my book.

[42]. *String Theory - volume I, volume II* Joseph Polchinski,
Cambridge University Press (2001).
Advanced treatment of string theory by one of the outstanding researchers
in this field.

[43]. *String Theory* David Tong,
Lecture notes available at http://www.damtp.cam.ac.uk/user/tong/string.html
A comprehensive introduction.

On The Mathematical Formalism:

[44]. *Advanced Calculus* Murray R. Spiegel,
McGraw-Hill Book Company (1974).
A comprehensive treatment of the subject matter expressed by the title,
supplemented by a wonderful collection of solved problems. Knowledge of
elementary Calculus is a prerequisite.

[45]. *Mathematical Methods of Physics* Jon Mathews and R.L. Walker,
Addison-Wesley Publishing Company, Inc. (1970).
A book which covers several topics in advanced mathematics essential to
theoretical physics in general and to quantum theory in particular.

[46]. *Methods of Mathematical Physics - volume I, volume II*
R. Courant and D. Hilbert, Wiley Classics Library (1989).
Outstanding treatment of the subject matter which stresses the deep
relation between physical content and mathematical formalism.

www.ingramcontent.com/pod-product-compliance
Lightning Source LLC
LaVergne TN
LVHW051058180726
843512LV00020B/1518